Radon Series on Computational and Applied Mathematics 5

Radon Series on Computational and Applied Mathematics

1 *Lectures on Advanced Computational Methods in Mechanics*
Johannes Kraus and Ulrich Langer (eds.), 2007

2 *Gröbner Bases in Symbolic Analysis*
Markus Rosenkranz and Dongming Wang (eds.), 2007

3 *Gröbner Bases in Control Theory and Signal Processing*
Hyungju Park and Georg Regensburger (eds.), 2007

4 *A Posteriori Estimates for Partial Differential Equations*
Sergey Repin, 2008

5 *Robust Algebraic Multilevel Methods and Algorithms*
Johannes Kraus and Svetozar Margenov, 2009

6 *Iterative Regularization Methods for Nonlinear Ill-Posed Problems*
Barbara Kaltenbacher, Andreas Neubauer and Otmar Scherzer, 2008

7 *Robust Static Super-Replication of Barrier Options*
Jan H. Maruhn, 2009

8 *Advanced Financial Modelling*
Hansjörg Albrecher, Wolfgang J. Runggaldier and Walter Schachermayer (eds.), 2009

Johannes Kraus
Svetozar Margenov

Robust Algebraic Multilevel Methods and Algorithms

Walter de Gruyter · Berlin · New York

Authors

Johannes Kraus
Johann Radon Institute for Computational and Applied Mathematics
Austrian Academy of Sciences
Altenberger Straße 69
4040 Linz
Austria
E-mail: johannes.kraus@oeaw.ac.at

Svetozar Margenov
Institute for Parallel Processing
Bulgarian Academy of Sciences
Acad. G. Bontchev str. 25A
1113 Sofia
Bulgaria
E-mail: margenov@parallel.bas.bg

Keywords
Algebraic multilevel iteration, algebraic multigrid method, preconditioning, conjugate gradient method, finite element method, discretization, partial differential equation

Mathematics Subject Classification 2000
65-01, 65-02, 65F10, 65N30, 65N50, 65N55

∞ Printed on acid-free paper which falls within the guidelines of the ANSI to ensure permanence and durability.

ISBN 978-3-11-019365-7

Bibliographic information published by the Deutsche Nationalbibliothek

The Deutsche Nationalbibliothek lists this publication in the Deutsche Nationalbibliografie; detailed bibliographic data are available in the Internet at http://dnb.d-nb.de.

Printed in Germany
Cover design: Martin Zech, Bremen.
Printing and binding: Hubert & Co. GmbH & Co. KG, Göttingen.

Preface

The iterative methods play an important role in solving linear equations that arise in real-world applications. Numerous properties of the problem may affect the efficiency of the solution. This book deals with algorithms for the solution of linear systems of algebraic equations with large-scale sparse matrices, with a focus on problems that are obtained after discretization of partial differential equations using finite element methods.

The monograph provides a comprehensive presentation of the recent advances in robust algebraic multilevel methods and algorithms including, e.g., the preconditioned conjugate gradient method, the algebraic multilevel iteration (AMLI) preconditioners, some relations to the classical algebraic multigrid (AMG) method and its recent modifications.

The first five chapters can serve as a short introductory course on the theory of AMLI methods and algorithms. The next part of the monograph is devoted to more advanced topics, including related issues of AMG methods, AMLI methods for discontinuous Galerkin systems, locking-free algorithms for coupled problems etc., ending with important aspects of implementation and one challenging application. This second part is addressed to some more experienced students and practitioners and can be used to complete a more advanced course on robust AMLI methods and their efficient application.

During the years, each of us cooperated with several coauthors on topics included in this volume. They definitely influenced and enriched our understanding of the field. Special thanks are due to them. Working on the monograph, we had a lot of fruitful discussions with Ludmil Zikatanov. We highly appreciate his suggestions and remarks. We thank also Petia Boyanova, and Ivan Georgiev for their careful reading of parts of preliminary drafts of the book.

This volume is intended for mathematicians, engineers, natural scientists etc. The monograph is partly based on, and initially stimulated by, the lecture course on Robust Parallel Algebraic Multigrid and Multilevel Techniques given in the frame of the Special Radon Semester on Computational Mechanics – Linz, October 3 – December 16, 2005.

We gratefully acknowledge the support by the Austrian Academy of Sciences. The work on this monograph has been also partially supported by the Bulgarian Academy of Sciences as well as by Austrian Science Foundation FWF Project P19170-N18, and the Bulgarian NSF Grants DO 02-147/08 and DO 02-338/08.

Johannes Kraus and Svetozar Margenov

Contents

1 Introduction

1.1 Finite element method (FEM)

We will start the presentation with the formulation of a boundary value problem for a second-order elliptic partial differential equation (PDE), which will serve then as a model problem in a brief introduction to the finite element method (FEM).

1.1.1 The elliptic model problem

Let us consider the elliptic boundary value problem

$$-\nabla \cdot (\mathbf{a}(\boldsymbol{x})\nabla u(\boldsymbol{x})) = f(\boldsymbol{x}) \quad \text{in} \quad \Omega, \tag{1.1a}$$

$$u = 0 \quad \text{on} \quad \Gamma_D, \tag{1.1b}$$

$$(\mathbf{a}(\boldsymbol{x})\nabla u(\boldsymbol{x})) \cdot \boldsymbol{n} = 0 \quad \text{on} \quad \Gamma_N, \tag{1.1c}$$

for an unknown scalar function $u(\boldsymbol{x})$ where Ω is a polygonal domain in two- or three-dimensional space $\mathbb{R}^d$, $d = 2, 3$, and $f(\boldsymbol{x})$ is a given squared Lebesgue integrable function, i.e.,

$$f \in L_2(\Omega) := \{v : v \text{ is defined on } \Omega \text{ and } \int_\Omega v^2 \, d\boldsymbol{x} < \infty\}. \tag{1.2}$$

The coefficient matrix $\mathbf{a}(\boldsymbol{x})$ in (1.1) is assumed to be symmetric positive definite (SPD) and uniformly bounded in Ω, i.e.,

$$c_1\|\boldsymbol{v}\|^2 \leq \boldsymbol{v}^T \mathbf{a}(\boldsymbol{x})\, \boldsymbol{v} \leq c_2\|\boldsymbol{v}\|^2 \quad \forall \boldsymbol{v} \in \mathbb{R}^d, \forall \boldsymbol{x} \in \Omega, \tag{1.3}$$

for some positive constants c_1 and c_2, and $\boldsymbol{n}$ is the outward unit vector normal to the boundary $\Gamma = \partial\Omega$. The disjoint parts of $\Gamma = \overline{\Gamma}_D \cup \overline{\Gamma}_N$ on which Dirichlet and Neumann boundary conditions are imposed are denoted by Γ_D and Γ_N, respectively.

Let $\mathcal{V}$ denote a linear (vector) space. Then we shall use the following notation:

Definition 1.1. A mapping $\mathcal{L} : \mathcal{V} \to \mathbb{R}$ is called a linear form if $\mathcal{L}(\cdot)$ is linear, i.e., for all $v, w \in \mathcal{V}$ and $\alpha, \beta \in \mathbb{R}$

$$\mathcal{L}(\alpha v + \beta w) = \alpha\mathcal{L}(v) + \beta\mathcal{L}(w).$$

Definition 1.2. If $\mathcal{V}$ is a linear space, then we say that $\mathcal{A} : \mathcal{V}\times\mathcal{V} \to \mathbb{R}$ is a bilinear form if $\mathcal{A}(\cdot,\cdot)$ is linear in each argument, i.e., for all $u, v, w \in \mathcal{V}$ and $\alpha, \beta \in \mathbb{R}$

$$\begin{aligned}\mathcal{A}(u, \alpha v + \beta w) &= \alpha\mathcal{A}(u, v) + \beta\mathcal{A}(u, w),\\ \mathcal{A}(\alpha u + \beta v, w) &= \alpha\mathcal{A}(u, w) + \beta\mathcal{A}(v, w).\end{aligned}$$

A bilinear form $\mathcal{A}(\cdot,\cdot)$ is said to be symmetric if $\mathcal{A}(v, w) = \mathcal{A}(w, v)$ for all $v, w \in \mathcal{V}$. Furthermore, a symmetric bilinear form $\mathcal{A}(\cdot,\cdot)$ on $\mathcal{V} \times \mathcal{V}$ is called a scalar product on $\mathcal{V}$ if $\mathcal{A}(v, v) > 0$ for all $v \in \mathcal{V}$, $v \neq 0$. The norm $\|\cdot\|_{\mathcal{A}}$ induced by the scalar product $\mathcal{A}(\cdot,\cdot)$ is defined by $\|v\|_{\mathcal{A}} = (\mathcal{A}(v, v))^{1/2}$.

Definition 1.3. A linear (vector) space $\mathcal{V}$ that is equipped with a scalar product $(\cdot,\cdot)_{\mathcal{V}}$ and the corresponding norm $\|\cdot\|_{\mathcal{V}}$ is called a Hilbert space if $\mathcal{V}$ is complete, i.e., if every Cauchy sequence in $\mathcal{V}$ is convergent and its limit also belongs to $\mathcal{V}$.

Starting point for the finite element solution of the basic problem (1.1) is the following weak formulation: Given $f \in L_2(\Omega)$ find $u \in \mathcal{V}$, satisfying

$$\mathcal{A}(u, v) = \mathcal{L}(v) \quad \forall v \in \mathcal{V}, \tag{1.4a}$$

$$\mathcal{A}(u, v) := \int_\Omega \mathbf{a}(\boldsymbol{x})\nabla u(\boldsymbol{x}) \cdot \nabla v(\boldsymbol{x})\, d\boldsymbol{x}, \tag{1.4b}$$

$$\mathcal{L}(v) := \int_\Omega f(\boldsymbol{x})\, v(\boldsymbol{x})\, d\boldsymbol{x}, \tag{1.4c}$$

$$\mathcal{V} := H^1_D(\Omega) \equiv \{v \in H^1(\Omega) : v = 0 \text{ on } \Gamma_D\}. \tag{1.4d}$$

The spaces $L_2(\Omega)$ and

$$H^1(\Omega) := \{v : v \text{ is defined on } \Omega \text{ and } \int_\Omega v^2 + \nabla v \cdot \nabla v\, d\boldsymbol{x} < \infty\} \tag{1.5}$$

are Hilbert spaces equipped with the scalar products

$$(v, w)_{L_2(\Omega)} \equiv (v, w) := \int_\Omega vw\, d\boldsymbol{x}, \tag{1.6}$$

$$(v, w)_{H^1(\Omega)} := \int_\Omega vw + \nabla v \cdot \nabla w\, d\boldsymbol{x}, \tag{1.7}$$

and the induced norms

$$\|v\|_{L_2(\Omega)} = (v, v)^{1/2}, \tag{1.8}$$

and

$$\|v\|_{H^1(\Omega)} = (v, v)^{1/2}_{H^1(\Omega)}, \tag{1.9}$$

respectively.

The weak formulation (1.4) of (1.1) is obtained from multiplying (1.1a) by an arbitrary test function $v \in \mathcal{V}$, integrating over Ω, and applying Green's formula (integration by parts). Any solution u of (1.4) is then called a weak solution of (1.1). If u is a weak solution of (1.1) it is not immediately clear that u is also a classical solution of (1.1) since the latter in general requires sufficient (stronger) regularity assumptions. Closely related to this subject matter is the fact that usually it is much easier to prove the existence (and uniqueness) of a (the) weak solution. This kind of existence results follow from the Lax–Milgram theorem, which is a variant of the Riesz' representation theorem in Hilbert space theory, see, e.g., [47], under the following assumptions[1]:

(A1) $\mathcal{A}(\cdot,\cdot)$ is continuous on $\mathcal{V} \times \mathcal{V}$, i.e.,

$$|\mathcal{A}(v,w)| \le \bar{c}\, \|v\|_{\mathcal{V}} \|w\|_{\mathcal{V}} \quad \forall v,w \in \mathcal{V}, \tag{1.10}$$

(A2) $\mathcal{A}(\cdot,\cdot)$ is $\mathcal{V}$-elliptic, i.e.,

$$\mathcal{A}(v,v) \ge \underline{c}\, \|v\|_{\mathcal{V}}^2 \quad \forall v \in \mathcal{V}, \tag{1.11}$$

(A3) $\mathcal{L}(\cdot)$ is continuous on $\mathcal{V}$, i.e.,

$$|\mathcal{L}(v)| \le c\, \|v\|_{\mathcal{V}} \quad \forall v \in \mathcal{V}. \tag{1.12}$$

Note that choosing $v = u$ in (1.4) and using (A2)–(A3) one obtains the relations $\underline{c}\, \|u\|_{\mathcal{V}}^2 \le \mathcal{A}(u,u) = \mathcal{L}(u) \le c\, \|u\|_{\mathcal{V}}$ and thus the stability estimate

$$\|u\|_{\mathcal{V}} \le \frac{c}{\underline{c}} \tag{1.13}$$

holds. Moreover, if we assume that u_1 and u_2 are two solutions of (1.4) we conclude that

$$\mathcal{A}(u_1 - u_2, v) = 0 \quad \forall v \in \mathcal{V}, \tag{1.14}$$

and the estimate (1.13) can be applied to the solution $(u_1 - u_2)$ of (1.14) for $c = 0$, i.e., $\|u_1 - u_2\|_{\mathcal{V}} = 0$, showing the uniqueness of the solution of (1.4).

Remark 1.4. It is not difficult to prove that the weak formulation (1.4) is equivalent to the following abstract minimization problem, see, e.g., [68]: Find $u \in \mathcal{V}$ such that

$$\mathcal{F}(u) = \min_{v \in \mathcal{V}} \mathcal{F}(v), \tag{1.15a}$$

$$\mathcal{F}(v) := \frac{1}{2}\mathcal{A}(v,v) - \mathcal{L}(v). \tag{1.15b}$$

[1] In this book, if not mentioned otherwise, it will be assumed that the considered problem is self-adjoint, which means that the bilinear form in (1.4) is symmetric.

1.1.2 Conforming discretizations

The numerical solution of variational problems like (1.4) or (1.15) involves their reformulation using proper finite-dimensional subspaces of the Hilbert spaces that provide the abstract framework for a comprehensive analysis.[2] The finite element method (FEM) has become an established technique for choosing such subspaces. This is due to its elegant mathematical description/formulation and due to the fact that the FEM methodology is highly practicable for solving many important (classes of) problems in engineering, medical, or life sciences.

Let us assume that Ω is a polygonal domain which is partitioned into finitely many subdomains, called elements. The elements e (which we will sometimes also denote by T) of this partition (triangulation) $\mathcal{T}_h$, which is also called a finite-element *mesh*, usually have a simple shape, e.g., triangles or quadrilaterals for two-dimensional domains, tetrahedra or hexahedra for three-dimensional domains. In the case when all elements e are congruent, the mesh is called regular. The quantity $h := \max_{e \in \mathcal{T}_h} h_e$ is called the mesh size (or mesh parameter).

The (most commonly used) subspaces $\mathcal{V}_h$ in the FEM are composed of piecewise polynomial functions. Let $C^0(\overline{\Omega})$ denote the space of continuous functions defined on $\overline{\Omega}$ and

$$C^1(\overline{\Omega}) = \{v \in C^0(\overline{\Omega}) : \mathrm{D}^\alpha v \in C^0(\overline{\Omega}),\ |\alpha| = 1\}$$

the space of $C^0(\overline{\Omega})$-functions whose first derivatives are continuous. Then, since (we assume that)

$$\begin{aligned} \mathcal{V}_h &:= \{v : v|_e \in P_r(e) \quad \forall e \in \mathcal{T}_h\}, \\ P_r(e) &:= \{w : w \text{ is a polynomial of degree less or equal to } r \text{ on } e\} \quad \forall e \in \mathcal{T}_h, \end{aligned}$$

the following continuity condition holds, see, e.g., [68]:

$$\mathcal{V}_h \subset H^1(\Omega) \text{ if and only if } \mathcal{V}_h \subset C^0(\overline{\Omega}).$$

The idea is now to replace the infinite-dimensional space $\mathcal{V}$ in (1.4) by $\mathcal{V}_h$, i.e., to consider the problem: Find $u_h \in \mathcal{V}_h$ such that

$$\mathcal{A}_h(u_h, v_h) = \mathcal{L}_h(v_h) \quad \forall v_h \in \mathcal{V}_h, \tag{1.16}$$

where in case of the scalar elliptic model problem $\mathcal{A}_h(\cdot,\cdot)$ and $\mathcal{L}_h(\cdot)$ are defined by

$$\mathcal{A}_h(u_h, v_h) \equiv \mathcal{A}(u_h, v_h) := \int_\Omega \mathbf{a}(x) \nabla u_h(x) \cdot \nabla v_h(x)\, dx, \tag{1.17a}$$

$$\mathcal{L}_h(v_h) \equiv \mathcal{L}(v_h) := \int_\Omega f(x)\, v_h(x)\, dx. \tag{1.17b}$$

[2] In a more general context these are Sobolev spaces.

Any FEM with piecewise polynomial functions that lie in the Sobolev space in which the variational problem is posed, i.e., $\mathcal{V}_h \subset \mathcal{V}$, is called a *conforming* method [35]. If the trial functions v_h are taken from the same space $\mathcal{V}_h$ in which the (approximate) solution u_h is sought for, the FEM is called a standard *Galerkin* method.

By subtracting (1.16) from (1.4a), thereby assuming that $\mathcal{V}_h \subset \mathcal{V}$, we find

$$\mathcal{A}(u - u_h, v_h) = 0 \quad \text{for all } v_h \in \mathcal{V}_h, \tag{1.18}$$

which is often referred to as the *Galerkin orthogonality*. This means that the FEM solution u_h is the projection with respect to the scalar product $\langle v, w\rangle_{\mathcal{A}} := \mathcal{A}(v, w)$ (defined on $\mathcal{V}$) of the exact solution u onto $\mathcal{V}_h$. In other words, u_h is the element of $\mathcal{V}_h$ closest to u in $\mathcal{V}$ with respect to the norm induced by the scalar product $\mathcal{A}(\cdot,\cdot)$, i.e.,

$$\|u - u_h\|_{\mathcal{A}} \leq \|u - v_h\|_{\mathcal{A}} \quad \text{for all } v_h \in \mathcal{V}_h. \tag{1.19}$$

Moreover, by using the $\mathcal{V}$-ellipticity (A2) and the continuity (A1) of the bilinear form $\mathcal{A}(\cdot,\cdot)$, it immediately follows from (1.18) that

$$\begin{aligned} \underline{c}\,\|u - u_h\|_{\mathcal{V}}^2 &\leq \mathcal{A}(u - u_h, u - u_h) \\ &= \mathcal{A}(u - u_h, u - v_h) + \mathcal{A}(u - u_h, v_h - u_h) \\ &\leq \overline{c}\,\|u - u_h\|_{\mathcal{V}}\,\|u - v_h\|_{\mathcal{V}}. \end{aligned}$$

After dividing by $\|u - u_h\|_{\mathcal{V}}$ we arrive at the following result, known as Céa's lemma, which plays an important role in deriving error estimates for various (conforming) FE approximations:

Lemma 1.5. *Let $u \in \mathcal{V}$ be the solution of (1.4), $u_h \in \mathcal{V}_h$ the solution of (1.16), where $\mathcal{V}_h \subset \mathcal{V}$ and let the assumptions (A1)–(A3) be satisfied. Then*

$$\|u - u_h\|_{\mathcal{V}} \leq \frac{\overline{c}}{\underline{c}} \inf_{v_h \in \mathcal{V}_h} \|u - v_h\|_{\mathcal{V}}. \tag{1.20}$$

1.1.3 Nonconforming discretizations

In certain real-life problems the condition that the finite element space is a subspace of the function space in which the variational problem is posed is too restrictive. In particular, this might be the case if the boundary conditions cannot be satisfied exactly, or if the bilinear form $\mathcal{A}(\cdot,\cdot)$ cannot be computed exactly on the finite element space $\mathcal{V}_h$. If the space $\mathcal{V}_h$, being used to solve a $\mathcal{V}$-elliptic problem, is not contained in $\mathcal{V}$ the FEM is called a *nonconforming* method, i.e., a discretization based on nonconforming elements. Then, in addition to the approximation error there is a second source of error of the numerical solution, which is called *consistency error*.

Violating the conformity condition $\mathcal{V}_h \subset \mathcal{V}$ the $\mathcal{V}$-norm in general is no longer well-defined on $\mathcal{V}_h$. A remedy for this shortcoming is to use mesh-dependent norms in the convergence analysis. The following *broken norm* is such an example: Given a partition $\mathcal{T}_h$ of Ω, we define

$$\|v\|_{m,h} := \sqrt{\sum_{T \in \mathcal{T}_h} \|v\|_{m,T}^2} \tag{1.21}$$

where $\|\cdot\|_{m,T}$ is the induced norm on the space $H^m(T)$ and thus $\|v\|_{m,h} = \|v\|_{m,\Omega}$ for all v in the Sobolev space $H^m(\Omega)$.

Let us consider now the weak formulation (1.16) where elements v_h of the FE space $\mathcal{V}_h$ do not necessarily have to satisfy any continuity conditions. In case of our elliptic model problem the bilinear form $\mathcal{A}_h(\cdot,\cdot)$ can be defined by

$$\mathcal{A}_h(u_h, v_h) := \sum_{T \in \mathcal{T}_h} \int_T \mathbf{a}(T)\, \nabla u_h(\boldsymbol{x}) \cdot \nabla v_h(\boldsymbol{x})\, d\boldsymbol{x}. \tag{1.22}$$

Here $\mathbf{a}(T)$ is a piecewise constant, symmetric positive definite (SPD) matrix, defined by the integral averaged values of $\mathbf{a}(\boldsymbol{x})$ over each element T from the triangulation $\mathcal{T}_h$, i.e.,

$$\mathbf{a}(T) = \frac{1}{|T|} \int_T \mathbf{a}(\boldsymbol{x})\, d\boldsymbol{x} \quad \forall T \in \mathcal{T}_h. \tag{1.23}$$

In this way strong coefficient jumps across the boundaries between adjacent finite elements from $\mathcal{T}_h$ are allowed.

Remark 1.6. Note that the positive definiteness of $\mathbf{a}(T)$ is a consequence of $\mathbf{a}(\boldsymbol{x})$ being SPD pointwise for all $\boldsymbol{x} \in T$: If $\boldsymbol{w}^T \mathbf{a}(\boldsymbol{x}) \boldsymbol{w} > 0$ for all $\boldsymbol{w} \neq \mathbf{0}$ for all $\boldsymbol{x} \in T$ then $\boldsymbol{w}^T \mathbf{a}(T) \boldsymbol{w} = \frac{1}{|T|} \boldsymbol{w}^T \left(\int_T \mathbf{a}(\boldsymbol{x})\, d\boldsymbol{x}\right) \boldsymbol{w} = \frac{1}{|T|} \int_T \boldsymbol{w}^T \mathbf{a}(\boldsymbol{x}) \boldsymbol{w}\, d\boldsymbol{x} > 0$.

We relax the continuity and ellipticity assumptions on the bilinear form $\mathcal{A}_h(\cdot,\cdot)$, which is assumed to be well-defined on both spaces $\mathcal{V}$ and $\mathcal{V}_h$ in the following way:

(A4) $\mathcal{A}_h(\cdot,\cdot)$ is continuous on $(\mathcal{V} + \mathcal{V}_h) \times \mathcal{V}_h$, i.e.,

$$|\mathcal{A}_h(v, w_h)| \le \bar{c}\, \|v\|_h \|w_h\|_h \quad \forall v \in \mathcal{V} + \mathcal{V}_h, \ \forall w_h \in \mathcal{V}_h, \tag{1.24}$$

(A5) $\mathcal{A}_h(\cdot,\cdot)$ is $\mathcal{V}_h$-elliptic, i.e.,

$$\mathcal{A}_h(v_h, v_h) \ge \underline{c}\, \|v_h\|_h^2 \quad \forall v_h \in \mathcal{V}_h. \tag{1.25}$$

Then the error of the numerical solution in the broken semi-norm

$$\|v\|_h := \sqrt{\mathcal{A}_h(v,v)} \quad \forall v \in \mathcal{V} + \mathcal{V}_h \tag{1.26}$$

can be bounded in terms of the approximation error (first term in (1.27)) and the consistency error (second term in (1.27)):

Lemma 1.7. *Under the above hypotheses there exists a constant c independent of h such that*

$$\|u - u_h\|_h \le c \left(\inf_{v_h \in \mathcal{V}_h} \|u - v_h\|_h + \sup_{w_h \in \mathcal{V}_h} \frac{|\mathcal{A}_h(u, w_h) - \mathcal{L}_h(w_h)|}{\|w_h\|_h} \right). \tag{1.27}$$

Proof. First we note that by the triangle inequality we have

$$\|u - u_h\|_h \le \|u - v_h\|_h + \|u_h - v_h\|_h. \tag{1.28}$$

Moreover, from (A4) and (A5), using also (1.16), it follows that

$$\begin{aligned} \underline{c}\,\|u_h - v_h\|_h^2 &\le \mathcal{A}_h(u_h - v_h, u_h - v_h) \\ &= \mathcal{A}_h(u - v_h, u_h - v_h) - \mathcal{A}_h(u, u_h - v_h) + \mathcal{L}_h(u_h - v_h) \\ &\le \overline{c}\,\|u - v_h\|_h\,\|u_h - v_h\|_h + |\mathcal{A}_h(u, u_h - v_h) - \mathcal{L}_h(u_h - v_h)| \end{aligned}$$

and thus, substituting $w_h = u_h - v_h$, the last term in (1.28) can be estimated by

$$\|u_h - v_h\|_h \le \frac{1}{\underline{c}} \left(\overline{c}\,\|u - v_h\|_h + \frac{|\mathcal{A}_h(u, w_h) - \mathcal{L}_h(w_h)|}{\|w_h\|_h} \right). \tag{1.29}$$

Together, (1.28) and (1.29) imply the estimate (1.27). □

Example. A simple nonconforming element for the discretization of second-order elliptic boundary value problems is the Crouzeix–Raviart element, which is also called *nonconforming P_1 element*, see, e.g., [35]. The related FE space is defined by

$$\begin{aligned} \mathcal{V}_h \equiv \mathcal{M}_*^1 := \{v \in L_2(\Omega) : \; & v|_T \text{ is linear for every } T \text{ in } \mathcal{T}_h, \\ & v \text{ is continuous at the midpoints of the edges (faces) of } T\} \end{aligned} \tag{1.30}$$

where $\mathcal{T}_h$ is a partition of a 2D (3D) polygonal (polyhedral) domain Ω into triangles (tetrahedra) T.

1.1.4 Structure and properties of the stiffness matrix

The numerical solution of boundary value problems like (1.1) by the finite element method typically leads to the problem of solving large systems of linear algebraic equations. The general procedure is to consider a basis of the finite element space $\mathcal{V}_h$, which we denote by $\Phi = \{\phi_1, \phi_2, \ldots, \phi_N\}$. Then any function v_h in $\mathcal{V}_h$ has the unique representation

$$v_h = \sum_{i=1}^{N} v_i \phi_i$$

where the real numbers v_i are the expansion coefficients of v_h (with respect to the basis Φ). Representing the solution u_h of (1.16) as $u_h = \sum_{i=1}^{N} u_i \phi_i$ it can easily be seen that (1.16) is equivalent to

$$\sum_{i=1}^{N} \mathcal{A}_h(\phi_i, \phi_j) u_i = \mathcal{L}_h(\phi_j), \quad j = 1, 2, \ldots, N, \tag{1.31}$$

which in matrix form reads as

$$A\mathbf{u} = \mathbf{b}. \tag{1.32}$$

Here $\mathbf{u} = (u_i) \in \mathbb{R}^N$ is the vector of unknowns, and the right-hand side vector $\mathbf{b} = (b_j) \in \mathbb{R}^N$ is defined by

$$b_j = \mathcal{L}_h(\phi_j), \quad 1 \le j \le N. \tag{1.33}$$

Since the basis functions $\phi_i \in \Phi$ have a local support by construction only few (typically $\mathcal{O}(1)$) nonzero entries

$$a_{ij} = \mathcal{A}_h(\phi_i, \phi_j), \quad 1 \le i, j \le N, \tag{1.34}$$

occur in each row of the *stiffness matrix* $A = (a_{ij}) \in \mathbb{R}^{N \times N}$, which is then called a *sparse* matrix. Another important observation is that for a $\mathcal{V}$-elliptic bilinear form $\mathcal{A}_h(\cdot,\cdot)$, cf. (A2), we find

$$\begin{aligned} \mathbf{v}^T A \mathbf{v} &= \sum_{i,j=1}^{N} v_i \mathcal{A}_h(\phi_i, \phi_j) v_j = \mathcal{A}_h\Big(\sum_{i=1}^{N} v_i \phi_i, \sum_{j=1}^{N} v_j \phi_j\Big) \\ &= \mathcal{A}_h(v_h, v_h) \ge \underline{c}\, \|v_h\|_{\mathcal{V}}^2 > 0 \quad \forall \mathbf{v} \ne \mathbf{0}; \end{aligned} \tag{1.35}$$

Moreover, the symmetry of $\mathcal{A}_h(\cdot,\cdot)$ obviously implies the symmetry of A, which together with (1.35) shows that A is SPD in the present context.

If $h := \max_{T\in\mathcal{T}_h} h_T$ is the mesh size of the triangulation $\mathcal{T}_h$, and the mesh is assumed to be quasi-uniform, i.e., for all $T \in \mathcal{T}_h$ the conditions

$$h_T \geq \delta_1 h, \tag{1.36}$$

$$\frac{\rho_T}{h_T} \geq \delta_2, \tag{1.37}$$

are satisfied, then the spectral condition number of the stiffness matrix arising from a second-order elliptic boundary value problem can be estimated by

$$\kappa(A) = \frac{\lambda_{\max}(A)}{\lambda_{\min}(A)} = \mathcal{O}(h^{-2}). \tag{1.38}$$

This can easily be seen from the following estimates, see, e.g., [68]: There are constants c_1, c_2, c_3 only depending on δ_1 and δ_2 in (1.36) and (1.37) but not on h such that for all $v_h = \sum_{i=1}^{N} v_i \phi_i \in \mathcal{V}_h$ we have

$$c_1 h^2 \|\mathbf{v}\|^2 \leq \|v_h\|^2_{L_2(\Omega)} \leq c_2 h^2 \|\mathbf{v}\|^2, \tag{1.39}$$

$$\mathcal{A}_h(v_h, v_h) \leq c_3 h^{-2} \|v_h\|^2_{L_2(\Omega)}. \tag{1.40}$$

Remark 1.8. The inequality (1.40) is a so-called inverse estimate: In case of the elliptic model problem this gives rise to bound the L_2-norm of the gradient of v_h by the L_2-norm of v_h itself, which is possible for functions $v_h \in \mathcal{V}_h$ at the price of a factor h^{-1} in the standard setting.[3]

Now, since for an arbitrary vector $\mathbf{v} \in \mathbb{R}^N$, $\mathbf{v} \neq \mathbf{0}$, there exists a uniquely determined function $v_h \in \mathcal{V}_h$ such that $\mathbf{v}^T A\mathbf{v} = \mathcal{A}_h(v_h, v_h)$, we deduce from (1.39) and (1.40)

$$\frac{\mathbf{v}^T A\mathbf{v}}{\|\mathbf{v}\|^2} = \frac{\mathcal{A}_h(v_h, v_h)}{\|\mathbf{v}\|^2} \leq c_3 h^{-2} \frac{\|v_h\|^2_{L_2(\Omega)}}{\|\mathbf{v}\|^2} \leq c_2 c_3, \tag{1.41}$$

and, further using the $\mathcal{V}$-ellipticity of $\mathcal{A}_h(\cdot,\cdot)$

$$\frac{\mathbf{v}^T A\mathbf{v}}{\|\mathbf{v}\|^2} = \frac{\mathcal{A}_h(v_h, v_h)}{\|\mathbf{v}\|^2} \geq \underline{c} \frac{\|v_h\|^2_{\mathcal{V}}}{\|\mathbf{v}\|^2} \geq \underline{c} \frac{\|v_h\|^2_{L_2(\Omega)}}{\|\mathbf{v}\|^2} \geq \underline{c}\, c_1 h^2. \tag{1.42}$$

Thence, in view of (1.41) and (1.42), there exist constants c and C that are independent of h such that

$$\lambda_{\max}(A) \leq C, \quad \lambda_{\min}(A) \geq c\, h^2,$$

which shows that $\kappa(A) \leq \frac{C}{c} h^{-2}$.

[3]For a proof of estimates like (1.39) and (1.40) we refer the reader to [47, 68].

1.2 Ill-conditioned problems

In this section we summarize certain classes of boundary value problems that will be subject to a more detailed elaboration later in this book. The reason for including this short overview here is to touch also on some frequently met difficulties regarding the solution of the arising systems of linear algebraic equation, i.e., to discuss some computational issues that typically result in ill-conditioned problems.

1.2.1 Anisotropic problems

Let us comment first on the model boundary value problem for one simple scalar second-order elliptic partial differential equation, i.e., problem (1.1) where the coefficient $\mathbf{a}(\boldsymbol{x})$ is a $d \times d$ SPD matrix.

Then, depending on the specification of the matrix $\mathbf{a}(\boldsymbol{x})$ any standard (conforming or nonconforming) finite element method can easily result in a highly ill-conditioned problem even when using a quasi-uniform mesh. One such example, where one typically meets an additional increase of the condition number of the stiffness matrix A, is the case in which the matrix $\mathbf{a}(\boldsymbol{x})$ is ill-conditioned. For instance, if

$$\mathbf{a}(\boldsymbol{x}) = \begin{bmatrix} 1 & 0 \\ 0 & \epsilon \end{bmatrix},$$

that is, considering the simplest anisotropic (orthotropic) problem, the condition number of A in general will be of order $\epsilon^{-1}h^{-2}$, which for extremely strong anisotropy, i.e., for $\epsilon \to 0$, results in highly ill-conditioned problems. This phenomenon we call *coefficient anisotropy*.

Another similar effect is caused by elements with very large aspect ratio. In case of such meshes that contain extremely stretched elements the aspect ratio typically appears as a factor in the condition number as well. This phenomenon of *mesh anisotropy* is therefore another source of an increase of the condition number, which of course in general interferes with coefficient anisotropy if present. In fact, for piecewise linear elements it is not difficult to show (see Section 3.1) that coefficient anisotropy can be described in terms of mesh anisotropy and vice versa such that it suffices to study either of these two phenomena.

1.2.2 Problems with highly varying coefficients

In the setting of the scalar elliptic problem (1.1) we will study another configuration that badly affects the condition number of A. This difficulty is caused by a *rough* coefficient $\mathbf{a}(\boldsymbol{x})$, i.e., the case in which the entries of $\mathbf{a}(\boldsymbol{x})$ are no longer smooth functions over the whole domain but are of *high variation*. In particular, we will be concerned with discontinuous coefficients where the (largest) jump discontinuity typically is present as a factor in the condition number of the stiffness matrix.

In the (so far) well established theory of robust multilevel methods a standard assumption is that the coefficient variation can be resolved on the coarsest mesh partition in the sense that the coefficient may have arbitrary large jumps (jump discontinuities) between coarse elements but varies mildly on each element of the coarse(st) mesh. Most of the robustness results that will be stated in this book cover exactly this case. Beyond that we will also present some very recent developments addressing problems with *high-frequency-high-contrast coefficients*. Robust multilevel methods for discontinuous Galerkin discretizations of elliptic problems of this kind will be discussed in Chapter 8.

1.2.3 Elastic deformation of almost incompressible materials

Another class of ill-conditioned problems we will target at is related to the elastic deformation of almost incompressible materials. Let Ω be a bounded open subset of $\mathbb{R}^d$, $d = 2$ or $d = 3$, which is associated with the reference configuration of an elastic body. It is well known from linear elasticity theory (see, e.g., [35]) that the governing equations describing the deformation of the body under the influence of applied forces (taking into account only first order terms in the displacement $\boldsymbol{u}$) are given by

$$-\operatorname{div} \boldsymbol{\sigma} = \boldsymbol{f} \quad \text{in } \Omega, \tag{1.43a}$$

$$\boldsymbol{u} = \boldsymbol{0} \quad \text{on } \Gamma_D, \tag{1.43b}$$

$$\sum_{j=1}^{d} \sigma_{ij} n_j = g_i \quad \text{on } \Gamma_N, \quad 1 \le i \le d, \tag{1.43c}$$

where $\boldsymbol{\sigma}$ denotes the stress tensor, $\boldsymbol{f}$ the body force, $\boldsymbol{u}$ the displacement field, $\boldsymbol{n}$ is the outwards pointing unit normal vector and $\boldsymbol{g}$ is a surface traction on the part Γ_N of the boundary $\Gamma = \overline{\Gamma}_D \cup \overline{\Gamma}_N$. Moreover, Γ_D denotes the part of the boundary on which the displacement is given. Writing stress and strain in vector form, i.e., $\boldsymbol{\sigma} = (\sigma_{11}, \sigma_{22}, \sigma_{12})^T$, $\boldsymbol{\varepsilon} = (\varepsilon_{11}, \varepsilon_{22}, \varepsilon_{12})^T$ in the 2D model, and $\boldsymbol{\sigma} = (\sigma_{11}, \sigma_{22}, \sigma_{33}, \sigma_{12}, \sigma_{13}, \sigma_{23})^T$, $\boldsymbol{\varepsilon} = (\varepsilon_{11}, \varepsilon_{22}, \varepsilon_{33}, \varepsilon_{12}, \varepsilon_{13}, \varepsilon_{23})^T$ in the 3D model, the stress-strain relation (for St. Venant–Kirchhoff materials) is given by Hooke's law, i.e.,

$$\boldsymbol{\sigma} = C \cdot \boldsymbol{\varepsilon},$$

where

$$C := \frac{E}{(1+\nu)(1-2\nu)} \begin{bmatrix} 1-\nu & \nu & 0 \\ \nu & 1-\nu & 0 \\ 0 & 0 & 1-2\nu \end{bmatrix},$$

for two-dimensional problems, and

$$C := \frac{E}{(1+\nu)(1-2\nu)} \begin{bmatrix} 1-\nu & \nu & \nu & 0 & 0 & 0 \\ \nu & 1-\nu & \nu & 0 & 0 & 0 \\ \nu & \nu & 1-\nu & 0 & 0 & 0 \\ 0 & 0 & 0 & 1-2\nu & 0 & 0 \\ 0 & 0 & 0 & 0 & 1-2\nu & 0 \\ 0 & 0 & 0 & 0 & 0 & 1-2\nu \end{bmatrix}$$

in three space dimensions. Here E denotes Young's modulus of elasticity and ν is the Poisson ratio.

Introducing the Lamé coefficients

$$\lambda = \frac{\nu E}{(1+\nu)(1-2\nu)}, \qquad \mu = \frac{E}{2(1+\nu)},$$

and the symmetric gradient $\nabla^{(s)}\boldsymbol{u} := \boldsymbol{\varepsilon} = \boldsymbol{\varepsilon}(\boldsymbol{u})$,

$$\varepsilon_{ij} = \frac{1}{2}\left(\frac{\partial u_i}{\partial x_j} + \frac{\partial u_j}{\partial x_i}\right),$$

equation (1.43a) yields the classical Lamé differential equation

$$-2\,\mu \operatorname{div} \nabla^{(s)}\boldsymbol{u} - \lambda \operatorname{grad} \operatorname{div} \boldsymbol{u} = \boldsymbol{f} \tag{1.44}$$

for the displacements u_i, $1 \le i \le d$.

When the Poisson ratio ν tends to $1/2$ the material becomes incompressible and in the limiting case when $\nu = \frac{1}{2}$ the boundary value problem is ill-posed. We will pay special attention to the case when ν is very close to the incompressible limit and study the robustness of the presented multilevel methods with respect to the Poisson ratio.

1.2.4 High-Reynolds-number flow

A second class of vector field problems that will be subject to our considerations in Chapter 9 is related to the Dirichlet initial-boundary value problem for the Navier–Stokes equations

$$\frac{\partial \boldsymbol{u}}{\partial t} + (\boldsymbol{u}\cdot\nabla)\boldsymbol{u} = -\nabla p + \frac{1}{Re}\nabla^2\boldsymbol{u} + \boldsymbol{f} \quad (\boldsymbol{x},t) \in \Omega \times (0,T) \tag{1.45a}$$

$$\nabla \cdot \boldsymbol{u} = 0 \quad (\boldsymbol{x},t) \in \Omega \times (0,T) \tag{1.45b}$$

$$\boldsymbol{u} = \boldsymbol{0} \quad (\boldsymbol{x},t) \in \Gamma \times (0,T) \tag{1.45c}$$

$$\boldsymbol{u} = \boldsymbol{0} \quad (\boldsymbol{x},t) \in \Omega \times \{0\} \tag{1.45d}$$

where Ω is a bounded and connected domain in $\mathbb{R}^d$, and $\Gamma = \partial\Omega$. The linearized form of equation (1.45a)

$$\frac{\partial \boldsymbol{u}}{\partial t} - \frac{1}{Re}\nabla^2 \boldsymbol{u} + (\boldsymbol{w}\cdot\nabla)\boldsymbol{u} + \nabla p = \boldsymbol{f}$$

is called the Oseen equation. Here $\boldsymbol{u}$ denotes the (unknown) velocity field, $\boldsymbol{w}$ is an (old) approximation of the velocity, p denotes the pressure, $\boldsymbol{f}$ some outer force, and Re the dimensionless Reynolds number. We assume also that Ω is such that the H^2-regularity property holds for the steady Stokes problem

$$-\frac{1}{Re}\nabla^2 \boldsymbol{u} + \nabla p = \boldsymbol{f} \tag{1.46a}$$

$$\nabla\cdot\boldsymbol{u} = 0 \quad \text{in } \Omega \tag{1.46b}$$

$$\boldsymbol{u} = \boldsymbol{0} \quad \text{on } \Gamma. \tag{1.46c}$$

Employing a stable time discretization and a mixed finite element discretization in space the Oseen problem (arising from fixed-point linearization of the Navier–Stokes problem) results in a saddle-point problem of the form

$$\begin{bmatrix} A(\mathbf{w}_h) & B^T \\ B & -C \end{bmatrix}\begin{bmatrix} \mathbf{u}_h \\ \mathbf{p}_h \end{bmatrix} = \begin{bmatrix} \mathbf{f}_h \\ \mathbf{g}_h \end{bmatrix}.$$

Note that the condition number of the coupled matrix depends on a problem parameter, namely the Reynolds number. High Reynolds numbers result in very ill-conditioned problems. As it will be shown in Chapter 9, this indefinite problem can be decoupled into two SPD systems by means of certain projection schemes and a proper choice of the underlying FE spaces. Then the robustness of the numerical method is achieved by applying robust linear solvers to the decoupled SPD systems.

1.3 Preliminaries on iterative methods

1.3.1 Stationary methods

Starting with an initial guess $\mathbf{x}_{(0)}$ for the exact solution, an iterative method for solving (1.32), which we denote as

$$A\mathbf{x} = \mathbf{b}, \tag{1.47}$$

here, is defined by a sequence of functions $(\psi_k)_{k\geq 1}$ generating a sequence of approximations $(\mathbf{x}_{(k)})_{k\geq 1}$. That is,

$$\mathbf{x}_{(k)} = \psi_k(\mathbf{x}_{(0)}, \mathbf{x}_{(1)}, \ldots, \mathbf{x}_{(k-1)}; A, \mathbf{b}) \quad \text{for} \quad k \geq 1. \tag{1.48}$$

Definition 1.9 ([5, 116]).[4] If for some integer $n > 0$, ψ_k is independent of k for all $k \geq n$, then the iterative method (1.48) is said to be stationary. Otherwise it is nonstationary. If for each k, ψ_k is a linear function of $\mathbf{x}_{(0)}, \mathbf{x}_{(1)}, \dots, \mathbf{x}_{(k-1)}$, then the method is said to be linear. Otherwise it is nonlinear. A stationary method of the form (1.48) is called of order s, $s \leq n$, if for $k \geq n$ the approximation $\mathbf{x}_{(k)}$ depends on $\mathbf{x}_{(k-s)}, \mathbf{x}_{(k-s+1)}, \dots, \mathbf{x}_{(k-1)}$ but not on $\mathbf{x}_{(l)}$ for $l < k - s$.

In case of a linear stationary iterative method of first order (1.48) can be represented as

$$\mathbf{x}_{(k)} = G\mathbf{x}_{(k-1)} + \mathbf{d}, \quad k = 1, 2, 3, \dots, \tag{1.49}$$

with an iteration matrix G and a constant vector $\mathbf{d}$.

Classical iterative methods are based on a splitting of A,

$$A = C - R, \tag{1.50}$$

where the $N \times N$ matrix C, which is also called the preconditioner (or the splitting matrix), is assumed to be nonsingular.

The basic iteration then is given by

$$\mathbf{x}_{(k)} = \mathbf{x}_{(k-1)} + C^{-1}\left(\mathbf{b} - A\mathbf{x}_{(k-1)}\right), \quad k = 1, 2, 3, \dots, \tag{1.51}$$

or, equivalently,

$$C\mathbf{x}_{(k)} = R\mathbf{x}_{(k-1)} + \mathbf{b}, \quad k = 1, 2, 3, \dots. \tag{1.52}$$

Remark 1.10. If $A = L + D + U$, where L is the strictly lower and U the strictly upper triangular part of A, $D = \operatorname{diag}(A)$ the diagonal part, and $T = \operatorname{tridiag}(A)$ the tridiagonal part of A, then some popular schemes can be represented by: $C = I$ (Richardson), $C = D$ (Jacobi), $C = D + L$ (Gauss–Seidel), $C = T$ (line Jacobi) or $C = \frac{1}{\omega}(D + \omega L)$, $\omega \neq 0$ (SOR).

Let $\mathbf{e}_{(k)} = \mathbf{x}_{(k)} - \mathbf{x}$ denote the error of the k-th vector iterate $\mathbf{x}_{(k)}$ and $\mathbf{r}_{(k)} = \mathbf{b} - A\mathbf{x}_{(k)}$ the k-th residual. Then we have

$$\begin{aligned} \mathbf{e}_{(k)} &= \mathbf{x}_{(k)} - \mathbf{x} \\ &= \mathbf{x}_{(k-1)} + C^{-1}\left(\mathbf{b} - A\mathbf{x}_{(k-1)}\right) - \mathbf{x} - C^{-1}\left(\mathbf{b} - A\mathbf{x}\right) \\ &= \left(I - C^{-1}A\right)\mathbf{e}_{(k-1)} = \left(I - C^{-1}A\right)^k \mathbf{e}_{(0)} \end{aligned} \tag{1.53}$$

[4] In some books a method of order s (in our terminology) is referred to as a method of degree s, see, e.g., [116].

and

$$\begin{aligned}\mathbf{r}_{(k)} &= \mathbf{b} - A\mathbf{x}_{(k)} = \mathbf{b} - A\left(\mathbf{x}_{(k-1)} + C^{-1}\left(\mathbf{b} - A\mathbf{x}_{(k-1)}\right)\right) \\ &= \left(I - AC^{-1}\right)\mathbf{r}_{(k-1)} = \left(I - AC^{-1}\right)^k \mathbf{r}_{(0)}. \end{aligned} \tag{1.54}$$

We may guess from (1.53) and (1.54) that the convergence behavior of the iteration (1.51) will depend on the approximation properties of the preconditioner, i.e., the better the matrices AC^{-1} and $C^{-1}A$ resemble the identity matrix I, the faster the method will converge. The following necessary and sufficient condition for convergence in terms of the spectral radius ρ is a classical result. A proof can be found in reference [5].

Theorem 1.11 ([5]). *The sequence of vectors $(\mathbf{x}_{(k)})_{k\geq 1}$ in (1.52) converges to the solution of (1.47) for any $\mathbf{x}_{(0)}$ if and only if $\rho(C^{-1}R) = \rho(I - C^{-1}A) < 1$.*

1.3.2 Polynomial acceleration

The basic iterative method (1.49) can be accelerated by generating a new sequence of iterates $(\tilde{\mathbf{x}}_{(n)})_{n\geq 0}$, using proper linear combinations of the first $(n+1)$ basic iterates $\mathbf{x}_{(0)}, \mathbf{x}_{(1)}, \ldots, \mathbf{x}_{(n)}$, i.e.,

$$\tilde{\mathbf{x}}_{(n)} = \sum_{k=0}^{n} \tau_{n,k}\mathbf{x}_{(k)}, \quad n = 0, 1, 2, \ldots. \tag{1.55}$$

Then, if one imposes the condition

$$\sum_{k=0}^{n} \tau_{n,k} = 1, \ n \geq 0 \tag{1.56}$$

on the coefficients $\tau_{n,k}$, the accelerated iteration is regular in the sense that $\tilde{\mathbf{x}}_{(n)} = \mathbf{x}$ for all $n \geq 0$ whenever the initial guess $\mathbf{x}_{(0)}$ equals the exact solution $\mathbf{x}$ of (1.47).

Denoting the error of $\tilde{\mathbf{x}}_{(n)}$ by $\tilde{\mathbf{e}}_{(n)} = \tilde{\mathbf{x}}_{(n)} - \mathbf{x}$, we obtain from (1.53), (1.55) and (1.56) the identity

$$\begin{aligned}\tilde{\mathbf{e}}_{(n)} &= \sum_{k=0}^{n} \tau_{n,k}\mathbf{x}_{(k)} - \mathbf{x} = \sum_{k=0}^{n} \tau_{n,k}\left(\mathbf{x}_{(k)} - \mathbf{x}\right) \\ &= \sum_{k=0}^{n} \tau_{n,k}\mathbf{e}_{(k)} = \sum_{k=0}^{n} \tau_{n,k}G^k\mathbf{e}_{(0)} \\ &= \sum_{k=0}^{n} \tau_{n,k}G^k\tilde{\mathbf{e}}_{(0)} = Q_n(G)\tilde{\mathbf{e}}_{(0)}, \end{aligned} \tag{1.57}$$

where $Q_n(G)$ is the matrix polynomial associated with the algebraic polynomial

$$Q_n(x) = \tau_{n,0} + \tau_{n,1}x + \tau_{n,2}x^2 + \ldots + \tau_{n,n}x^n. \tag{1.58}$$

Obviously, (1.56) is equivalent to

$$Q_n(1) = 1. \tag{1.59}$$

Definition 1.12. We call any combined algorithm that produces a sequence of approximations $(\tilde{\mathbf{x}}_{(n)})_{n\geq 0}$ a Krylov subspace method or polynomial method if the n-th error vector $\tilde{\mathbf{e}}_{(n)}$ can be written as in (1.57).

Sometimes this procedure is also referred to as a semi-iterative method with respect to the basic iteration (1.49), see, e.g., [110].

It follows from (1.56)–(1.57) that the errors of the accelerated iterates (for which from now on we skip again the tilde symbol for the sake of convenience), satisfy

$$\mathbf{e}_{(n)} = Q_n(G)\mathbf{e}_{(0)} = Q_n(I - C^{-1}A)\mathbf{e}_{(0)} = P_n(C^{-1}A)\mathbf{e}_{(0)}, \tag{1.60}$$

where the associated algebraic polynomials $Q_n(x)$ and $P_n(x) = Q_n(1-x)$ have the representation

$$\begin{aligned} Q_n(x) &= \prod_{k=1}^{n}(\alpha_k x + 1 - \alpha_k), \\ P_n(x) &= \prod_{k=1}^{n}(1 - \alpha_k x). \end{aligned} \tag{1.61}$$

Hence, the regularity condition (1.59) for $P_n(x)$, which is $P_n(0) = 1$, is fulfilled.

In order to make the accelerated scheme practicable, we aim at minimizing the (virtual) spectral radius of $P_n(C^{-1}A)$. In other words, we look for a polynomial $P_n(x)$ of the form (1.61) that minimizes

$$\max_{\lambda_1 \leq x \leq \lambda_N} |P_n(x)|. \tag{1.62}$$

The classical solution to this problem is given in Theorem 1.13 (see, e.g., [5, 110, 116]).

Theorem 1.13. *Let Π_n^1 denote the set of polynomials of degree n that take the value 1 at the origin. For $0 < a < b$, the transformed Chebyshev polynomial $\tilde{P}_n(x)$ of degree $n \geq 0$ associated with the interval $[a,b]$ is defined by*

$$\tilde{P}_n(x) = \frac{T_n\left(\frac{2x-(b+a)}{b-a}\right)}{T_n\left(\frac{-(b+a)}{b-a}\right)}. \tag{1.63}$$

Herein $T_n(z)$ is the n-th Chebyshev polynomial of the first kind, which can be obtained via the recursion (*see, e.g.,* [1]):

$$\begin{aligned} T_0(z) &= 1, \\ T_1(z) &= z, \\ T_{n+1}(z) &= 2zT_n(z) - T_{n-1}(z), \quad n = 1, 2, 3, \ldots, \quad z \in \mathbb{R}. \end{aligned} \tag{1.64}$$

Under these assumptions, we have

$$\max_{a \le x \le b} \left| T_n\left(\frac{2x - (b+a)}{b-a}\right)\right| = 1, \tag{1.65}$$

$$\max_{a \le x \le b} |\tilde{P}_n(x)| = \min_{P_n \in \Pi_n^1} \max_{a \le x \le b} |P_n(x)|, \tag{1.66}$$

$$\tilde{P}_n(x) = \prod_{k=1}^{n} (1 - \alpha_k x) \quad \textit{for all } x \in [a, b], \tag{1.67}$$

$$\begin{aligned} \textit{where} \quad \alpha_k &= \frac{2}{(b-a)\cos\theta_k + (b+a)} \\ \textit{and} \quad \theta_k &= \frac{2(k+1)\pi}{2n}. \end{aligned} \tag{1.68}$$

Proof. Using the trigonometric identity

$$\cos((k+1)\phi) = 2\cos\phi\cos(k\phi) - \cos((k-1)\phi), \quad -\pi \le \phi \le \pi,$$

we deduce

$$T_n(\cos\phi) = \cos(n\phi), \quad -\pi \le \phi \le \pi.$$

Therefore,

$$\begin{aligned} &\max_{a \le x \le b} \left| T_n\left(\frac{2x - (b+a)}{b-a}\right)\right| \\ &= \max_{-1 \le z \le 1} |T_n(z)| = \max_{-\pi \le \phi \le \pi} |T_n(\cos\phi)| = \max_{-\pi \le \phi \le \pi} |\cos(n\phi)| = 1. \end{aligned}$$

In order to prove (1.66), we assume that there exists a polynomial $P_n \in \Pi_n^1$ that satisfies

$$\max_{a \le x \le b} |P_n(x)| < \max_{a \le x \le b} |\tilde{P}_n(x)|.$$

Consider now

$$R_n(x) = T_n\left(\frac{2x - (b+a)}{b-a}\right) - P_n(x) T_n\left(\frac{-(b+a)}{b-a}\right),$$

which is a polynomial of degree n. Since

$$\max_{a \le x \le b} \left| P_n(x) T_n \left(\frac{-(b+a)}{b-a} \right) \right| < \max_{a \le x \le b} |\tilde{P}_n(x)| \left| T_n \left(\frac{-(b+a)}{b-a} \right) \right| = 1$$

and

$$T_n \left(\frac{2x_i - (b+a)}{b-a} \right) = (-1)^i$$

for

$$x_i = \frac{(b-a)\cos\left(\frac{i\pi}{n}\right) + (b+a)}{2}, \qquad i = 0, 1, 2, \dots, n,$$

$R_n(x)$ changes the sign in each interval (x_i, x_{i+1}). Thus, additionally to the root $x = 0$, $R_n(x)$ has n roots, which is in contradiction to its degree n.

Finally, comparing the roots of $\prod_{k=1}^{n}(1 - \alpha_k x)$ with those of $\tilde{P}_n(x)$, and taking notice of $\tilde{P}_n(0) = 1$, it follows that the two polynomials are identical if

$$\frac{1}{\alpha_k} = \frac{(b-a)\theta_k + (b+a)}{2},$$

proving (1.67)–(1.68). □

In the next section we will discuss one particular Krylov subspace method, namely the method of conjugate gradients (CG), and some of its modifications related to the inclusion of preconditioning techniques.

1.4 Conjugate gradients (CG)

1.4.1 From steepest descent to conjugate gradients

As a starting point, we note that solving the linear system (1.47) and minimizing the functional ϕ, defined by

$$\phi(\mathbf{x}) = \frac{1}{2}\mathbf{x}^T A \mathbf{x} - \mathbf{x}^T \mathbf{b}, \tag{1.69}$$

are equivalent problems if the matrix A is symmetric and positive definite (SPD). Under this assumption, $\phi(\mathbf{x})$ has the uniquely determined minimum $-\frac{1}{2}\mathbf{b}^T A^{-1} \mathbf{b}$, attained when $\mathbf{x} = A^{-1}\mathbf{b}$.

A simple iterative minimization procedure for the quadratic functional ϕ is the following one: Given the current approximation $\mathbf{x}_{(n-1)}$, we calculate

$$\mathbf{r}_{(n-1)} = \mathbf{b} - A\mathbf{x}_{(n-1)} = -\nabla\phi(\mathbf{x}_{(n-1)}) \tag{1.70}$$

in order to minimize $\phi(\mathbf{x}_{(n-1)} + \alpha_n \mathbf{r}_{(n-1)})$ with respect to α_n. Following these lines, we deduce

$$\alpha_n = \frac{\langle \mathbf{r}_{(n-1)}, \mathbf{r}_{(n-1)} \rangle}{\langle A\mathbf{r}_{(n-1)}, \mathbf{r}_{(n-1)} \rangle}. \tag{1.71}$$

The vector $\mathbf{x}_{(n-1)}$ is then actualized according to

$$\mathbf{x}_{(n)} = \mathbf{x}_{(n-1)} + \alpha_n \mathbf{r}_{(n-1)}, \quad n = 1, 2, 3, \ldots. \tag{1.72}$$

Because $\mathbf{r}_{(n-1)}$ is the negative gradient of ϕ at $\mathbf{x}_{(n-1)}$, the iteration (1.71)–(1.72) is referred to as the method of steepest descent. From

$$\begin{aligned}
\|\mathbf{e}_{(n)}\|_A^2 &= \|\mathbf{r}_{(n)}\|_{A^{-1}}^2 = \|\mathbf{r}_{(n-1)} - \alpha_n A\mathbf{r}_{(n-1)}\|_{A^{-1}}^2 \\
&= \langle A^{-1}\mathbf{r}_{(n-1)} - \alpha_n \mathbf{r}_{(n-1)}, \mathbf{r}_{(n-1)} - \alpha_n A\mathbf{r}_{(n-1)} \rangle \\
&= \langle A^{-1}\mathbf{r}_{(n-1)}, \mathbf{r}_{(n-1)} \rangle - 2\alpha_n \langle \mathbf{r}_{(n-1)}, \mathbf{r}_{(n-1)} \rangle + \alpha_n^2 \langle \mathbf{r}_{(n-1)}, A\mathbf{r}_{(n-1)} \rangle \\
&= \langle A^{-1}\mathbf{r}_{(n-1)}, \mathbf{r}_{(n-1)} \rangle - \frac{\langle \mathbf{r}_{(n-1)}, \mathbf{r}_{(n-1)} \rangle^2}{\langle \mathbf{r}_{(n-1)}, A\mathbf{r}_{(n-1)} \rangle} \\
&= \|\mathbf{r}_{(n-1)}\|_{A^{-1}}^2 \left(1 - \frac{\langle \mathbf{r}_{(n-1)}, \mathbf{r}_{(n-1)} \rangle^2}{\langle A^{-1}\mathbf{r}_{(n-1)}, \mathbf{r}_{(n-1)} \rangle \langle \mathbf{r}_{(n-1)}, A\mathbf{r}_{(n-1)} \rangle} \right) \\
&\le \|\mathbf{r}_{(n-1)}\|_{A^{-1}}^2 \left(1 - \frac{1}{\frac{\lambda_N(A)}{\lambda_1(A)}} \right) = \|\mathbf{e}_{(n-1)}\|_A^2 \left(1 - \frac{1}{\kappa(A)} \right)
\end{aligned} \tag{1.73}$$

we conclude its global convergence.

The bound (1.73) can be improved by expanding the residuals into linear combinations of the eigenvectors $\{_i : 1 \le i \le N\}$ of A. Taking into account

$$\begin{aligned}
&\frac{\langle \mathbf{r}_{(n-1)}, \mathbf{r}_{(n-1)} \rangle^2}{\langle A^{-1}\mathbf{r}_{(n-1)}, \mathbf{r}_{(n-1)} \rangle \langle \mathbf{r}_{(n-1)}, A\mathbf{r}_{(n-1)} \rangle} \\
&\ge \min_{\mathbf{c} \neq \mathbf{0}} \frac{\langle \sum_i c_i{}_i, \sum_i c_i{}_i \rangle^2}{\langle A^{-1} \sum_i c_i{}_i, \sum_i c_i{}_i \rangle \langle \sum_i c_i{}_i, A \sum_i c_i{}_i \rangle} = \min_{\tilde{\mathbf{c}} \neq \mathbf{0}} \frac{\sum_i \tilde{c}_i \sum_i \tilde{c}_i}{\sum_i \frac{\tilde{c}_i}{\lambda_i} \sum_i \tilde{c}_i \lambda_i}
\end{aligned}$$

and using Lagrange's method to calculate $\max_{\tilde{\mathbf{c}} \neq \mathbf{0}} \sum_i \frac{\tilde{c}_i}{\lambda_i} \sum_i \tilde{c}_i \lambda_i$, subject to the constraint $\sum_i \tilde{c}_i = 1$, the following sharp bound can be established:

$$\|\mathbf{e}_{(n)}\|_A^2 \le \|\mathbf{e}_{(n-1)}\|_A^2 \left(1 - \frac{4}{2 + \kappa(A) + 1/\kappa(A)} \right). \tag{1.74}$$

Unfortunately, the steepest descent method makes slow progress, especially if the gradients occurring in (1.72) have very similar directions. A way to overcome this difficulty is to demand the search directions to be mutually orthogonal with respect to the (energy) inner product $\langle\cdot,\cdot\rangle_A$, that is,

$$\langle \mathbf{p}_{(i)}, \mathbf{p}_{(j)}\rangle_A = \langle \mathbf{p}_{(i)}, A\mathbf{p}_{(j)}\rangle = \mathbf{p}_{(i)}^T A\mathbf{p}_{(j)} = 0 \quad \text{for} \quad i \neq j. \tag{1.75}$$

The conjugate gradient algorithm, stated below, has this essential feature.

Algorithm 1.14 (Conjugate Gradients, see, e.g., [63]).

$n = 0;\quad \mathbf{r}_{(0)} = \mathbf{b} - A\mathbf{x}_{(0)};$
while (termination criterion is false) **do**
  $n = n + 1$
  if $(n = 1)$ **then**

$$\mathbf{p}_{(1)} = \mathbf{r}_{(0)} \tag{1.76}$$

  else

$$\beta_n = -\frac{\langle \mathbf{r}_{(n-1)}, \mathbf{p}_{(n-1)}\rangle_A}{\langle \mathbf{p}_{(n-1)}, \mathbf{p}_{(n-1)}\rangle_A} \tag{1.77}$$

$$\mathbf{p}_{(n)} = \mathbf{r}_{(n-1)} + \beta_n \mathbf{p}_{(n-1)} \tag{1.78}$$

  end

$$\alpha_n = \frac{\langle \mathbf{r}_{(n-1)}, \mathbf{r}_{(n-1)}\rangle}{\langle \mathbf{p}_{(n)}, \mathbf{p}_{(n)}\rangle_A} \tag{1.79}$$

$$\mathbf{x}_{(n)} = \mathbf{x}_{(n-1)} + \alpha_n \mathbf{p}_{(n)} \tag{1.80}$$

$$\mathbf{r}_{(n)} = \mathbf{b} - A\mathbf{x}_{(n)} \tag{1.81}$$

end

1.4.2 Convergence analysis of the CG method

Let us start the analysis of the CG method with proving the following fundamental lemma characterizing the Krylov subspaces induced by Algorithm 1.14.

Lemma 1.15. *After n iterations of the conjugate gradient method we have*

$$\begin{aligned} \operatorname{span}\{\mathbf{p}_{(1)}, \mathbf{p}_{(2)}, \dots, \mathbf{p}_{(n)}\} &= \operatorname{span}\{\mathbf{r}_{(0)}, \mathbf{r}_{(1)}, \dots, \mathbf{r}_{(n-1)}\} && (1.82)\\ &= \operatorname{span}\{\mathbf{r}_{(0)}, A\mathbf{r}_{(0)}, \dots, A^{n-1}\mathbf{r}_{(0)}\} \\ \langle \mathbf{r}_{(i)}, \mathbf{r}_{(j)}\rangle &= 0 \quad \textit{for} \quad i \neq j, \quad 0 \le i, j \le n-1, && (1.83)\\ \langle \mathbf{p}_{(i)}, \mathbf{p}_{(j)}\rangle_A &= 0 \quad \textit{for} \quad i \neq j, \quad 1 \le i, j \le n. && (1.84) \end{aligned}$$

Proof. The proof is based on mathematical induction. Using the fact $\mathbf{p}_{(1)} = \mathbf{r}_{(0)}$, one can easily see that the lemma holds for $n = 1$ and $n = 2$. Suppose now, that all three statements (1.82)–(1.84) are true for $n = k$, $k \geq 2$.

In particular,

$$\mathbf{p}_{(k)} \in \operatorname{span}\{\mathbf{r}_{(0)}, A\mathbf{r}_{(0)}, \ldots, A^{k-1}\mathbf{r}_{(0)}\}.$$

Writing $\mathbf{r}_{(k)}$ in the form

$$\mathbf{r}_{(k)} = \mathbf{b} - A\mathbf{x}_{(k)} = \mathbf{b} - A\mathbf{x}_{(k-1)} - \alpha_k A\mathbf{p}_{(k)} = \mathbf{r}_{(k-1)} - \alpha_k A\mathbf{p}_{(k)} \tag{1.85}$$

yields

$$\mathbf{r}_{(k)} \in \operatorname{span}\{\mathbf{r}_{(0)}, A\mathbf{r}_{(0)}, \ldots, A^{k}\mathbf{r}_{(0)}\}.$$

On the other hand, the induction hypothesis implies

$$A^{k-1}\mathbf{r}_{(0)} \in \operatorname{span}\{\mathbf{p}_{(1)}, \mathbf{p}_{(2)}, \ldots, \mathbf{p}_{(k)}\},$$

which together with (1.85) gives

$$A^{k}\mathbf{r}_{(0)} \in \operatorname{span}\{\mathbf{r}_{(0)}, \mathbf{r}_{(1)}, \ldots, \mathbf{r}_{(k)}\}.$$

Hence, $\operatorname{span}\{\mathbf{r}_{(0)}, \mathbf{r}_{(1)}, \ldots, \mathbf{r}_{(k)}\} = \operatorname{span}\{\mathbf{r}_{(0)}, A\mathbf{r}_{(0)}, \ldots, A^{k}\mathbf{r}_{(0)}\}$. From (1.78) we conclude in a similar way

$$\operatorname{span}\{\mathbf{p}_{(1)}, \mathbf{p}_{(2)}, \ldots, \mathbf{p}_{(k+1)}\} = \operatorname{span}\{\mathbf{r}_{(0)}, \mathbf{r}_{(1)}, \ldots, \mathbf{r}_{(k)}\},$$

completing the induction step for (1.82).

Next we prove $\langle \mathbf{r}_{(k)}, \mathbf{r}_{(j)} \rangle = 0$ for $j < k$. For $j = k-1$ we find

$$\begin{aligned}
\langle \mathbf{r}_{(k)}, \mathbf{r}_{(k-1)} \rangle &= \langle \mathbf{r}_{(k-1)}, \mathbf{r}_{(k-1)} \rangle - \alpha_k \langle \mathbf{p}_{(k)}, \mathbf{r}_{(k-1)} \rangle_A \\
&= \langle \mathbf{r}_{(k-1)}, \mathbf{r}_{(k-1)} \rangle \left(1 - \frac{\langle \mathbf{p}_{(k)}, \mathbf{r}_{(k-1)} \rangle_A}{\langle \mathbf{p}_{(k)}, \mathbf{p}_{(k)} \rangle_A} \right) \\
&= \langle \mathbf{r}_{(k-1)}, \mathbf{r}_{(k-1)} \rangle \left(1 - \frac{\langle \mathbf{p}_{(k)}, \mathbf{r}_{(k-1)} \rangle_A}{\langle \mathbf{p}_{(k)}, \mathbf{r}_{(k-1)} \rangle_A + \beta_k \langle \mathbf{p}_{(k)}, \mathbf{p}_{(k-1)} \rangle_A} \right) \\
&= 0,
\end{aligned}$$

and, similarly,

$$\begin{aligned}
\langle \mathbf{r}_{(k)}, \mathbf{r}_{(j)} \rangle &= \langle \mathbf{r}_{(k-1)}, \mathbf{r}_{(j)} \rangle - \frac{\langle \mathbf{r}_{(k-1)}, \mathbf{r}_{(k-1)} \rangle}{\langle \mathbf{p}_{(k)}, \mathbf{p}_{(k)} \rangle_A} \langle \mathbf{p}_{(k)}, \mathbf{r}_{(j)} \rangle_A \\
&= -\frac{\langle \mathbf{r}_{(k-1)}, \mathbf{r}_{(k-1)} \rangle}{\langle \mathbf{p}_{(k)}, \mathbf{p}_{(k)} \rangle_A} \langle \mathbf{p}_{(k)}, \sum_{i=1}^{j+1} \eta_i \mathbf{p}_{(i)} \rangle_A \\
&= 0
\end{aligned}$$

for $j < k-1$. Thus, (1.82) and (1.83) are established and we may use (1.83) in the induction step for (1.84). Let us remind the induction assumption $\langle \mathbf{p}_{(i)}, \mathbf{p}_{(j)} \rangle_A = 0$ for $i \neq j,\ 1 \leq i, j \leq k$. Then

$$\begin{aligned}\langle \mathbf{p}_{(k+1)}, \mathbf{p}_{(k)} \rangle_A &= \langle \mathbf{r}_{(k)} + \beta_{k+1} \mathbf{p}_{(k)}, \mathbf{p}_{(k)} \rangle_A \\ &= \langle \mathbf{r}_{(k)}, \mathbf{p}_{(k)} \rangle_A - \langle \mathbf{r}_{(k)}, \mathbf{p}_{(k)} \rangle_A = 0.\end{aligned}$$

Finally, since

$$\begin{aligned}\langle \mathbf{p}_{(k+1)}, \mathbf{p}_{(j)} \rangle_A &= \langle \mathbf{r}_{(k)}, \mathbf{p}_{(j)} \rangle_A + \beta_{k+1} \langle \mathbf{p}_{(k)}, \mathbf{p}_{(j)} \rangle_A \\ &= \langle \mathbf{r}_{(k)}, A\mathbf{p}_{(j)} \rangle = \frac{1}{\alpha_j} \left(\langle \mathbf{r}_{(k)}, \mathbf{r}_{(j-1)} \rangle - \langle \mathbf{r}_{(k)}, \mathbf{r}_{(j)} \rangle \right) = 0\end{aligned}$$

for $1 \leq j < k$, the proof of the lemma is completed. □

An elementary computation shows that $\frac{\partial \phi(\mathbf{x}_{(n)})}{\partial \alpha_n} = 0$ implies

$$\alpha_n = \frac{\langle \mathbf{r}_{(n-1)}, \mathbf{p}_{(n)} \rangle}{\langle \mathbf{p}_{(n)}, \mathbf{p}_{(n)} \rangle_A} = \frac{\langle \mathbf{r}_{(n-1)}, \mathbf{r}_{(n-1)} \rangle}{\langle \mathbf{p}_{(n)}, \mathbf{p}_{(n)} \rangle_A} \tag{1.86}$$

so that α_n defined by (1.79) minimizes ϕ along the direction $\mathbf{p}_{(n)}$ passing through $\mathbf{x}_{(n-1)}$.

The following Theorem 1.16 tells us that in exact arithmetic the CG methods yields the exact solution of (1.47) after at most N iteration steps, where N is the dimension of the linear system.

Theorem 1.16. *Consider Algorithm 1.14 with the termination criterion* $\mathbf{r}_{(n)} = \mathbf{0}$. *Then, (in exact arithmetic) we have* $A\mathbf{x}_{(n)} = \mathbf{b}$ *for some* $n \leq N$.

Proof. The theorem follows from Lemma 1.15, because there exist at most N mutually orthogonal nonzero vectors $\mathbf{r}_{(n)}$, $n \geq 0$. □

However, this result is only of theoretical interest when using the CG method as an iterative solver for very large systems of equations. That is why we will now study the convergence behavior of Algorithm 1.14.

First of all, the recurrence (1.80) can be represented as

$$\mathbf{x}_{(n)} - \mathbf{x}_{(0)} = \sum_{i=1}^{n} \alpha_i \mathbf{p}_{(i)}. \tag{1.87}$$

Next, we observe that

$$\begin{aligned}\langle \mathbf{r}_{(j-1)}, \mathbf{p}_{(j)} \rangle &= \langle \mathbf{x} - \mathbf{x}_{(j-1)}, \mathbf{p}_{(j)} \rangle_A = \langle \mathbf{x} - \mathbf{x}_{(0)} - \sum_{i=1}^{j-1} \alpha_i \mathbf{p}_{(i)}, \mathbf{p}_{(j)} \rangle_A \\ &= \langle \mathbf{x} - \mathbf{x}_{(0)}, \mathbf{p}_{(j)} \rangle_A \quad \text{for} \quad 1 \le j \le n.\end{aligned}$$

Therefore,

$$\alpha_j = \frac{\langle \mathbf{r}_{(j-1)}, \mathbf{p}_{(j)} \rangle}{\langle \mathbf{p}_{(j)}, \mathbf{p}_{(j)} \rangle_A} = \frac{\langle \mathbf{x} - \mathbf{x}_{(0)}, \mathbf{p}_{(j)} \rangle_A}{\langle \mathbf{p}_{(j)}, \mathbf{p}_{(j)} \rangle_A}$$

and thus

$$\langle \mathbf{x}_{(n)} - \mathbf{x}_{(0)}, \mathbf{p}_{(j)} \rangle_A = \alpha_j \langle \mathbf{p}_{(j)}, \mathbf{p}_{(j)} \rangle_A = \langle \mathbf{x} - \mathbf{x}_{(0)}, \mathbf{p}_{(j)} \rangle_A \quad \text{for} \quad 1 \le j \le n.$$

In other words, $\mathbf{x}_{(n)} - \mathbf{x}_{(0)}$ is the projection of the initial error with respect to $\langle \cdot, \cdot \rangle_A$ onto the space W_n spanned by the first n search directions $\mathbf{p}_{(1)}, \mathbf{p}_{(2)}, \ldots, \mathbf{p}_{(n)}$. On account of this fact,

$$\begin{aligned}\|\mathbf{e}_{(n)}\|_A &= \|\mathbf{x} - \mathbf{x}_{(0)} + \mathbf{x}_{(0)} - \mathbf{x}_{(n)}\|_A \\ &\le \|\mathbf{e}_{(0)} - \mathbf{q}\|_A \quad \forall \mathbf{q} \in W_n = \operatorname{span}\{\mathbf{p}_{(1)}, \mathbf{p}_{(2)}, \ldots, \mathbf{p}_{(n)}\}.\end{aligned} \tag{1.88}$$

Remark 1.17. The linear vector spaces $\mathbb{R}^N$ and $W_n \subset \mathbb{R}^N$ are Hilbert spaces for the inner product $\langle \cdot, \cdot \rangle_A$. This implies that for any $\mathbf{x}' \in \mathbb{R}^N$ there exists a uniquely determined element $\mathbf{q}' \in W_n$ providing the best approximation of $\mathbf{x}'$ in the norm induced by the inner product. The vector $\mathbf{q}'$ is the projection of $\mathbf{x}'$ onto W_n with respect to this inner product.

Recalling Lemma 1.15 we notice that

$$\begin{aligned}W_n &= \operatorname{span}\{\mathbf{r}_{(0)}, \ldots, A^{n-1}\mathbf{r}_{(0)}\} \\ &= \operatorname{span}\{A(\mathbf{x} - \mathbf{x}_{(0)}), \ldots, A^n(\mathbf{x} - \mathbf{x}_{(0)})\}.\end{aligned}$$

Remark 1.18. The spaces W_n are called the Krylov subspaces related to the CG method.

Finally, from (1.88) and Theorem 1.13 we conclude the following error estimate.

Theorem 1.19. *The error $\mathbf{e}_{(n)} = \mathbf{x}_{(n)} - \mathbf{x}$ of the n-th approximation vector $\mathbf{x}_{(n)}$ computed via the CG method (Algorithm 1.14) satisfies*

$$\begin{aligned}\|\mathbf{e}_{(n)}\|_A &= \|\mathbf{x} - \mathbf{x}_{(n)}\|_A = \min_{P_n \in \Pi_n^1} \|P_n(A)(\mathbf{x} - \mathbf{x}_{(0)})\|_A \\ &\le \min_{P_n \in \Pi_n^1} \max_{1 \le i \le N} |P_n(\lambda_i)| \|\mathbf{x} - \mathbf{x}_{(0)}\|_A \\ &\le \overline{\rho}(\tilde{P}_n(A)) \|\mathbf{e}_{(0)}\|_A = \frac{2\sigma^n}{1 + \sigma^{2n}} \|\mathbf{e}_{(0)}\|_A\end{aligned} \tag{1.89}$$

where $\tilde{P}_n$ is the n-th shifted Chebyshev polynomial, associated with the interval $[\lambda_1(A), \lambda_N(A)]$, and σ is defined by $\sigma = \frac{\sqrt{\kappa(A)}-1}{\sqrt{\kappa(A)}+1}$.

Proof. In view of (1.60) the bound (1.89) follows by calculating the virtual spectral radius of the matrix polynomial (1.67)–(1.68), which is given by

$$\overline{\rho}(\tilde{P}_n(A)) = \frac{1}{\left|T_n\left(\frac{-(\lambda_N+\lambda_1)}{\lambda_N-\lambda_1}\right)\right|} = \frac{2\sigma^n}{1+\sigma^{2n}}, \tag{1.90}$$

where

$$\sigma = \frac{\sqrt{\kappa}-1}{\sqrt{\kappa}+1} = \frac{1-\frac{1}{\sqrt{\kappa}}}{1+\frac{1}{\sqrt{\kappa}}} \approx 1 - \frac{2}{\sqrt{\kappa}} \quad \text{and} \quad \kappa = \kappa(A). \tag{1.91}$$

□

1.4.3 Preconditioned conjugate gradient (PCG) method

The preconditioned conjugate gradient (PCG) method has established itself as the method of choice for iteratively solving large sparse symmetric positive definite (SPD) systems of linear algebraic equations. It is obtained by applying Algorithm 1.14 to the transformed system[5]

$$\tilde{A}\tilde{\mathbf{x}} = \tilde{\mathbf{b}}, \tag{1.92}$$

where $\tilde{A} = C^{-\frac{1}{2}}AC^{-\frac{1}{2}}, \tilde{\mathbf{x}} = C^{\frac{1}{2}}\mathbf{x}$ and $\tilde{\mathbf{b}} = C^{-\frac{1}{2}}\mathbf{b}$. The decomposition $C = C^{\frac{1}{2}}C^{\frac{1}{2}}$ of the preconditioner is not needed explicitly in the algorithm[6] (see, e.g., [63]), so that it is possible to work with the matrix C itself.

Algorithm 1.20 works with the pseudoresiduals $\mathbf{z}_{(n)} = C^{-1}\mathbf{r}_{(n)}$ instead of $\mathbf{r}_{(n)}$, thereby minimizing the quadratic functional

$$\phi(\tilde{\mathbf{x}}) = \frac{1}{2}\tilde{\mathbf{x}}^T\tilde{A}\tilde{\mathbf{x}} - \tilde{\mathbf{x}}^T\tilde{\mathbf{b}}, \tag{1.93}$$

i.e., solving the system $C^{-\frac{1}{2}}A\mathbf{x} = C^{-\frac{1}{2}}\mathbf{b}$, which is consistent with $A\mathbf{x} = \mathbf{b}$. Thus, the preconditioned CG method will converge towards the exact solution of (1.47) as well.

[5]Under the assumption that C is symmetric and positive definite, the preconditioned system matrix $\tilde{A}$ is SPD as well.

[6]Its steps can be rewritten avoiding an explicit reference to the matrix $C^{\frac{1}{2}}$.

Algorithm 1.20 (Preconditioned Conjugate Gradients, see, e.g., [25]).

$n = 0; \quad \mathbf{r}_{(0)} = \mathbf{b} - A\mathbf{x}_{(0)};$

while (termination criterion is false) **do**

solve $\quad C\mathbf{z}_{(n)} = \mathbf{r}_{(n)}$ (1.94)

$n = n + 1$

$\gamma_{n-1} = \langle \mathbf{r}_{(n-1)}, \mathbf{z}_{(n-1)} \rangle$

if ($n = 1$) **then**

$\mathbf{p}_{(1)} = \mathbf{z}_{(0)}$ (1.95)

else

$$\beta_n = \frac{\gamma_{n-1}}{\gamma_{n-2}} \tag{1.96}$$

$$\mathbf{p}_{(n)} = \mathbf{z}_{(n-1)} + \beta_n \mathbf{p}_{(n-1)} \tag{1.97}$$

end

$$\mathbf{q}_{(n)} = A\mathbf{p}_{(n)} \tag{1.98}$$

$$\alpha_n = \frac{\gamma_{n-1}}{\langle \mathbf{p}_{(n)}, \mathbf{q}_{(n)} \rangle} \tag{1.99}$$

$$\mathbf{x}_{(n)} = \mathbf{x}_{(n-1)} + \alpha_n \mathbf{p}_{(n)} \tag{1.100}$$

$$\mathbf{r}_{(n)} = \mathbf{r}_{(n-1)} - \alpha_n \mathbf{q}_{(n)} \tag{1.101}$$

end

Remark 1.21. The Krylov subspaces involved in Algorithm 1.20 are generated by the preconditioned matrix, that is,

$$W_n^\star = \operatorname{span}\{\mathbf{r}_{(0)}, (AC^{-1})\mathbf{r}_{(0)}, \ldots, (AC^{-1})^{n-1}\mathbf{r}_{(0)}\}. \tag{1.102}$$

Similar arguments as in the derivation of Theorem 1.19 result in the same (virtual) rate of convergence except that the bound (1.89) now involves the spectral condition number of the preconditioned matrix, i.e., $\kappa = \kappa(C^{-1}A)$.

Hence, the PCG method combines several favorable properties, which are summarized below:

(a) It is optimal in the sense of minimizing in the n-th step the A-norm of the error $\mathbf{e}_{(n)} = \mathbf{e}_{(0)} - \mathbf{q}^\star$, where $\mathbf{q}^\star$ denotes an arbitrary element of the related Krylov space $W_n^\star$, cf. (1.88).

(b) Its rate of convergence can be improved by appropriate preconditioning.

(c) The algorithm is free from any parameters to be estimated.

(d) The memory requirements for an implementation are low due to the short (three-term) recurrences: only two sparse matrices (including the preconditioner C) and five vectors have to be stored when implementing Algorithm 1.20.

(e) The number of arithmetic operations per iteration is low: Algorithm 1.20 involves one matrix-vector product, three vector updates (SAXPYs) and two inner products (SDOTs) per iteration.

A more detailed convergence analysis of the PCG method that takes into account the eigenvalue distribution of the preconditioned matrix sometimes yields sharper estimates, for instance, in the case of isolated or clustered eigenvalues (see, e.g., [12, 13]). For a survey of iterative solution methods we refer the reader to [103].

1.4.4 Generalized conjugate gradient (GCG) method

In the following we will consider a generalization of the PCG method, known as generalized conjugate gradient (GCG) algorithm that is well adapted for so-called variable-step preconditioning, see e.g., [4, 18, 19, 96]. In this case the preconditioner is no longer given by a linear mapping (an SPD matrix) but it can be defined via an iterative process itself. We will come back to this method in the next chapter in order to formulate and analyze a powerful combined algorithm known as nonlinear algebraic multilevel iteration (nonlinear AMLI) method.

The main difference to the standard PCG method is that the option of a more general preconditioner requires an explicit orthogonalization of the search directions $\mathbf{p}_{(n)}$. Let $M[\cdot]$ denote the preconditioner, i.e., a mapping from $\mathbb{R}^N$ to $\mathbb{R}^N$, which can also be nonlinear. Hence, $\mathbf{z}_{(n)} = M^{-1}[\mathbf{r}_{(n)}]$ replaces the step (1.94) in Algorithm 1.20. Then, assuming that A is SPD the n-th search direction is orthonormalized with respect to the A-inner product $\langle \cdot, A\,\cdot \rangle$ against the m_n previous directions, i.e.,

$$\mathbf{p}_{(n)} = \mathbf{z}_{(n-1)} - \sum_{j=n-m_n}^{n-1} \frac{\langle \mathbf{z}_{(n-1)}, A\mathbf{p}_{(j)} \rangle}{\langle \mathbf{p}_{(j)}, A\mathbf{p}_{(j)} \rangle} \mathbf{p}_{(j)} \tag{1.103}$$

where $\{m_n\}_{n=1,2,\dots}$ is a sequence of truncation parameters. The untruncated version corresponds to $m_n = n-1$ for all n, whereas $m_n = \min(n-1, m_{\max})$ results in a pure truncation. The advised truncation strategy we are using (if not mentioned otherwise) is to restart the (untruncated) algorithm at every $m_{\max}$ iterations, which corresponds to the choice $m_n = \mathrm{mod}(n-1, m_{\max})$.

Algorithm 1.22 (Generalized Conjugate Gradients, see, e.g., [19, 96]).

$n = 0; \quad \mathbf{r}_{(0)} = \mathbf{b} - A\mathbf{x}_{(0)};$

while (termination criterion is false) **do**

$$\mathbf{z}_{(n)} = M^{-1}[\mathbf{r}_{(n)}] \tag{1.104}$$

$n = n + 1$

$\mathbf{p}_{(n)} = \mathbf{z}_{(n-1)}$

$$\mathbf{q}_{(n)} = A\mathbf{p}_{(n)} \tag{1.105}$$

for $j = n - m_n$ **to** $n - 1$ (1.106)

$\beta = \dfrac{\langle \mathbf{q}_{(n)}, \mathbf{p}_{(j)} \rangle}{\gamma_j}$

$\mathbf{p}_{(n)} = \mathbf{p}_{(n)} - \beta \mathbf{p}_{(j)}$

$\mathbf{q}_{(n)} = \mathbf{q}_{(n)} - \beta \mathbf{q}_{(j)}$

end

$\gamma_n = \langle \mathbf{p}_{(n)}, \mathbf{q}_{(n)} \rangle$

$\alpha_n = \dfrac{\langle \mathbf{r}_{(n)}, \mathbf{p}_{(n)} \rangle}{\gamma_n}$

$\mathbf{x}_{(n)} = \mathbf{x}_{(n-1)} + \alpha_n \mathbf{p}_{(n)}$

$\mathbf{r}_{(n)} = \mathbf{r}_{(n-1)} - \alpha_n \mathbf{q}_{(n)}$

end

Note that the matrix vector products required for (1.103) are hidden in the step (1.105) and thus they do not appear explicitly in the loop (1.106), i.e., performing the step (1.105) before entering (1.106) results in A-orthogonal search directions. This particular construction is a natural choice because for a fixed linear mapping $M^{-1}[\cdot] = C^{-1}$, i.e., for an SPD preconditioner C, the above (untruncated) GCG method reduces to the standard preconditioned CG algorithm. Moreover, if $0 \le m_n \le m_{n-1} + 1$ for all n, it can be shown ([5, 18]) that

$$\begin{aligned} \langle \mathbf{p}_{(k)}, A\mathbf{p}_{(j)} \rangle &= 0 \qquad \forall j, k \text{ such that } n - m_n \le j < k < n, \\ \langle \mathbf{r}_{(k)}, \mathbf{p}_{(j)} \rangle &= 0 \qquad \forall j, k \text{ such that } n - m_n \le j < k \le n \end{aligned}$$

and that the following optimality property holds:

$$\|\mathbf{x} - \mathbf{x}_{(n)}\|_A = \min_{\mathbf{p} \in \operatorname{span}\{\mathbf{p}_{(n-m_n)}, \ldots, \mathbf{p}_{(n-1)}\}} \|\mathbf{x} - \mathbf{x}_{(n-m_n)} - \mathbf{p}\|_A. \tag{1.107}$$

However, it should be mentioned that it is possible to choose different inner products (not only $\langle \cdot, A\,\cdot \rangle$ as described before) in the orthogonalization procedure, e.g., $\langle A\cdot, A\cdot\rangle$ or $\langle\cdot,\cdot\rangle$, cf. [18]. This makes the GCG algorithm also applicable to indefinite problems thus providing an alternative to the well-known Uzawa algorithm [82].

Regarding the convergence of the GCG iteration a very general result can be found in reference [4]. A more specific but improved bound on the local decrease of the error in GCG-type iterations holds true when the nonlinear preconditioner becomes close to a linear operator, see [96].[7] Let us recall this result from reference [96] that will be used later in the analysis of the nonlinear algebraic multilevel iteration method.

Theorem 1.23 ([96]). *Let A, C be SPD matrices of size $N\times N$ and $M^{-1}[\cdot]$ a mapping from $\mathbb{R}^N$ to $\mathbb{R}^N$. Let $\mathbf{d}$, $\mathbf{x}_{(0)}$ be vectors of $\mathbb{R}^N$ and let $\mathbf{r}_{(n)}$, $\mathbf{x}_{(n)}$ be the sequences of iterates and residuals generated by applying the GCG (FCG) algorithm with preconditioner $M[\cdot]$ to the linear system $A\mathbf{x} = \mathbf{d}$. If for any n,*

$$\frac{\|M^{-1}[\mathbf{r}_{(n)}] - C^{-1}\mathbf{r}_{(n)}\|_C}{\|C^{-1}\mathbf{r}_{(n)}\|_C} \le \epsilon_n < 1$$

then

$$\frac{\|\mathbf{x}-\mathbf{x}_{(n+1)}\|_A}{\|\mathbf{x}-\mathbf{x}_{(n)}\|_A} \le \sqrt{1 - \frac{4\kappa(1-\epsilon_n)^2}{(\kappa + \epsilon_n^2(\kappa-1) + (1-\epsilon_n)^2)^2}}$$

where $\kappa = \kappa(C^{-1}A)$.

[7] where the studied variant of the GCG algorithm is referred to as Flexible Conjugate Gradients (FCG).

2 Algebraic multilevel iteration methods

The material selected in this chapter follows the spirit of the AMLI methods as presented originally in [16, 17]. At the same time, some needed generalizations are made to support the cases of nonconforming FEM and/or discontinuous Galerkin discretizations as well as the nonlinear AMLI.

2.1 Block-factorization: Schur complement

Consider now a general matrix A, which is assumed to be symmetric positive definite and partitioned in a two-by-two block form

$$A = \begin{bmatrix} A_{11} & A_{12} \\ A_{21} & A_{22} \end{bmatrix}. \tag{2.1}$$

The following LU factorization holds true:

$$A = \begin{bmatrix} A_{11} & \\ A_{21} & S \end{bmatrix} \begin{bmatrix} I_1 & A_{11}^{-1}A_{12} \\ & I_2 \end{bmatrix}, \tag{2.2}$$

where S stands for the Schur complement, i.e., $S = A_{22} - A_{21}A_{11}^{-1}A_{12}$. Various preconditioning techniques are based on approximations of (2.2).

Lemma 2.1. *Let* A *be a symmetric positive definite matrix,* $\mathbf{x} = \left[\begin{smallmatrix} \mathbf{x}_1 \\ \mathbf{x}_2 \end{smallmatrix}\right]$ *be a block-vector corresponding to the two-by-two representation (2.1) of* A*, and* $\mathbf{x}_2$ *be fixed. Then, the following extremal property of the Schur complement holds:*

$$\mathbf{x}_2^T S \mathbf{x}_2 = \min_{\mathbf{x}_1} \mathbf{x}^T A \mathbf{x}. \tag{2.3}$$

Proof.

$$\begin{aligned}
\mathbf{x}^T A \mathbf{x} &= [\mathbf{x}_1^T\ \mathbf{x}_2^T] \begin{bmatrix} A_{11} & A_{12} \\ A_{21} & A_{22} \end{bmatrix} \begin{bmatrix} \mathbf{x}_1 \\ \mathbf{x}_2 \end{bmatrix} \\
&= \mathbf{x}_1^T A_{11}\mathbf{x}_1 + \mathbf{x}_1^T A_{12}\mathbf{x}_2 + \mathbf{x}_2^T A_{21}\mathbf{x}_1 + \mathbf{x}_2^T A_{22}\mathbf{x}_2 \\
&\quad + \mathbf{x}_2^T A_{21}A_{11}^{-1}A_{12}\mathbf{x}_2 - \mathbf{x}_2^T A_{21}A_{11}^{-1}A_{12}\mathbf{x}_2 \\
&= \mathbf{x}_2^T S \mathbf{x}_2 + \left(\mathbf{x}_1 + A_{11}^{-1}A_{12}\mathbf{x}_2\right)^T A_{11}\left(\mathbf{x}_1 + A_{11}^{-1}A_{12}\mathbf{x}_2\right).
\end{aligned}$$

Since A is symmetric positive definite, the same obviously holds for A_{11}, and therefore

$$\mathbf{x}^T A\mathbf{x} \geq \mathbf{x}_2^T S\mathbf{x}_2. \tag{2.4}$$

The equality in (2.4) is reached for

$$\mathbf{x}_1 = -A_{11}^{-1} A_{12} \mathbf{x}_2^T$$

which completes the proof. □

Corollary 2.2. *The Schur complement S of any symmetric and positive definite matrix A is also symmetric and positive definite.*

The above statement follows directly from the definition of S and the relation (2.3).

The following alternative forms of block two-by-two factorizations are also used in some cases:

1. LDL^T block-factorization:

$$A = \begin{bmatrix} I_1 & \\ A_{21}A_{11}^{-1} & I_2 \end{bmatrix} \begin{bmatrix} A_{11} & \\ & S \end{bmatrix} \begin{bmatrix} I_1 & A_{11}^{-1}A_{21}^T \\ & I_2 \end{bmatrix}.$$

2. LL^T block-factorization:

$$A = \begin{bmatrix} L_{11} & \\ L_{21} & L_{22} \end{bmatrix} \begin{bmatrix} L_{11} & L_{21}^T \\ & L_{22}, \end{bmatrix},$$

where

$$L_{11} = A_{11}^{1/2},$$

$$L_{21} = A_{21}A_{11}^{-1/2},$$

$$L_{22}^2 = A_{22} - A_{21}A_{11}^{-1}A_{12}, \qquad \text{i.e.,} \qquad L_{22} = S^{1/2}.$$

The considered two-by-two block factorizations provide a general framework for the construction of preconditioners. In this context, it is important to note that the Schur complement of a given sparse matrix is not sparse in general. This leads to the problem how to construct a proper sparse approximation of S. Some advanced answers to this question are given in the present book.

The efficiency of the preconditioners based on block factorization strongly depends on the coupling of the partitioning (2.1). It is characterized by the corresponding constant in the strengthened Cauchy–Bunyakowski–Schwarz (CBS) inequality.

2.2 Local estimates of the CBS constant

Let $W = V_1 \times V_2$ be a splitting of the vector space, which is consistent with the partitioning (2.1). We will use also the notations $\mathbf{v}_i \in V_i$, $i = 1, 2$, and $W_1 = \{\mathbf{v} = [\mathbf{v}_1^T, \mathbf{0}^T]^T\}$, $W_2 = \{\mathbf{v} = [\mathbf{0}^T, \mathbf{v}_2^T]^T\}$.

The CBS constant measures the strength of the off-diagonal blocks of $A_{12} = A_{21}^T$ in relation to the diagonal blocks and can be defined as the minimal γ satisfying the strengthened Cauchy–Bunyakowski–Schwarz inequality

$$|\mathbf{v}_1^T A_{12}\mathbf{v}_2| \leq \gamma \left\{\mathbf{v}_1^T A_{11}\mathbf{v}_1 \ \mathbf{v}_2^T A_{22}\mathbf{v}_2\right\}^{1/2}. \tag{2.5}$$

The following three lemmas (see, e.g., [5, 53]) form the theoretical background of the CBS constant estimates.

Lemma 2.3. *Let A be symmetric positive semidefinite, A_{11} be positive definite and γ be the smallest constant satisfying (2.5). Then:*

(a) $\gamma \leq 1$.

(b) $\gamma = 1$ *if there exists* $\mathbf{w} = \left[\begin{smallmatrix}\mathbf{v}_1\\ \mathbf{v}_2\end{smallmatrix}\right] \in \ker(A)$ *for which* $\mathbf{v}_2 \notin \ker(A_{22})$.

(c) $\gamma < 1$ *if for any* $\mathbf{w} = \left[\begin{smallmatrix}\mathbf{v}_1\\ \mathbf{v}_2\end{smallmatrix}\right] \in \ker(A)$ *it holds that* $\mathbf{v}_2 \in \ker(A_{22})$.

(d) *Under the assumption of (c),*

$$\gamma = \sup_{\mathbf{v}_i \in V_i \setminus \ker(A_{ii}),\, i=1,2} \frac{\mathbf{v}_1{}^T A_{12}\mathbf{v}_2}{(\mathbf{v}_1{}^T A_{11}\mathbf{v}_1 \ \mathbf{v}_2{}^T A_{22}\mathbf{v}_2)^{1/2}}.$$

Proof. Let $\mathbf{w} = \left[\begin{smallmatrix}\mathbf{v}_1\\ \mathbf{v}_2\end{smallmatrix}\right]$ where $\mathbf{v}_1, \mathbf{v}_2 \neq \mathbf{0}$. By assumption, the matrix A is symmetric positive semidefinite, i.e.,

$$\mathbf{w}^T A\mathbf{w} = \mathbf{v}_1^T A_{11}\mathbf{v}_1 + \mathbf{v}_2^T A_{22}\mathbf{v}_2 + 2\mathbf{v}_1{}^T A_{12}\mathbf{v}_2 \geq 0. \tag{2.6}$$

Let us denote by $\gamma(\mathbf{v}_1, \mathbf{v}_2)$ the CBS constant corresponding to the vectors $\mathbf{v}_1, \mathbf{v}_2$. It is readily seen that $\gamma(\mathbf{v}_1, \mathbf{v}_2) = \gamma(\alpha\mathbf{v}_1, \beta\mathbf{v}_2)$ for any $\alpha, \beta \neq 0$. Consequently, we can always assume that the vectors $\mathbf{v}_1, \mathbf{v}_2$ are properly scaled, which does not change the value of γ.

To prove (a), we consider first the case $\mathbf{v}_2{}^T A_{22}\mathbf{v}_2 \neq 0$. The pivot block A_{11} is positive definite by assumption, that is $\mathbf{v}_1{}^T A_{11}\mathbf{v}_1 > 0$, and we can scale $\mathbf{v}_1$ such that $\mathbf{v}_1{}^T A_{11}\mathbf{v}_1 = \mathbf{v}_2{}^T A_{22}\mathbf{v}_2$. After substitution of this equality in (2.6) we get

$$|\mathbf{v}_1{}^T A_{12}\mathbf{v}_2| \leq \mathbf{v}_1{}^T A_{11}\mathbf{v}_1,$$

and therefore $\gamma(\mathbf{v}_1, \mathbf{v}_2) \leq 1$.

Let us continue with the alternative case ${\mathbf{v}_2}^T A_{22}\mathbf{v}_2 = 0$, and let $\mathbf{v}_1 = \tau\hat{\mathbf{v}}_1$. Then

$$\tau\hat{\mathbf{v}}_1^T A_{11}\hat{\mathbf{v}}_1 + 2\hat{\mathbf{v}}_1^T A_{12}\mathbf{v}_2 \geq 0$$

for any $\tau > 0$, which obviously holds if and only if

$$\hat{\mathbf{v}}_1^T A_{12}\mathbf{v}_2 \geq 0.$$

If we assume that $A_{12}\mathbf{v}_2 \neq \mathbf{0}$, and if we choose $\hat{\mathbf{v}}_1 = -A_{12}\mathbf{v}_2$, we obtain the contradiction

$$-\|A_{12}\mathbf{v}_2\| > 0.$$

Therefore

$$ {\mathbf{v}_2}^T A_{22}\mathbf{v}_2 = 0, \qquad \text{i.e.,} \qquad A_{12}\mathbf{v}_2 = \mathbf{0}. \tag{2.7}$$

In this case the CBS inequality is satisfied for any γ which completes the proof of statement (a).

Now, let us assume that $\left[\begin{smallmatrix}\mathbf{v}_1\\ \mathbf{v}_2\end{smallmatrix}\right] \in \ker(A)$, such that $\mathbf{v}_2 \notin \ker(A_{22})$. As in the proof of (a), we scale $\mathbf{v}_1$ to get ${\mathbf{v}_1}^T A_{11}\mathbf{v}_1 = {\mathbf{v}_2}^T A_{22}\mathbf{v}_2$, substitute in (2.6) and obtain

$$-{\mathbf{v}_1}^T A_{12}\mathbf{v}_2 = {\mathbf{v}_1}^T A_{11}\mathbf{v}_1.$$

Therefore $\gamma(\mathbf{v}_1, \mathbf{v}_2) = 1$ which is the statement of (b).

Let the condition from (c) be satisfied and let us consider the case $A\mathbf{w} = \mathbf{0}$ which leads to $A_{22}\mathbf{v}_2 = \mathbf{0}$. Then $A_{12}\mathbf{v}_2 = \mathbf{0}$ (see (2.7)) and therefore $A_{11}\mathbf{v}_1 = \mathbf{0}$ (see (2.6) in the case of equality). Since A_{11} is symmetric positive definite, we obtain $\mathbf{v}_1 = \mathbf{0}$. But this case is obviously not interesting for the analysis of the CBS constant because (2.5) is satisfied for any γ. In the case $A\mathbf{w} \neq \mathbf{0}$, the inequality (2.6) is strict, all arguments in the proof of (a) hold with strict inequalities, and the result is $\gamma < 1$ which completes the proof of (c). At the end, the statement (d) follows directly from the definition of γ. □

Lemma 2.4. *Let A be a symmetric positive semidefinite matrix satisfying condition (c) from Lemma 2.3. Then*

(a)

$$\gamma^2 = \sup_{\mathbf{v}_2 \in V_2 \setminus \ker(A_{22})} \frac{{\mathbf{v}_2}^T A_{21} A_{11}^{-1} A_{12}\mathbf{v}_2}{{\mathbf{v}_2}^T A_{22}\mathbf{v}_2}. \tag{2.8}$$

(b) *for any* $\mathbf{v}_2 \in V_2 \setminus \ker(A_{22})$

$$1 - \gamma^2 \leq \frac{{\mathbf{v}_2}^T S\mathbf{v}_2}{\mathbf{v}_2^T A_{22}\mathbf{v}_2} < 1, \tag{2.9}$$

where the left-hand side inequality is sharp and the right-hand side inequality is sharp if $\ker(A_{12}) \neq \{0\}$.

Proof. Equality (d) from the previous lemma is applied to get

$$\gamma = \sup_{\mathbf{v}_i \in V_i \setminus \ker(A_{ii}),\ i=1,2} \frac{\mathbf{v}_1{}^T A_{12} \mathbf{v}_2}{\{\mathbf{v}_1{}^T A_{11} \mathbf{v}_1\ \mathbf{v}_2{}^T A_{22} \mathbf{v}_2\}^{1/2}}$$

$$= \sup_{\mathbf{v}_1 \neq \mathbf{0},\ \mathbf{v}_2 \in V_2 \setminus \ker(A_{22})} \frac{\mathbf{v}_1{}^T A_{11}^{-1/2} A_{12} \mathbf{v}_2}{\{\mathbf{v}_1{}^T \mathbf{v}_1\ \mathbf{v}_2{}^T A_{22} \mathbf{v}_2\}^{1/2}},$$

where the supremum is reached for $\mathbf{v}_1 = A_{11}^{-1/2} A_{12} \mathbf{v}_2$ and therefore

$$\gamma = \left\{ \sup_{\mathbf{v}_2 \in V_2 \setminus \ker(A_{22})} \frac{\mathbf{v}_2{}^T A_{21} A_{11}^{-1} A_{12} \mathbf{v}_2}{\mathbf{v}_2{}^T A_{22} \mathbf{v}_2} \right\}^{1/2},$$

which is exactly (a). The inequalities (b) follow directly from (a) and the definition of the Schur complement $S = A_{22} - A_{21} A_{11}^{-1} A_{12}$. □

Now, we are ready to show how the CBS constant can be estimated locally (see also [83]). Let us assume that

$$A = \sum_{E \in \mathcal{E}} R_E^T A_E R_E, \qquad \mathbf{v} = \sum_{E \in \mathcal{E}} R_E^T \mathbf{v}_E, \tag{2.10}$$

where A_E are symmetric positive semidefinite local matrices, $\mathcal{E}$ is some index set, and the summation is understood as assembling, i.e., the matrices R_E^T represent the natural inclusions. The global splitting naturally induces the two-by-two block representation of the local matrix A_E and the related vector $\mathbf{v}_E$, namely,

$$A_E = \begin{bmatrix} A_{E:11} & A_{E:12} \\ A_{E:21} & A_{E:22} \end{bmatrix}, \quad \mathbf{v}_E = \begin{bmatrix} \mathbf{v}_{E:1} \\ \mathbf{v}_{E:2} \end{bmatrix}. \tag{2.11}$$

Lemma 2.5. *Let the local matrices A_E, $E \in \mathcal{E}$, satisfy the condition (c) from Lemma 2.3. Let also $V_{E:i}$, $i = 1, 2$, be the natural restriction of V_i induced by the local matrix A_E. Then*

$$\gamma \le \max_{E \in \mathcal{E}} \gamma_E < 1 \tag{2.12}$$

where γ_E stands for the local CBS constant corresponding to A_E, that is

$$\gamma_E^2 = \sup_{\mathbf{v}_{E:2} \in V_{E:2} \setminus \ker(A_{E:22})} \frac{\mathbf{v}_{E:2}^T A_{E:21} A_{E:11}^{-1} A_{E:12} \mathbf{v}_{E:2}}{\mathbf{v}_{E:2}^T A_{E:22} \mathbf{v}_{E:2}}. \tag{2.13}$$

Proof. It is important to note, that the assumption (c) from Lemma 2.3 ensures the strong estimates $\gamma_E < 1$. Then,

$$\begin{aligned}
\left|\mathbf{v}_1^T A_{12}\mathbf{v}_2\right| &= \left|\sum_{E\in\mathcal{E}} \mathbf{v}_{E:1}^T A_{E:12}\mathbf{v}_{E:2}\right| \le \sum_{E\in\mathcal{E}} \left|\mathbf{v}_{E:1}^T A_{E:12}\mathbf{v}_{E:2}\right| \\
&\le \sum_{E\in\mathcal{E}} \gamma_E \sqrt{\mathbf{v}_{E:1}^T A_{E:11}\mathbf{v}_{E:1}}\sqrt{\mathbf{v}_{E:2}^T A_{E:22}\mathbf{v}_{E:2}} \\
&\le \max_{E\in\mathcal{E}} \gamma_E \sum_{E\in\mathcal{E}} \sqrt{\mathbf{v}_{E:1}^T A_{E:11}\mathbf{v}_{E:1}}\sqrt{\mathbf{v}_{E:2}^T A_{E:22}\mathbf{v}_{E:2}} \\
&\le \max_{E\in\mathcal{E}} \gamma_E \sqrt{\sum_{E\in\mathcal{E}} \mathbf{v}_{E:1}^T A_{E:11}\mathbf{v}_{E:1}}\sqrt{\sum_{E\in\mathcal{E}} \mathbf{v}_{E:2}^T A_{E:22}\mathbf{v}_{E:2}} \\
&= \max_{E\in\mathcal{E}} \gamma_E \sqrt{\mathbf{v}_1^T A_{11}\mathbf{v}_1}\sqrt{\mathbf{v}_2^T A_{22}\mathbf{v}_2}
\end{aligned}$$

which completes the proof. □

2.3 Two-level preconditioning methods

2.3.1 Algebraic two-level methods

We consider now the additive (M_A) and the multiplicative (M_F) two-level preconditioners in a purely algebraic setting, starting with the following simplified variants:

$$M_A = \begin{bmatrix} A_{11} & \\ & A_{22} \end{bmatrix}, \tag{2.14}$$

$$M_F = \begin{bmatrix} A_{11} & \\ A_{21} & A_{22} \end{bmatrix} \begin{bmatrix} I & A_{11}^{-1}A_{12} \\ & I \end{bmatrix}. \tag{2.15}$$

In other words, the additive preconditioner M_A consists of the block-diagonal part of the original matrix, while the multiplicative preconditioner M_F is obtained by substituting with A_{22} the Schur complement S in the exact factorization (2.2).

The relative condition number of the introduced preconditioners can be estimated in terms of the CBS constant.

Theorem 2.6. *Let* $\mathbf{w} = \left[\begin{smallmatrix}\mathbf{v}_1\\ \mathbf{v}_2\end{smallmatrix}\right]$ *be a block-vector which is consistent with the two-by-two representation of* A*, see (2.1). Then*

$$\begin{aligned}
(1-\gamma)\left(\mathbf{v}_1^T A_{11}\mathbf{v}_1 + \mathbf{v}_2^T A_{22}\mathbf{v}_2\right) &\le \mathbf{w}^T A\mathbf{w} \\
&\le (1+\gamma)\left(\mathbf{v}_1^T A_{11}\mathbf{v}_1 + \mathbf{v}_2^T A_{22}\mathbf{v}_2\right), \quad (2.16)
\end{aligned}$$

and therefore

$$(1-\gamma)\mathbf{w}^T M_A \mathbf{w} \le \mathbf{w}^T A \mathbf{w} \le (1+\gamma)\mathbf{w}^T M_A \mathbf{w}. \qquad (2.17)$$

Proof. The definition of the CBS constant (2.5) is combined with the inequality $2ab \le a^2 + b^2$ to get the right hand side of (2.16):

$$\begin{aligned}
\mathbf{w}^T A\mathbf{w} &= \mathbf{v}_1^T A_{11}\mathbf{v}_1 + \mathbf{v}_2^T A_{22}\mathbf{v}_2 + \mathbf{v}_2^T A_{21}\mathbf{v}_1 + \mathbf{v}_1^T A_{12}\mathbf{v}_2 \\
&\le \mathbf{v}_1^T A_{11}\mathbf{v}_1 + \mathbf{v}_2^T A_{22}\mathbf{v}_2 + 2\gamma\sqrt{\mathbf{v}_1^T A_{11}\mathbf{v}_1}\sqrt{\mathbf{v}_2^T A_{22}\mathbf{v}_2} \\
&\le (1+\gamma)\left(\mathbf{v}_1^T A_{11}\mathbf{v}_1 + \mathbf{v}_2^T A_{22}\mathbf{v}_2\right).
\end{aligned}$$

Similarly, the inequality $-2ab \ge -a^2 - b^2$ is used to finalize the proof, namely

$$\begin{aligned}
\mathbf{w}^T A\mathbf{w} &= \mathbf{v}_1^T A_{11}\mathbf{v}_1 + \mathbf{v}_2^T A_{22}\mathbf{v}_2 + \mathbf{v}_2^T A_{21}\mathbf{v}_1 + \mathbf{v}_1^T A_{12}\mathbf{v}_2 \\
&\ge \mathbf{v}_1^T A_{11}\mathbf{v}_1 + \mathbf{v}_2^T A_{22}\mathbf{v}_2 - 2\gamma\sqrt{\mathbf{v}_1^T A_{11}\mathbf{v}_1}\sqrt{\mathbf{v}_2^T A_{22}\mathbf{v}_2} \\
&\ge (1-\gamma)\left(\mathbf{v}_1^T A_{11}\mathbf{v}_1 + \mathbf{v}_2^T A_{22}\mathbf{v}_2\right).
\end{aligned}$$

□

Theorem 2.7. *The following estimates hold for the multiplicative two-level preconditioner*

$$(1-\gamma^2)\mathbf{w}^T M_F \mathbf{w} \le \mathbf{w}^T A \mathbf{w} \le \mathbf{w}^T M_F \mathbf{w}. \qquad (2.18)$$

Proof. The statement follows directly from (2.9) and the relations

$$A = \begin{bmatrix} I & \\ A_{21}A_{11}^{-1} & I \end{bmatrix} \begin{bmatrix} A_{11} & \\ & S \end{bmatrix} \begin{bmatrix} I & A_{11}^{-1}A_{12} \\ & I \end{bmatrix}$$

and

$$M_F = \begin{bmatrix} I & \\ A_{21}A_{11}^{-1} & I \end{bmatrix} \begin{bmatrix} A_{11} & \\ & A_{22} \end{bmatrix} \begin{bmatrix} I & A_{11}^{-1}A_{12} \\ & I \end{bmatrix}.$$

Then, the following chain of relations proves the left hand side of (2.18)

$$\begin{aligned}
\mathbf{w}^T A\mathbf{w} &= \hat{\mathbf{w}}^T \begin{bmatrix} A_{11} & \\ & S \end{bmatrix} \hat{\mathbf{w}} \\
&= \hat{\mathbf{v}}_1^T A_{11}\hat{\mathbf{v}}_1 + \hat{\mathbf{v}}_2^T S\hat{\mathbf{v}}_2 \\
&\ge \hat{\mathbf{v}}_1^T A_{11}\hat{\mathbf{v}}_1 + (1-\gamma^2)\hat{\mathbf{v}}_2^T A_{22}\hat{\mathbf{v}}_2 \\
&\ge (1-\gamma^2)\hat{\mathbf{w}}^T \begin{bmatrix} A_{11} & \\ & A_{22} \end{bmatrix} \hat{\mathbf{w}} \\
&= (1-\gamma^2)\mathbf{w}^T M_F \mathbf{w},
\end{aligned}$$

where $\mathbf{w} = \left[\begin{smallmatrix}\mathbf{v}_1\\ \mathbf{v}_2\end{smallmatrix}\right]$ and $\hat{\mathbf{w}} = \left[\begin{smallmatrix}\hat{\mathbf{v}}_1\\ \hat{\mathbf{v}}_2\end{smallmatrix}\right] = \left[\begin{smallmatrix} I & A_{11}^{-1}A_{12}\\ & I\end{smallmatrix}\right]\mathbf{w}$. The right hand side of (2.18) follows similarly, applying the right hand side of (2.9). □

Corollary 2.8. *The results of the last two theorems, namely (2.17) and (2.18), are summarized in the following relative condition number estimates of the considered additive and multiplicative two-level preconditioners*

$$\kappa\left(M_A^{-1}A\right) \le \frac{1+\gamma}{1-\gamma}, \tag{2.19}$$

$$\kappa\left(M_F^{-1}A\right) \le \frac{1}{1-\gamma^2}. \tag{2.20}$$

We end this section with a brief presentation of a more general setting of the basic two-level preconditioners, which are defined under the assumptions

$$\alpha_1 A_{11} \le C_{11} \le \beta_1 A_{11} \tag{2.21}$$

and

$$\alpha_2 A_{22} \le C_{22} \le \beta_2 A_{22} \tag{2.22}$$

respectively

$$\tilde{\alpha}_2 A_{22} \le C_{22} + A_{21}C_{11}^{-1}A_{12} \le \tilde{\beta}_2 A_{22}, \tag{2.23}$$

where we assume that the approximations C_{11} and C_{22} are scaled such that

$$\beta_2 \le \beta_1 \tag{2.24}$$

and

$$\gamma^2 < \alpha_1 \le 1 \le \beta_1, \tag{2.25}$$

$$\gamma^2 \le \tilde{\alpha}_2 \le 1 \le \tilde{\beta}_2. \tag{2.26}$$

The inequalities (2.21)–(2.23) are in a positive semidefinite sense where C_{ii} are symmetric and positive definite matrices.

(a) The generalized additive preconditioner is introduced as

$$M_A = \begin{bmatrix} C_{11} & 0\\ 0 & C_{22}\end{bmatrix}. \tag{2.27}$$

Then, under the assumptions (2.21), (2.22) and (2.24), for $\varkappa_A = \varkappa(M_A^{-1}A)$, we have

$$\begin{aligned}\varkappa_A \;\le\; & \frac{\beta_1}{\alpha_1(1-\gamma^2)}\left\{\frac{1}{2}\left(1+\frac{\alpha_1}{\alpha_2}\right)+\sqrt{\left[\frac{1}{2}\left(1-\frac{\alpha_1}{\alpha_2}\right)\right]^2+\frac{\alpha_1}{\alpha_2}\gamma^2}\right\}\\ & \times\left\{\frac{1}{2}\left(1+\frac{\beta_2}{\beta_1}\right)+\sqrt{\left[\frac{1}{2}\left(1-\frac{\beta_2}{\beta_1}\right)\right]^2+\frac{\beta_2}{\beta_1}\gamma^2}\right\}.\end{aligned} \tag{2.28}$$

(b) The generalized multiplicative preconditioner (or of block Gauss–Seidel form) is then defined by

$$M_F = \begin{bmatrix} C_{11} & 0 \\ A_{21} & C_{22} \end{bmatrix} \begin{bmatrix} I_1 & C_{11}^{-1}A_{12} \\ 0 & I_2 \end{bmatrix}. \tag{2.29}$$

Under the assumptions (2.21), (2.23), (2.25), and (2.26) the relative condition number $\varkappa_F = \varkappa(M_F^{-1}A)$ satisfies the estimate

$$\varkappa_F \leq \frac{\beta_1 + \tilde{\beta}_2 - 2\gamma^2 + \sqrt{(\beta_1 - \tilde{\beta}_2)^2 + 4(1-\beta_1)(1-\tilde{\beta}_2)\gamma^2}}{\alpha_1 + \tilde{\alpha}_2 - 2\gamma^2 + \sqrt{(\alpha_1 - \tilde{\alpha}_2)^2 + 4(1-\alpha_1)(1-\tilde{\alpha}_2)\gamma^2}}. \tag{2.30}$$

Detailed proofs of (2.28) and (2.30), and an analysis of other versions of constructing M_A and M_F are found, for instance, in [5] and [11].

2.3.2 Two-level preconditioners for FEM systems

The classical theory of the optimal order two-level preconditioners for FEM systems was first developed in [11, 24], see also [6]. The crucial question is how to construct a two-by-two splitting of the stiffness matrix such that the related CBS constant is far away from the upper limit of one.

The general framework requires to define two nested finite element spaces $\mathcal{V}_H \subset \mathcal{V}_h$, that correspond to two consecutive (regular) mesh refinements. The well-studied case of conforming linear finite elements is the starting point in the theory of two-level and multilevel methods. Let $\mathcal{T}_H$ and $\mathcal{T}_h$ be two successive mesh refinements of the domain Ω, which correspond to $\mathcal{V}_H$ and $\mathcal{V}_h$. Let $\{\phi_H^{(k)}, k = 1, 2, \ldots, N_H\}$ and $\{\phi_h^{(k)}, k = 1, 2, \ldots, N_h\}$ be the standard finite element nodal basis functions. We split the meshpoints $\mathbf{N}_h$ from $\mathcal{T}_h$ into two groups: the first group contains the nodes $\mathbf{N}_H$ from $\mathcal{T}_H$ and the second one consists of the rest, where the latter are the newly added node-points $\mathbf{N}_{h\backslash H}$ from $\mathcal{T}_h \backslash \mathcal{T}_H$. Next we define the so-called hierarchical basis functions

$$\{\tilde{\phi}_h^{(k)}, k = 1, 2, \ldots, N_h\} = \{\phi_H^{(l)} \text{ on } \mathcal{T}_H\} \cup \{\phi_h^{(m)} \text{ on } \mathcal{T}_h \backslash \mathcal{T}_H\}. \tag{2.31}$$

Let then $\tilde{A}_h$ be the corresponding hierarchical stiffness matrix . Under the splitting (2.31) both matrices A_h and $\tilde{A}_h$ admit in a natural way a two-by-two block structure

$$A_h = \begin{bmatrix} A_{11} & A_{12} \\ A_{21} & A_{22} \end{bmatrix} \begin{matrix} \}N_{h\backslash H} \\ \}N_H \end{matrix} \tag{2.32}$$

and

$$\tilde{A}_h = \begin{bmatrix} \tilde{A}_{11} & \tilde{A}_{12} \\ \tilde{A}_{21} & \tilde{A}_{22} \end{bmatrix} = \begin{bmatrix} A_{11} & \tilde{A}_{12} \\ \tilde{A}_{21} & A_H \end{bmatrix} \begin{matrix} \}N_{h\backslash H} \\ \}N_H \end{matrix} . \tag{2.33}$$

It is well known that for the considered linear conforming finite elements, the transformation matrix which relates the nodal point vectors for the standard and the hierarchical basis functions has the form $J_h = \left[\begin{smallmatrix} I & J_{12} \\ 0 & I \end{smallmatrix}\right]$, i.e.,

$$\mathbf{v} = \begin{bmatrix} \mathbf{v}_1 \\ \mathbf{v}_2 \end{bmatrix} = J_h \begin{bmatrix} \tilde{\mathbf{v}}_1 \\ \tilde{\mathbf{v}}_2 \end{bmatrix}, \qquad \begin{array}{l} \mathbf{v}_1 = \tilde{\mathbf{v}}_1 + J_{12}\tilde{\mathbf{v}}_2 \\ \mathbf{v}_2 = \tilde{\mathbf{v}}_2 \end{array}.$$

Clearly, the hierarchical stiffness matrix $\tilde{A}_h$ is more dense than A_h and therefore its action on a vector is computationally more expensive. The transformation matrix J_h, however, enables us in practical implementations to work with A_h, since $\tilde{A}_h = J_h^T A_h J_h$. It is also important to note that the transformation of the nodal basis matrix A_h to the hierarchical basis matrix $\tilde{A}_h$ does not change the Schur complement, i.e.,

$$S = A_{22} - A_{21}A_{11}^{-1}A_{12} = \tilde{A}_{22} - \tilde{A}_{21}\tilde{A}_{11}^{-1}\tilde{A}_{12} = \tilde{S}. \tag{2.34}$$

It is easy to check that in the case of conforming linear finite elements the hierarchical two-level splitting (2.33) satisfies the conditions of Lemma 2.5. The index set $\mathcal{E}$ consists of the standard macroelements $E \in \mathcal{T}_h$, and A_E are the macroelement stiffness matrices. Then Lemma 2.4 implies the following simple rule to compute the local CBS constant:

$$\gamma_E^2 = 1 - \mu_1, \tag{2.35}$$

where μ_1 is the minimal eigenvalue of the generalized eigenproblem

$$\tilde{S}_E \mathbf{v}_{E:2} = \mu A_e \mathbf{v}_{E:2}, \qquad \mathbf{v}_{E:2} \neq \mathbf{c}, \tag{2.36}$$

$\mathbf{c}^T = (c, c, \ldots, c)$, c is a real constant. Here, the macroelement $E \in \mathcal{T}_h$ is a (usually uniform) refinement of the current coarser grid element $e \in \mathcal{T}_H$.

In the case of nonconforming finite elements, the local analysis follows the same scheme, where the definition of a two-level hierarchical basis is a key problem (see, e.g., [31, 60, 90]). The techniques in the recently studied cases of discontinuous Galerkin (DG) methods (see [81, 83]) are more specific, see Chapter 8. The examples there well illustrate the potential for novel applications provided by the generalized setting of Lemma 2.5. What is important to note is the leading general principle of local analysis, which will be demonstrated in the next chapters for various advanced nonconforming FE and DG discretizations.

2.4 Linear AMLI methods

In what follows we will denote by $M^{(k)}$ a preconditioner for a finite element (stiffness) matrix $A^{(k)}$ corresponding to a k times refined mesh ($0 \leq k \leq \ell$). We will also make use of the corresponding k-th level hierarchical matrix $\tilde{A}^{(k)}$, which is related to $A^{(k)}$ via a two-level hierarchical basis (HB) transformation $J^{(k)}$, i.e.,

$$\tilde{A}^{(k)} = (J^{(k)})^T A^{(k)} J^{(k)}. \tag{2.37}$$

By $A_{ij}^{(k)}$ and $\tilde{A}_{ij}^{(k)}$, $1 \leq i, j \leq 2$, we denote the blocks of $A^{(k)}$ and $\tilde{A}^{(k)}$ that correspond to the fine-coarse partitioning of degrees of freedom (DOF) where the DOF associated with the coarse mesh are numbered last.

The multilevel methods have evolved from two-level methods. The straightforward recursive extension leads to the class of HB methods for which the condition number grows in general exponentially with the number of levels ℓ. Therefore, in order to obtain multilevel preconditioners (of both additive or multiplicative type) with an optimal order condition number, i.e.,

$$\varkappa(M^{(\ell)^{-1}} A^{(\ell)}) = O(1),$$

and optimal computational complexity (linearly proportional to the number of degrees of freedom N_ℓ at the finest discretization level), HB preconditioners are combined with various types of stabilization techniques.

One particular purely algebraic stabilization technique is the so-called Algebraic Multilevel Iteration (AMLI) method, where a specially constructed matrix polynomial P_{υ_k} of degree υ_k is used at some or all of the levels $k = k_0 + 1, \ldots, \ell$. The AMLI methods have originally been introduced and studied in a multiplicative form, see [16, 17]. The presentation of the next three sections follows reference [72].

The task is to build a preconditioner $M^{(\ell)}$ for the coefficient matrix $A^{(\ell)} := A_h$, cf. (2.33), at the level of the finest mesh.

Starting at level 0 (associated with the coarsest mesh) on which a complete LU factorization of the matrix $A^{(0)}$ is performed, we define

$$M^{(0)} := A^{(0)}. \tag{2.38}$$

Given the preconditioner $M^{(k-1)}$ at level $k-1$ the preconditioner $M^{(k)}$ at level k is defined by

$$M^{(k)^{-1}} := U^{(k)} D^{(k)} L^{(k)} \tag{2.39}$$

where

$$U^{(k)} := \begin{bmatrix} I & -C_{11}^{(k)^{-1}} \tilde{A}_{12}^{(k)} \\ 0 & I \end{bmatrix}, \quad L^{(k)} := \begin{bmatrix} I & 0 \\ -\tilde{A}_{21}^{(k)} C_{11}^{(k)^{-1}} & I \end{bmatrix} \tag{2.40}$$

and

$$D^{(k)} := \begin{bmatrix} C_{11}^{(k)^{-1}} & 0 \\ 0 & Z^{(k-1)^{-1}} \end{bmatrix}. \tag{2.41}$$

Here we use the approximation

$$Z^{(k-1)^{-1}} := \left(I - P^{(k)}(M^{(k-1)^{-1}}A^{(k-1)})\right)\left(A^{(k-1)}\right)^{-1} \tag{2.42}$$

to the Schur complement $S = A^{(k-1)} - \tilde{A}_{21}^{(k)} C_{11}^{(k)^{-1}} \tilde{A}_{12}^{(k)}$ where $A^{(k-1)} := A_H = \tilde{A}_{22}^{(k)}$ is the coarse-level stiffness matrix (stiffness matrix at level $k-1$), which can be obtained from the two-level hierarchical basis representation (2.33) at level k, and $P^{(k)}$ is a polynomial of degree υ_k satisfying[1]

$$P^{(k)}(0) = 1. \tag{2.43}$$

Then one finds that (2.42) is equivalent to

$$Z^{(k-1)^{-1}} = M^{(k-1)^{-1}} Q^{(k)}(A^{(k-1)} M^{(k-1)^{-1}}) \tag{2.44}$$

where the polynomial $Q^{(k)}$ is given by

$$Q^{(k)}(t) = \frac{1 - P^{(k)}(t)}{t}. \tag{2.45}$$

The preconditioning step (1.94) in the PCG method, see Algorithm 1.20, requires the solution of a system $M^{(\ell)}\mathbf{v}^{(\ell)} = \mathbf{d}^{(\ell)}$ with the preconditioner $M^{(\ell)}$ where the right-hand side vector $\mathbf{d}^{(\ell)}$ is the actual residual of the linear system to solve. In the present context the action of the inverse of the preconditioner is implemented using a certain AMLI cycle. In order to describe different cycles formally, let $\boldsymbol{\upsilon} = (\upsilon_1, \upsilon_2, \dots, \upsilon_l)^T$ be a vector whose k-th component υ_k defines the degree of the stabilization polynomial at level $k = 1, 2, \dots, \ell$. Using first-order stabilization at all intermediate levels, i.e., $\upsilon_k = 1$, $P^{(k)}(t) = P_1(t) = 1 - p_{11}t$, corresponds to the so-called V-cycle AMLI and pure second-order stabilization, i.e., $\upsilon_k = 2$, $P^{(k)}(t) = P_2(t) = 1 - p_{21}t - p_{22}t^2$, results in the W-cycle iteration, $1 \leq k < \ell$. A more detailed algorithmic presentation of general (linear and nonlinear) AMLI preconditioning will be given in the last chapter of this book.

Let us now study the spectral condition number $\kappa(M^{(\ell)^{-1}}A^{(\ell)})$ where $M^{(\ell)}$ denotes the recursively defined linear AMLI preconditioner, cf. (2.38)–(2.42). The presented results are in the spirit of [10, 16, 17] and have the same recursive structure, that is, an estimate at level k involves the same type of estimate at level $k-1$.

Since $M^{(0)} = A^{(0)}$ implies

$$\lambda_{\min}(M^{(0)^{-1}}A^{(0)}) = \lambda_{\max}(M^{(0)^{-1}}A^{(0)}) = 1$$

[1] Further assumptions on $P^{(k)}$ will follow later in this section.

the bound[2]

$$\rho_0^{(k-1)} \leq \frac{\mathbf{u}^T M^{(k-1)^{-1}} \mathbf{u}}{\mathbf{u}^T (A^{(k-1)})^{-1} \mathbf{u}} \leq \rho_1^{(k-1)} \leq \lambda \quad \forall \mathbf{u} \neq \mathbf{0} \tag{2.46}$$

holds for $k-1=0$ and $\rho_0^{(0)} = \rho_1^{(0)} = 1$. Next, we consider the matrix

$$C^{(k)} = \begin{bmatrix} C_{11}^{(k)} & \tilde{A}_{12}^{(k)} \\ \tilde{A}_{21}^{(k)} & A^{(k-1)} + \tilde{A}_{21}^{(k)} (C_{11}^{(k)})^{-1} \tilde{A}_{12}^{(k)} \end{bmatrix} \tag{2.47}$$

corresponding to the multiplicative two-level preconditioner at level k. Its inverse has the decomposition

$$C^{(k)^{-1}} = L^{(k)T} \overline{D}^{(k)} L^{(k)}$$

where

$$\overline{D}^{(k)} = \begin{bmatrix} C_{11}^{(k)^{-1}} & 0 \\ 0 & (A^{(k-1)})^{-1} \end{bmatrix}$$

and the matrix $L^{(k)}$ is given by (2.40).

The basic assumption in the analysis of the multilevel preconditioner is an approximation property of the form

$$1 \leq \frac{\mathbf{v}^T A^{(k)} \mathbf{v}}{\mathbf{v}^T C^{(k)} \mathbf{v}} \leq \vartheta_k \leq \vartheta \quad \forall \mathbf{v} \neq \mathbf{0}, \quad k = 1, 2, \ldots, \ell, \tag{2.48}$$

cf. (2.30). Moreover, let

$$Q^{(k)}(t) \geq 0 \quad \forall t \in I_\sigma^{(k-1)} := [\rho_0^{(k-1)}, \rho_1^{(k-1)}], \tag{2.49}$$

and

$$r_0^{(k-1)} := \min_{t \in I_\sigma^{(k-1)}} t\, Q^{(k)}(t), \tag{2.50a}$$

$$r_1^{(k-1)} := \max_{t \in I_\sigma^{(k-1)}} t\, Q^{(k)}(t). \tag{2.50b}$$

Then for any vector $\mathbf{y}_2 \neq \mathbf{0}$ and $\mathbf{x}_2 := (A^{(k-1)})^{-1/2} \mathbf{y}_2$ we have

$$\begin{aligned} & \frac{\mathbf{y}_2^T M^{(k-1)^{-1}} Q^{(k)}(A^{(k-1)} M^{(k-1)^{-1}}) \mathbf{y}_2}{\mathbf{y}_2^T (A^{(k-1)})^{-1} \mathbf{y}_2} \\ = & \frac{\mathbf{x}_2^T (A^{(k-1)})^{1/2} M^{(k-1)^{-1}} Q^{(k)}(A^{(k-1)} M^{(k-1)^{-1}}) (A^{(k-1)})^{1/2} \mathbf{x}_2}{\mathbf{x}_2^T \mathbf{x}_2} \\ = & \frac{\mathbf{x}_2^T X Q^{(k)}(X) \mathbf{x}_2}{\mathbf{x}_2^T \mathbf{x}_2} \end{aligned}$$

[2]or equivalently $\rho_0^{(k-1)} \leq \lambda_{\min}(M^{(k-1)^{-1}} A^{(k-1)}) \leq \lambda_{\max}(M^{(k-1)^{-1}} A^{(k-1)}) \leq \rho_1^{(k-1)}$

where $X = (A^{(k-1)})^{1/2} M^{(k-1)^{-1}} (A^{(k-1)})^{1/2}$. Hence we have the relation

$$r_0^{(k-1)} \mathbf{y}_2^T A^{(k-1)^{-1}} \mathbf{y}_2 \leq \mathbf{y}_2^T Z^{(k-1)^{-1}} \mathbf{y}_2 \leq r_1^{(k-1)} \mathbf{y}_2^T A^{(k-1)^{-1}} \mathbf{y}_2 \quad \forall k \tag{2.51}$$

for (2.44). Now, let $\mathbf{v}$ be a fixed nonzero vector, and $\mathbf{y} = (\mathbf{y}_1^T, \mathbf{y}_2^T)^T = L^{(k)}\mathbf{v}$. Then

$$\frac{\mathbf{v}^T M^{(k)^{-1}} \mathbf{v}}{\mathbf{v}^T C^{(k)^{-1}} \mathbf{v}} = \frac{\mathbf{y}^T D^{(k)} \mathbf{y}}{\mathbf{y}^T \overline{D}^{(k)} \mathbf{y}} = \frac{\mathbf{y}_1^T C_{11}^{(k)^{-1}} \mathbf{y}_1 + \mathbf{y}_2^T Z^{(k-1)^{-1}} \mathbf{y}_2}{\mathbf{y}_1^T C_{11}^{(k)^{-1}} \mathbf{y}_1 + \mathbf{y}_2^T A^{(k-1)^{-1}} \mathbf{y}_2}, \tag{2.52}$$

and thus, by using (2.51), we arrive at

$$\min\{1, r_0^{(k-1)}\} \leq \frac{\mathbf{v}^T M^{(k)^{-1}} \mathbf{v}}{\mathbf{v}^T C^{(k)^{-1}} \mathbf{v}} \leq \max\{1, r_1^{(k-1)}\} \quad \forall \mathbf{v} \neq \mathbf{0}. \tag{2.53}$$

Together with (2.48) this gives

$$\begin{aligned} \rho_0^{(k)} := \min\{1, r_0^{(k-1)}\} \quad &\leq \quad \frac{\mathbf{v}^T M^{(k)^{-1}} \mathbf{v}}{\mathbf{v}^T A^{(k)^{-1}} \mathbf{v}} \\ &\leq \quad \vartheta_k \max\{1, r_1^{(k-1)}\} =: \rho_1^{(k)} \quad \forall \mathbf{v} \neq \mathbf{0}. \end{aligned} \tag{2.54}$$

For $\rho_0^{(k-1)} \geq 1$, by choosing $Q^{(k)}(t) := Q(t)$ such that

$$t\,Q(t) \geq 1 \quad \forall t \in [1, \rho_1^{(k-1)}]$$

it follows that $\rho_0^{(k)} \geq 1$ and hence $1 \leq \lambda_{\min}(M^{(k)^{-1}} A^{(k)})$ for $k = 1, 2, \ldots, \ell$. On the other hand, λ becomes a uniform upper bound for $\lambda_{\max}(M^{(k)^{-1}} A^{(k)})$ if the condition

$$\rho_1^{(k)} = \vartheta_k \max_{t \in I_\sigma^{(k-1)}} t\,Q^{(k)}(t) \leq \lambda \tag{2.55}$$

can be met for all k. In case of a second-degree stabilization polynomial $P^{(k)}$, which implies that Q is a linear function, cf. (2.45), the best choice of Q is given by

$$Q(t) = \frac{\lambda + 1}{\lambda} - \frac{1}{\lambda} t, \tag{2.56}$$

which corresponds to

$$P^{(k)}(t) = P_2(t) := 1 - \left(1 + \frac{1}{\lambda}\right) t + \frac{1}{\lambda} t^2. \tag{2.57}$$

Note that in this case Q is the linear function with the smallest maximum of $t\,Q(t)$ subject to the condition that $t\,Q(t) \geq 1$ for all t in $[1, \lambda]$. The maximum is achieved for $t = (1 + \lambda)/2$ in which case (2.55) reduces to

$$\vartheta_k \leq \vartheta \leq \frac{4\lambda^2}{(\lambda + 1)^2} \quad \forall k.$$

Then

$$\lambda \leq \frac{\vartheta + 2\sqrt{\vartheta}}{4 - \vartheta}. \tag{2.58}$$

For the V-cycle method, i.e., for $P^{(k)}$ being a linear function and therefore Q being constant, the condition $t\,Q(t) \geq 1$ for $t \geq 1$ results in the optimal choice $Q(t) = 1$, which corresponds to

$$P^{(k)}(t) = P_1(t) := 1 - t. \tag{2.59}$$

We summarize the results for the V- and W-cycle of linear AMLI in a theorem.

Theorem 2.9. *Consider an SPD matrix $A^{(\ell)}$ and the preconditioner $M^{(\ell)}$ defined by (2.38)–(2.42) and assume that the approximation property (2.48) holds. Then the (linear) AMLI V-cycle preconditioner $M_V^{(\ell)}$, associated with the polynomial (2.59), satisfies*

$$\kappa(M_V^{(\ell)^{-1}} A^{(\ell)}) \leq \prod_{k=1}^{\ell} \vartheta_k \leq \vartheta^\ell. \tag{2.60}$$

Moreover, if (2.48) holds for some $\vartheta < 4$, then the relative condition number of the (linear) AMLI W-cycle preconditioner $M_W^{(\ell)}$, associated with the polynomial (2.57), is bounded by

$$\kappa(M_W^{(\ell)^{-1}} A^{(\ell)}) \leq \frac{\vartheta + 2\sqrt{\vartheta}}{4 - \vartheta} =: \lambda(\vartheta) \tag{2.61}$$

Proof. It remains to prove (2.60). Using (2.54) under the hypothesis (2.46) the bound follows immediately by induction. □

2.5 Nonlinear AMLI methods

As we observed in the analysis of the previous section, the W-cycle of linear AMLI is quite sensitive to a proper choice of the polynomial that is used to define the matrix $Z^{(k-1)}$. In practice, this demands tight bounds for the spectrum of the preconditioned matrix $M^{(k-1)^{-1}} A^{(k-1)}$, which are computationally expensive.

That is why we will also present a parameter-free algorithm in this section. It is based on inner GCG-type iterations resulting in variable-step preconditioners that define nonlinear mappings in general. The considered algorithm is similar to variable-step preconditioning methods as they were introduced in [19].

Typically, the nonlinear AMLI yields a preconditioner that is close to a linear mapping, and this property can be exploited to derive a theoretical bound on its rate of convergence.

Let us first define the nonlinear multilevel preconditioner $M^{(k)}[\cdot]$ at level k, $1 \leq k \leq \ell$, that is (in general) a nonlinear mapping from $\mathbb{R}^{N_k}$ to $\mathbb{R}^{N_k}$:

$$M^{(k)^{-1}}[\mathbf{y}] := U^{(k)} D^{(k)}[L^{(k)}\mathbf{y}] \tag{2.62}$$

The matrices $U^{(k)}$ and $L^{(k)}$ are given by (2.40) and

$$D^{(k)}[\mathbf{z}] = \left[\begin{array}{c} C_{11}^{(k)^{-1}} \mathbf{z}_1 \\ Z^{(k-1)^{-1}}[\mathbf{z}_2] \end{array} \right]. \tag{2.63}$$

The (nonlinear) mapping $Z^{(k-1)}[\cdot]$ is defined by

$$\begin{array}{lcll} Z^{(0)}[\cdot] & := & A^{(0)} & \\ Z^{(k)}[\cdot] & := & M^{(k)}[\cdot] & \text{if } \upsilon = 1 \text{ and } k > 0 \\ Z^{(k)}[\cdot] & := & M_{\upsilon}^{(k)}[\cdot] & \text{if } \upsilon > 1 \text{ and } k > 0 \end{array} \tag{2.64}$$

where

$$M_{\upsilon}^{(k)^{-1}}[\mathbf{d}] := \mathbf{x}_{(\upsilon)}$$

and $\mathbf{x}_{(\upsilon)}$ is the υ-th iterate obtained when applying the GCG algorithm, see Algorithm 1.22, to the linear system $A^{(k)}\mathbf{x} = \mathbf{d}$ thereby using $M^{(k)}[\cdot]$ as a preconditioner and starting with the initial guess $\mathbf{x}_{(0)} = \mathbf{0}$. In the setting of the nonlinear AMLI preconditioner the vector $\boldsymbol{\upsilon} = (\upsilon_1, \upsilon_2, \ldots, \upsilon_\ell)^T$ specifies how many inner GCG iterations are performed at each of the levels $k = 1, 2, \ldots, \ell - 1$, and $\upsilon_\ell > 0$ denotes the maximum number of orthogonal search directions at level ℓ (the fine-grid level). The additional GCG-type variable-step iterations on certain levels (those levels k for which $\upsilon_k > 1$) involve the use of again the same type of variable-step preconditioner. We restrict our analysis here to the case in which a fixed number υ of inner GCG-type iterations is performed at every intermediate level, that is, employing the vector

$$\boldsymbol{\upsilon} = \boldsymbol{\upsilon}_W = [\upsilon, \upsilon, \ldots, \upsilon, m_{\max}]^T \tag{2.65}$$

where the algorithm is restarted at level ℓ at every $m_{\max}$ iterations, cf. Section 1.4.4. We will refer to this choice as the (parameter-free) nonlinear υ-fold W-cycle

AMLI. Without any inner iterations (and without restart at level ℓ) the considered nonlinear AMLI reduces to the V-cycle of linear AMLI presented in the last section where also the GCG method at level ℓ reduces to the standard PCG algorithm, cf. Algorithm 1.20.

Let us now study the convergence properties of the nonlinear AMLI method. That means, we want to derive bounds on the (local) decrease of the error in the norm induced by the coefficient matrix $A := A^{(\ell)}$. If $\mathbf{x}_{(i)}$ denotes the i-th iterate generated by the nonlinear AMLI, we aim at deriving a bound of the form

$$\frac{\|\mathbf{x} - \mathbf{x}_{(i+1)}\|_A}{\|\mathbf{x} - \mathbf{x}_{(i)}\|_A} \leq \delta < 1. \tag{2.66}$$

A very general result of this kind has first been proven in [4] and has been applied in [18] to indefinite problems. Based on this result a convergence theory for variable-step multilevel preconditioning methods has been established in [19] with a focus on hierarchical basis matrices. However, the assumptions in [19] result in quite pessimistic estimates for the convergence rate of the nonlinear AMLI method. As it has been shown in [96] the bound on the local decrease of the error in GCG-type iterations can be improved considerably if the nonlinear preconditioner becomes close to a linear operator.

We will first formulate a useful corollary to Theorem 1.23, see [72].

Corollary 2.10. *Consider the matrices $A^{(k)}$ as well as the approximations $C^{(k)}$ defined by (2.47) where $A^{(\ell)}$ is assumed to be an SPD matrix. If υ is some positive integer, $M^{(k)}[\cdot]$ is the preconditioner defined by (2.62)–(2.64), and*

$$\frac{\|C^{(k)} {M^{(k)}}^{-1}[\mathbf{v}] - \mathbf{v}\|_{{C^{(k)}}^{-1}}}{\|\mathbf{v}\|_{{C^{(k)}}^{-1}}} \leq \epsilon_k < 1 \quad \forall \mathbf{v} \neq \mathbf{0} \tag{2.67}$$

then

$$\frac{\|A^{(k)} {M_\upsilon^{(k)}}^{-1}[\mathbf{v}] - \mathbf{v}\|_{(A^{(k)})^{-1}}}{\|\mathbf{v}\|_{(A^{(k)})^{-1}}} \leq \delta_k(\upsilon) \quad \forall \mathbf{v} \neq \mathbf{0} \tag{2.68}$$

where

$$\delta_k(\upsilon) = \left(1 - \frac{4\kappa(1-\epsilon_k)^2}{(1+\kappa-2\epsilon_k+\kappa\epsilon_k^2)^2}\right)^{\upsilon/2} \tag{2.69}$$

and $\kappa = \kappa({C^{(k)}}^{-1} A^{(k)})$.

Proof. The assumptions on $A^{(\ell)}$ imply that $A^{(k)}$ and $C^{(k)}$ are SPD. Let $\mathbf{v}$ be an arbitrary nonzero vector of dimension N_k. Then $\mathbf{x}_{(\upsilon)} := {M_\upsilon^{(k)}}^{-1}[\mathbf{v}]$ is the υ-th iterate of the GCG algorithm applied to the linear system $A^{(k)}\mathbf{x} = \mathbf{v}$ using the preconditioner $M^{(k)}[\cdot]$ and starting with the initial guess $\mathbf{x}_{(0)} = \mathbf{0}$. The corresponding residual is given by $\mathbf{r}_{(\upsilon)} = \mathbf{v} - A^{(k)}\mathbf{x}_{(\upsilon)}$. Taking into account (2.67), Theorem 1.23 shows that

$$\begin{aligned} &\|A^{(k)} M_{\upsilon}^{(k)^{-1}}[\mathbf{v}] - \mathbf{v}\|_{(A^{(k)})^{-1}} = \|\mathbf{r}_{(\upsilon)}\|_{(A^{(k)})^{-1}} \\ &= \|\mathbf{x} - \mathbf{x}_{(\upsilon)}\|_{A^{(k)}} \leq \delta_k(1)\|\mathbf{x} - \mathbf{x}_{(\upsilon-1)}\|_{A^{(k)}} \leq \ldots \leq \delta_k(\upsilon)\|\mathbf{x} - \mathbf{x}_{(0)}\|_{A^{(k)}} \\ &= \delta_k(\upsilon)\|\mathbf{r}_{(0)}\|_{(A^{(k)})^{-1}} = \delta_k(\upsilon)\|\mathbf{v}\|_{(A^{(k)})^{-1}}. \end{aligned}$$

□

The following lemma provides the key to the convergence analysis as presented in [72]. It relates the accuracy of the approximation of $A^{(k-1)}$ by the preconditioner $M_{\upsilon}^{(k-1)}[\cdot]$ at level $k-1$ to the accuracy of the approximation of $C^{(k)}$ by $M^{(k)}[\cdot]$.

Lemma 2.11. *Consider the same operators as in Corollary 2.10. If*

$$\frac{\|A^{(k-1)} M_{\upsilon}^{(k-1)^{-1}}[\mathbf{u}] - \mathbf{u}\|_{(A^{(k-1)})^{-1}}}{\|\mathbf{u}\|_{(A^{(k-1)})^{-1}}} \leq \delta_{k-1} \quad \forall \mathbf{u} \neq \mathbf{0}$$

then

$$\frac{\|C^{(k)} M^{(k)^{-1}}[\mathbf{v}] - \mathbf{v}\|_{C^{(k)^{-1}}}}{\|\mathbf{v}\|_{C^{(k)^{-1}}}} \leq \delta_{k-1} \quad \forall \mathbf{v} \neq \mathbf{0}.$$

Proof. Let $\mathbf{v}$ be an arbitrary (but fixed) nonzero vector. First we observe that

$$\frac{\|C^{(k)} M^{(k)^{-1}}[\mathbf{v}] - \mathbf{v}\|_{C^{(k)^{-1}}}}{\|\mathbf{v}\|_{C^{(k)^{-1}}}} = \frac{(C^{(k)} M^{(k)^{-1}}[\mathbf{v}] - \mathbf{v})^T (M^{(k)^{-1}}[\mathbf{v}] - C^{(k)^{-1}}\mathbf{v})}{\mathbf{v}^T C^{(k)^{-1}} \mathbf{v}}.$$

Let $\mathbf{y} = (\mathbf{y}_1^T, \mathbf{y}_2^T)^T = L^{(k)}\mathbf{v}$, where the partitioning of $\mathbf{y}$ is according to the splitting at level k. Then, since $Z^{(k-1)^{-1}}[\cdot] = M_{\upsilon}^{(k-1)^{-1}}[\cdot]$ we find

$$C^{(k)} M^{(k)^{-1}}[\mathbf{v}] - \mathbf{v} = L^{(k)^{-1}} \begin{bmatrix} \mathbf{0} \\ A^{(k-1)} M_{\upsilon}^{(k-1)^{-1}}[\mathbf{y}_2] - \mathbf{y}_2 \end{bmatrix},$$

$$M^{(k)^{-1}}[\mathbf{v}] - C^{(k)^{-1}}\mathbf{v} = L^{(k)^T} \begin{bmatrix} \mathbf{0} \\ M_{\upsilon}^{(k-1)^{-1}}[\mathbf{y}_2] - (A^{(k-1)})^{-1}\mathbf{y}_2 \end{bmatrix},$$

and

$$C^{(k)^{-1}}\mathbf{v} = L^{(k)^T} \begin{bmatrix} (C_{11}^{(k)})^{-1}\mathbf{y}_1 \\ (A^{(k-1)})^{-1}\mathbf{y}_2 \end{bmatrix}.$$

Thus,

$$\begin{aligned}
\frac{\|C^{(k)}M^{(k)^{-1}}[\mathbf{v}]-\mathbf{v}\|_{C^{(k)^{-1}}}}{\|\mathbf{v}\|_{C^{(k)^{-1}}}}
&= \frac{(A^{(k-1)}M_\upsilon^{(k-1)^{-1}}[\mathbf{y}_2]-\mathbf{y}_2)^T(M_\upsilon^{(k-1)^{-1}}[\mathbf{y}_2]-(A^{(k-1)})^{-1}\mathbf{y}_2)}{\mathbf{y}_1^T(C_{11}^{(k)})^{-1}\mathbf{y}_1+\mathbf{y}_2^T(A^{(k-1)})^{-1}\mathbf{y}_2} \\
&\le \frac{\|A^{(k-1)}M_\upsilon^{(k-1)^{-1}}[\mathbf{y}_2]-\mathbf{y}_2\|_{(A^{(k-1)})^{-1}}}{\|\mathbf{y}_2\|_{(A^{(k-1)})^{-1}}}.
\end{aligned}$$

□

The detailed algorithm of the nonlinear AMLI (including some implementation issues) will be presented in the last chapter. The main convergence result is stated now in the following theorem.

Theorem 2.12. *Consider the linear system $A^{(\ell)}\mathbf{x} = \mathbf{d}^{(\ell)}$ where $A^{(\ell)}$ is an SPD HB stiffness matrix, and, let $\mathbf{x}_{(i)}$ be the sequence of iterates generated by the nonlinear AMLI algorithm. Further, assume that the approximation property (2.48) holds. If υ, the number of inner GCG iterations at every coarse level (except the coarsest), is chosen such that*

$$\delta(\upsilon) := \left(1-\frac{4\vartheta(1-\epsilon)^2}{(1+\vartheta-2\epsilon+\vartheta\epsilon^2)^2}\right)^{\upsilon/2} \le \epsilon \tag{2.70}$$

for some positive $\epsilon < 1$ then

$$\frac{\|\mathbf{x}-\mathbf{x}_{(i+1)}\|_{A^{(\ell)}}}{\|\mathbf{x}-\mathbf{x}_{(i)}\|_{A^{(\ell)}}} \le \sqrt{1-\frac{4\vartheta(1-\epsilon)^2}{(1+\vartheta-2\epsilon+\vartheta\epsilon^2)^2}} = \delta(1) =: \delta < 1. \tag{2.71}$$

Proof. From the definition of $M^{(k)^{-1}}[\cdot]$, see (2.62)–(2.64), it follows that in the first step of recursion we have $M^{(1)^{-1}}[\cdot] = (C^{(1)})^{-1}$ and thus (2.67) holds for $k = 1$ and $\epsilon_1 = 0$. Now, Corollary 2.10 shows that the inequality (2.68) is valid for $\delta_1(\upsilon)$ given by (2.69) using $\epsilon_1 = 0$ and $\kappa = \vartheta$ where ϑ is the constant from the approximation property (2.48). Next, Lemma 2.11 yields $\epsilon_2 \le \delta_1(\upsilon)$. By induction we conclude

$$\epsilon_k \le \delta_{k-1}(\upsilon) \le \left(1-\frac{4\vartheta(1-\epsilon_{k-1})^2}{(1+\vartheta-2\epsilon_{k-1}+\vartheta\epsilon_{k-1}^2)^2}\right)^{\upsilon/2} \tag{2.72}$$

for any $k \geq 2$. Moreover, since the right-hand side of (2.72) approaches zero when υ increases, the sequence $(\epsilon_k)_{k=1,2,\ldots}$ is uniformly bounded by some $\epsilon < 1$ if υ is sufficiently large. Assuming that $\epsilon_{k-1} \leq \epsilon$ we find from (2.72) and (2.70) that $\epsilon_k \leq \delta(\upsilon) \leq \epsilon$ for all $k = 1, 2, \ldots$. Thus $\epsilon_\ell \leq \epsilon$ and the bound (2.71) follows from Theorem 1.23. □

In particular, for $\upsilon \geq 2$ condition (2.70) is satisfied if

$$1 - \epsilon \leq \frac{4\vartheta(1-\epsilon)^2}{(1 + \vartheta - 2\epsilon + \vartheta\epsilon^2)^2},$$

that is, if

$$\vartheta \leq f(\epsilon) = \frac{1 - \epsilon^2 + 2\epsilon^3 + 2\sqrt{\epsilon - 2\epsilon^2 + 3\epsilon^3 - 2\epsilon^4}}{(1+\epsilon^2)^2}. \tag{2.73}$$

Note that the condition number estimates for the two-level preconditioners (2.27) and (2.29) directly enter the approximation property (2.48). In the simplest case in which the multiplicative two-level preconditioner (2.47) is considered under the assumption $C_{11}^{(k)} = A_{11}^{(k)}$ this results in a direct relation between the CBS constant γ and the constant ϑ in (2.48), i.e., $\vartheta = 1/(1-\gamma^2)$, cf. Corollary 2.8.

We end this section with a remark on the comparison of the convergence factor of linear and nonlinear AMLI cycles.

Remark 2.13. The right-hand side of (2.73) has a unique maximum in the interval $(0, 1)$, which is achieved for $\epsilon = \bar{\epsilon} \approx 0.187248$. Using this value for ϵ in (2.70) shows that the sequence $(\epsilon_k)_{k=1,2,\ldots}$ is uniformly bounded by $\bar{\epsilon}$ if $\upsilon = 2$ and $\vartheta \leq 1.597$, or, $\upsilon = 3$ and $\vartheta \leq 2.298$, or, $\upsilon = 4$ and $\vartheta \leq 3.017$.

On the other hand, assuming for instance $\upsilon = 3$ and $\vartheta = 2$ we get $\epsilon < 0.0587$. Then, the bound (2.71) on the (local) error reduction factor in Theorem 2.12 is $\delta \approx 0.39$. For comparison, computing the condition number estimate (2.61) from Theorem 2.9 (using the same value of ϑ, i.e., $\kappa \leq (\vartheta + 2\sqrt{\vartheta})/(4-\vartheta) = 1 + \sqrt{2}$) the error reduction factor of the CG method preconditioned by the linear AMLI W-cycle can be bounded by $\delta \leq \frac{\kappa-1}{\kappa+1} \leq \sqrt{2} - 1 \approx 0.41$, cf. Theorem 1.19. It should be noted, however, that the derived estimates, especially for the nonlinear AMLI, are sometimes rather pessimistic, which can be seen from comparison with the available numerical tests.

2.6 Optimality conditions

For the case when $C_{11}^{(k)} = A_{11}^{(k)}$ at all levels $k = 1, 2, \ldots, \ell$ Theorem 2.9 shows that for the W-cycle of linear AMLI, i.e., for $\upsilon = 2$, the relative condition number can be stabilized by a second-order polynomial if

$$\sqrt{\vartheta} = \frac{1}{\sqrt{1-\gamma^2}} < 2 = \upsilon, \tag{2.74}$$

resulting in the bound (2.61). Thereby we assume that the approximation property (2.48) for the respective multiplicative two-level preconditioner holds, which in general allows also the usage of a proper approximation $C_{11}^{(k)} \neq A_{11}^{(k)}$. In fact, and this was shown in the original convergence analysis of linear AMLI methods [17], a stabilization of the condition number of the multiplicative multilevel preconditioner[3] can be achieved under the assumption

$$A_{11}^{(k)} \leq C_{11}^{(k)} \leq (1+\delta_1) A_{11}^{(k)} \tag{2.75}$$

on the approximation of the pivot block $A_{11}^{(k)}$ if

$$\frac{1}{\sqrt{1-\gamma^2}} < \upsilon. \tag{2.76}$$

Note that the spectral equivalence of $A_{11}^{(k)}$ and $C_{11}^{(k)}$ as formulated earlier in this chapter, see equation (2.21), can always be transformed into (2.75) by a proper scaling of $C_{11}^{(k)}$. It is also interesting to note that a stabilization of the condition number κ, i.e., a uniform bound on κ that is independent of the number of levels ℓ, is possible even for large(r) values of δ_1 in (2.75). It turns out that the effect of the constant in the spectral equivalence relation (2.75) enters almost linearly the relative condition number of the multilevel preconditioner in this case.

Assume now that we have a fully stabilized multilevel method. This means that our solution algorithm produces an approximation to the exact solution of the linear system arising from FE discretization of a given (boundary value) problem, thereby achieving a fixed prescribed reduction of the norm of the initial residual within a uniformly bounded number of iterations. Hence, this upper bound on the number of iterations does not depend on the meshsize h, i.e., the solutions for a repeatedly refined mesh (in principle for any number of regular refinement steps) are obtained at a constant number of iterations. Then the second condition to be fulfilled for an optimal-order solution process is that the computational cost of each single iteration is proportional to the total number of degrees of freedom (DOF). Finally, the third condition that is desirable in many situations is that the construction of the method, i.e., the setup of the preconditioner, is of optimal order of computational complexity. This last issue will be discussed in Chapter 10.

[3] by a properly shifted and scaled Chebyshev polynomial P_υ

The restriction on υ for optimal order of computational complexity per iteration is the following one. Consider a υ-fold W-cycle of either linear or nonlinear AMLI. Then every application of the recursively defined preconditioner at a given level k with a given number N_k of DOF involves υ applications of the preconditioner at level $k-1$ where the number N_{k-1} of DOF is smaller by some factor say ϱ. Hence, the computational work $w^{(k)}$ at level k can be estimated by

$$w^{(k)} \le c\, N_k + \upsilon\, w^{(k-1)}$$

for some constant c which depends on the (average) number of nonzero entries (per row) of the involved sparse matrices, e.g., the number of nonzeros per row of the incomplete triangular factors of the preconditioner $C_{11}^{(k)}$ if some incomplete factorization is used at this point. The work $w^{(\ell)}$ for one application of the preconditioner at level ℓ (associated with the finest mesh) can therefore be estimated by

$$\begin{aligned} w^{(\ell)} &\le c\,(N_\ell + \upsilon\, N_{\ell-1} + \ldots + \upsilon^\ell\, N_0) \\ &= c\, N_\ell \left(1 + \frac{\upsilon}{\varrho} + \left(\frac{\upsilon}{\varrho}\right)^2 + \ldots + \left(\frac{\upsilon}{\varrho}\right)^\ell\right) = c\, N_\ell \frac{1 - \left(\frac{\upsilon}{\varrho}\right)^{\ell+1}}{1 - \frac{\upsilon}{\varrho}}. \end{aligned}$$

Since the number of DOF at level $k-1$ is (assumed to be) $1/\varrho$ times the number of DOF at level k, each visit of level k must induce less than ϱ visits of level $k-1$ (at least in average). That means, if for example the coarsening ratio $\varrho = 4$ then two but also three inner GCG iterations, or, alternatively, the employment of second- but also third-degree matrix polynomials at every intermediate level, will result in a computational complexity $\mathcal{O}(N) = \mathcal{O}(N_\ell)$ of one (outer) iteration. The condition for optimal-order single iterations is thus

$$\upsilon < \varrho, \tag{2.77}$$

which combined with (2.76) results in the (combined) optimality conditions

$$\frac{1}{\sqrt{1-\gamma^2}} < \upsilon < \varrho. \tag{2.78}$$

In what follows we will assume that the default meaning of AMLI is the multiplicative one.

Remark 2.14. The optimality conditions for the symmetric preconditioner of block-diagonal (additive) form are given by

$$\sqrt{\frac{1+\gamma}{1-\gamma}} < \upsilon < \varrho. \tag{2.79}$$

Stabilization techniques for additive multilevel iteration methods and nearly optimal order parameter-free block-diagonal preconditioners of AMLI-type are discussed in references [6, 15, 97].

2.7 Robustness of the AMLI methods

2.7.1 Local analysis of the model problem

Let us consider the model problem corresponding to the bilinear form

$$\mathcal{A}(u,v) = \int_{\Omega} a(e)\left(\frac{\partial u}{\partial x_1}\frac{\partial v}{\partial x_1} + \frac{\partial u}{\partial x_2}\frac{\partial v}{\partial x_2}\right) d\Omega \tag{2.80}$$

where linear finite elements are used for discretization. Let us assume also that: a) the coarsest triangulation $\mathcal{T}_0$ of Ω consists of isosceles right triangles, the legs of which are aligned with the coordinate axes; b) the coefficient $a(\boldsymbol{x}) = a(e)$ is constant on the triangles $e \in \mathcal{T}_0$.

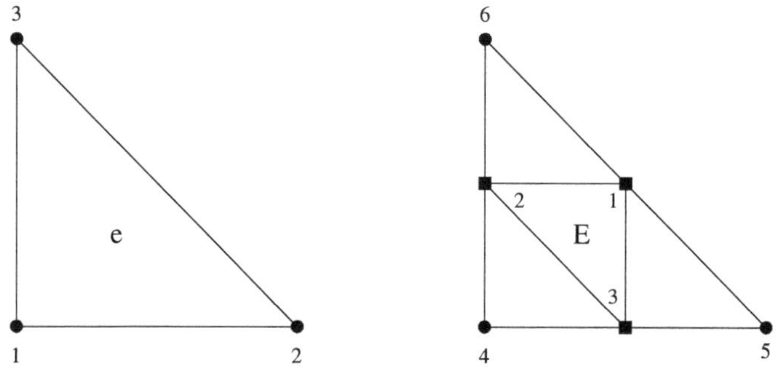

Figure 2.1: Element and macroelement for mesh of right triangles

Under the above assumptions, the local CBS constant γ_E is unique. The related element and macroelement stiffness matrices, and the Schur complement, which are needed for the local analysis, have the form

$$A_e = \frac{a(e)}{2}\begin{bmatrix} 2 & -1 & -1 \\ -1 & 1 & 0 \\ -1 & 0 & 1 \end{bmatrix},$$

$$A_E = \frac{a(e)}{2}\begin{bmatrix} 4 & -2 & -2 & 0 & 0 & 0 \\ -2 & 4 & 0 & -1 & 0 & -1 \\ -2 & 0 & 4 & -1 & -1 & 0 \\ 0 & -1 & -1 & 2 & 0 & 0 \\ 0 & 0 & -1 & 0 & 1 & 0 \\ 0 & -1 & 0 & 0 & 0 & 1 \end{bmatrix},$$

$$\tilde{S}_E = \frac{a(e)}{16} \begin{bmatrix} 8 & -4 & -4 \\ -4 & 5 & -1 \\ -4 & -1 & 5 \end{bmatrix},$$

where the node numbering is given in Figure 2.1. In this particular case, the 3×3 local eigenproblem (2.36) is reduced to the following 2×2 characteristic equation

$$\begin{vmatrix} 5 - 8\mu & -1 \\ -1 & 5 - 8\mu \end{vmatrix} = 0,$$

where the trivial eigenvalue, corresponding to the constant eigenvector, is excluded. Therefore, $1 - \gamma_E^2 = \mu_1 = 1/2$, that is $\gamma_E^2 = 1/2$, which implies the global estimate

$$\gamma^2 \leq \frac{1}{2}. \tag{2.81}$$

Now, the local approach will be applied to construct and analyze a simple approximation of the pivot block $A_{11} = A_{h:11}$ corresponding to the uniform triangulation $\mathcal{T}_h$, i.e.,

$$A_{11} = \sum_{E \in \mathcal{T}_h} R_E^T A_{E:11} R_E$$

where

$$A_{E:11} = \frac{a(e)}{2} \begin{bmatrix} 2 & -1 & -1 \\ -1 & 2 & 0 \\ -1 & 0 & 2 \end{bmatrix}.$$

We solve the local 3×3 eigenproblem

$$A_{E:11} \mathbf{v}_{E:1} = \lambda D_{E:11} \mathbf{v}_{E:1} \tag{2.82}$$

where $D_{E:11} = 2I$ is the diagonal part of $A_{E:11}$. The eigenvalues of (2.82) are as follows

$$\lambda_1 = \lambda_1(A_{E:11}) = \frac{a(e)}{4}(2 - \sqrt{2}), \qquad \lambda_3 = \lambda_3(A_{E:11}) = \frac{a(e)}{4}(2 + \sqrt{2}).$$

It is readily seen that

$$A_{E:11} \leq \lambda_3 D_{E:11} \leq \frac{\lambda_3}{\lambda_1} A_{E:11}. \tag{2.83}$$

Now, let us introduce the diagonal matrix

$$C_{11} = \sum_{E \in \mathcal{T}_h} \frac{a(e)}{4}(2 + \sqrt{2}) D_{E:11}. \tag{2.84}$$

The summation of the local estimates (2.83) in terms of (2.84) leads to the global estimate

$$A_{11} \leq C_{11} \leq \frac{2+\sqrt{2}}{2-\sqrt{2}} A_{11}. \tag{2.85}$$

Remark 2.15. For the considered model problem, the locally derived estimates (2.81) and (2.85) are uniform with respect to the size of the discrete problem (number of the mesh refinement levels), as well as to coefficient jumps on the coarsest mesh.

2.7.2 Robust preconditioning strategy

The assumption (2.75) is obviously satisfied for the diagonal matrix (2.84) with

$$\delta_1 = 2(1+\sqrt{2}),$$

see (2.85). Then, the next statement follows directly from (2.79), applying (2.81) and (2.75).

Corollary 2.16. *The AMLI preconditioner for the elliptic model problem (2.80) has optimal order of computational complexity. The optimality conditions (2.79) are satisfied for acceleration polynomials P_υ with a fixed degree $\upsilon \in \{2,3\}$. The related computational complexity estimates are robust with respect to coefficient jumps aligned with the coarsest triangulation $\mathcal{T}_0$.*

The general strategy for constructing efficient AMLI preconditioners is based on the assumption (2.75) and the optimality conditions (2.78) (or (2.79) when applicable). The following two conditions are fundamental for the robustness of AMLI algorithms:

- Proper uniform estimate of the CBS constant with respect to mesh and coefficient anisotropy, and/or other problem parameters.
- Optimal order preconditioning (approximation) of the pivot block A_{11} with respect to mesh and coefficient anisotropy, and/or other problem parameters.

"Proper" estimate of the CBS constant means that γ is far enough from one, so that the optimality conditions (2.79) or (2.78) can be satisfied. This requirement obviously depends on the space dimension $d \in \{2,3\}$ of the problem, which directly reflects on ϱ, the reduction factor of the number of degrees of freedom (DOF). A number of such estimates are presented in the next chapters, addressing different problems and different FE and DG discretizations. A key point in this respect is to combine the robustness of the AMLI algorithm with the robustness of the discretization which is governed by the problem under consideration. In the case of parameter-dependent problems, the goal is to get locking-free solution methods for systems arising after application of locking-free discretizations.

As we know, there is a general approach for deriving uniform estimates of the CBS constant, solving the related local eigenproblem (2.36). Contrary, the question how to construct robust preconditioners for the pivot block A_{11} is more complicated. Moreover, it is important to note that, unlike to the isotropic case, the condition number $\kappa(A_{11})$ deteriorates with the raise of anisotropy or/and when the related parameter tends to the limit case.

The commonly known theory of the optimal order solution methods for FEM elliptic problems is restricted to the case of coefficient jumps which are aligned with the coarse(st) geometric splitting, i.e. with $\mathcal{T}_0$. Such assumptions are usually made in the case of multilevel, multigrid and domain decomposition methods. There are many numerical tests confirming that the convergence of the related methods deteriorates if this condition is violated. At the same time, there are a lot of (multiscale and multiphysics) models of strongly heterogeneous media where the coefficient jumps can be resolved on the level of the finest mesh only. In the last part of this book we will show some pioneering results for such problems demonstrating robustness for classes of problems with extremely rough coefficients.

2.7.3 Hierarchical error estimators

The successive refinement of a finite element mesh provides a sequence of nested grids and finite element spaces for which a proper hierarchical decomposition is introduced. Here we briefly show how this sequence can be used for building not only multilevel preconditioners but also error estimates, see e.g. [23, 30, 41] and the references therein.

We will summarize some results from [30], with a restriction to the case of conforming linear finite elements, using the notations from Section 2.3.2. Let $u \in \mathcal{V}$ be the exact solution, and let $u_H \in \mathcal{V}_H$, $u_h \in \mathcal{V}_h$ be the finite element solutions of the elliptic boundary value problem (1.4) in $\mathcal{V}_H$ and $\mathcal{V}_h$, respectively. We will assume also that the *saturation condition* is satisfied, i.e., there exists a constant $\zeta < 1$ such that

$$\|u - u_h\|_{\mathcal{A}} \le \zeta \|u - u_H\|_{\mathcal{A}} . \tag{2.86}$$

Then the Galerkin orthogonality allows to show that

$$\|w_h\|_{\mathcal{A}} \le \|u - u_H\|_{\mathcal{A}} \le \frac{1}{1-\zeta^2} \|w_h\|_{\mathcal{A}} , \tag{2.87}$$

where $w_h = u_h - u_H$ and $\|\cdot\|_{\mathcal{A}}$ stands for the energy norm induced by the bilinear form $\mathcal{A}(\cdot,\cdot)$, see [23]. Thus $\eta = \|w_h\|_{\mathcal{A}}$ can serve as an *efficient and reliable error estimator*.

A cheaper error estimator $\overline{\eta}$ can be computed via the hierarchical decomposition

$$\mathcal{V}_h = \mathcal{V}_H \oplus \mathcal{V}_H^+ ,$$

associated with the hierarchical two-level splitting of the nodal basis, see (2.31). Let $\overline{\eta} = \|\overline{w}_h\|_{\mathcal{A}}$, where

$$\overline{w}_h \in \mathcal{V}_H^+ : \quad \mathcal{A}(\overline{w}_h, v_h) = \mathcal{L}(v_h) - \mathcal{A}(u_h, v_h) \qquad \forall v_h \in \mathcal{V}_H^+ . \tag{2.88}$$

Then the following estimates hold:

$$\|\overline{w}_h\|_{\mathcal{A}} \leq \|u - u_H\|_{\mathcal{A}} \leq \frac{1}{(1-\zeta^2)(1-\gamma^2)} \|\overline{w}_h\|_{\mathcal{A}} , \tag{2.89}$$

where γ is the related CBS constant.

Algebraically,

$$\overline{\eta}^2 = \langle A_{11}\overline{\mathbf{w}}_1, \overline{\mathbf{w}}_1 \rangle, \tag{2.90}$$

where $\overline{\mathbf{w}} = (\overline{\mathbf{w}}_1^T, \overline{\mathbf{w}}_2^T)^T$ is the hierarchical two-level basis vector of nodal unknowns, corresponding to $\overline{w}_h$, and

$$\overline{\mathbf{w}}_1 : \qquad A_{11}\overline{\mathbf{w}}_1 = \mathbf{l}_1 - \tilde{A}_{12}\overline{\mathbf{w}}_2, \tag{2.91}$$

$$\overline{\mathbf{w}}_2 : \qquad \tilde{A}_{22}\overline{\mathbf{w}}_2 = \mathbf{l}_2 \tag{2.92}$$

is the linear system arising from the FE discretization of (2.88).

The next lemma summarizes the construction of the hierarchical error estimators.

Lemma 2.17. *Let the saturation condition (2.86) be fulfilled, and let*

$$\overline{\mathbf{w}}_1 : \qquad A_{11}\overline{\mathbf{w}}_1 = \mathbf{l}_1 - \tilde{A}_{12}\overline{\mathbf{w}}_2, \tag{2.93}$$

$$\overline{\mathbf{w}}_2 : \qquad \tilde{A}_{22}\overline{\mathbf{w}}_2 = \mathbf{l}_2 , \tag{2.94}$$

where C_{11} is a preconditioner for A_{11} satisfying (2.75). Then the following hierarchical error estimate holds:

$$\|\overline{\mathbf{w}}_1\|_{C_{11}} \leq \|u - u_H\|_{\mathcal{A}} \leq \frac{1+\delta_1}{(1-\zeta^2)(1-\gamma^2)} \|\overline{\mathbf{w}}_1\|_{C_{11}} . \tag{2.95}$$

Proof. The inequalities follow directly from (2.89), the definition of $\overline{\mathbf{w}}_1$ (see (2.93)), and the inequality (2.75). □

Remark 2.18. The reliability of the hierarchical error estimator is controlled by the parameters ζ, γ and δ_1 while the efficiency is ensured by the optimal order computational complexity of the preconditioner C_{11}. This means that if the considered AMLI algorithm is robust, and if the saturation condition is satisfied, then the related hierarchical error estimator is robust.

Corollary 2.19. *Let us consider the model elliptic problem (2.80) with the locally defined (and properly scaled) diagonal preconditioner C_{11} of the pivot block A_{11}. Then, the estimates (2.81) and (2.85) lead to the hierarchical error estimate*

$$\|\overline{\mathbf{w}}_1\|_{C_{11}} \leq \|u - u_H\|_{\mathcal{A}} \leq \frac{2(3+\sqrt{2})}{1-\zeta^2}\|\overline{\mathbf{w}}_1\|_{C_{11}}, \tag{2.96}$$

which is robust with respect to the coefficient jumps.

3 Robust AMLI algorithms: Conforming linear finite elements

3.1 Some basic relations

A general important technique for finite element methods is to transform the arising integral over an arbitrary element to a standard reference element. We will use here the transformation method for a planar domain and triangular finite elements.

First we will show that the analysis for an arbitrary triangle (e) with coordinates (x_i, y_i), $i = 1, 2, 3$ can be done on the reference triangle $(\tilde{e})$, with coordinates $(0, 0)$ $(1, 0)$ and $(0, 1)$. Transforming the finite element functions between these triangles, the element bilinear form $\mathcal{A}_e(\cdot, \cdot)$ becomes (see e.g. [6, 14]):

$$\begin{aligned}
\mathcal{A}_e(u, v) &= \mathcal{A}_{\tilde{e}}(\tilde{u}, \tilde{v}) \\
&= \int_{\tilde{e}} \left[\frac{\partial \tilde{u}}{\partial \tilde{x}}, \frac{\partial \tilde{u}}{\partial \tilde{y}} \right] \left[\begin{array}{cc} (x_2 - x_1) & (y_2 - y_1) \\ (x_3 - x_1) & (y_3 - y_1) \end{array} \right]^{-1} \left[\begin{array}{cc} a_{11} & a_{12} \\ a_{21} & a_{22} \end{array} \right] \\
&\quad \times \left[\begin{array}{cc} (x_2 - x_1) & (x_3 - x_1) \\ (y_2 - y_1) & (y_3 - y_1) \end{array} \right]^{-1} \left[\frac{\partial \tilde{v}}{\partial \tilde{x}}, \frac{\partial \tilde{v}}{\partial \tilde{y}} \right]^T \left| \frac{\partial(x, y)}{\partial(\tilde{x}, \tilde{y})} \right| d\tilde{e},
\end{aligned}$$

where $0 < \tilde{x}, \tilde{y} < 1$, i.e., it takes the form

$$\mathcal{A}_{\tilde{e}}(\tilde{u}, \tilde{v}) = \int_{\tilde{e}} \sum_{i,j} \tilde{a}_{ij} \frac{\partial \tilde{u}}{\partial \tilde{x}_i} \frac{\partial \tilde{v}}{\partial \tilde{x}_j} d\tilde{e}, \tag{3.1}$$

where the coefficients $\tilde{a}_{ij}$ depend on both, the coordinates of the vertices of e (or, more precisely, the angles of e) and the coefficients a_{ij} of the differential operator. A similar form holds in case of $3D$ problems.

We conclude that it suffices for the analysis of uniform local bounds in the FEM setting to consider the (macro)element stiffness matrices for the reference triangle and arbitrary coefficients $[a_{ij}]$, or alternatively, for the Laplace operator, i.e.,

$$[a_{ij}] = \left[\begin{array}{cc} 1 & 0 \\ 0 & 1 \end{array} \right],$$

and an arbitrary triangle e.

Following the standard FEM assembling procedure we can write the global stiffness matrix A in the form

$$A = \sum_{e \in \mathcal{T}_k} R_e{}^T A_e R_e, \tag{3.2}$$

where A_e is the element stiffness matrix, R_e stands for the restriction mapping of the global vector of unknowns to the local vector corresponding to element $e \in \mathcal{T}_k$.

Consider now the Laplace operator and an arbitrary shaped linear triangular finite element.

Lemma 3.1 ([7, 85]). *The element stiffness matrix A_e for the Laplace operator can be written in the form*

$$A_e = \frac{1}{2} \begin{bmatrix} b+c & -c & -b \\ -c & a+c & -a \\ -b & -a & a+b \end{bmatrix}, \tag{3.3}$$

where a, b, and c equal the cotans of the angles in $e \in \mathcal{T}_h$.

Proof. We derive the element stiffness matrix corresponding to the bilinear form

$$\mathcal{A}_e(u, v) = \int_e (u_x v_x + u_y v_y) de$$

for a given arbitrary non-degenerate triangle e as shown in Figure 3.1. Let us introduce the notations $h = |OA|$, $p = |OB|$ and $q = |OC|$.

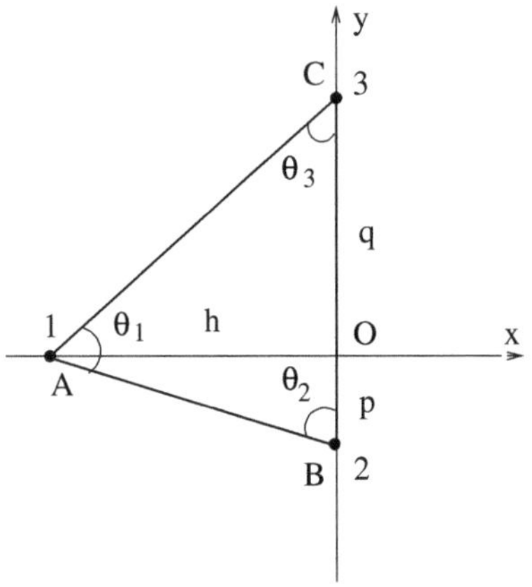

Figure 3.1: Derivation of the element stiffness matrix

Let θ_1, θ_2, θ_3 be the angles of the triangle, and let $a = \cot\theta_1$, $b = \cot\theta_2$, $c = \cot\theta_3$. Then the following relations are readily seen:

$$b = \frac{p}{h}, \quad c = \frac{q}{h},$$

$$a = \cot(\pi - (\theta_2 + \theta_3)) = -\frac{\cot\theta_2 \cot\theta_3 - 1}{\cot\theta_2 + \cot\theta_3} = \frac{h^2 - pq}{h(p+q)}.$$

For the element basis functions and their derivatives we obtain:

$$\phi_1 = -\frac{x}{h}, \quad \frac{\partial\phi_1}{\partial x} = -\frac{1}{h}, \quad \frac{\partial\phi_1}{\partial y} = 0,$$

$$\phi_2 = \frac{qx + h(q-y)}{h(p+q)}, \quad \frac{\partial\phi_2}{\partial x} = \frac{q}{h(p+q)}, \quad \frac{\partial\phi_2}{\partial y} = -\frac{1}{p+q},$$

$$\phi_3 = \frac{px + h(p+y)}{h(p+q)}, \quad \frac{\partial\phi_2}{\partial x} = \frac{p}{h(p+q)}, \quad \frac{\partial\phi_2}{\partial y} = \frac{1}{p+q}.$$

Taking into account that

$$|e| = \int de = \frac{h(p+q)}{2}$$

we substitute the obtained constants for the derivatives of the element basis functions in the formula

$$A_e(i,j) = \mathcal{A}(\phi_i, \phi_j)$$

which competes the proof. □

In the local analysis, without loss of generality we will assume that $|a| \le b \le c$. This can be concluded from the following lemma.

Lemma 3.2 ([14]). *Let $\theta_1, \theta_2, \theta_3$ be the angles in an arbitrary triangle. Then with $a = \cot\theta_1$, $b = \cot\theta_2$, $c = \cot\theta_3$ it holds:*

(i) $a = (1 - bc)/(b + c)$.

(ii) *If $\theta_1 \ge \theta_2 \ge \theta_3$ then $|a| \le b \le c$.*

(iii) $a + b > 0$.

Proof. As was already mentioned in the proof of the previous lemma,

$$a = \frac{1 - \cot\theta_2 \cot\theta_3}{\cot\theta_2 + \cot\theta_3},$$

which is part (i). To prove part (ii), note that if $\theta_1 \leq \frac{\pi}{2}$, then $\theta_1 \geq \theta_2 \geq \theta_3$ shows that $0 \leq a \leq b \leq c$. If the triangle is obtuse, i.e., $\theta_1 > \frac{\pi}{2}$, then $\theta_2 + \theta_3 < \frac{\pi}{2}$ and it follows that $a < 0$ and

$$|a| = \frac{\cot\theta_2 \cot\theta_3 - 1}{\cot\theta_2 + \cot\theta_3} = \cot\theta_2 \frac{\cot\theta_3 - 1/\cot\theta_2}{\cot\theta_3 + \cot\theta_2} < \cot\theta_2 = b.$$

Finally,

$$a + b = \frac{\sin(\theta_1 + \theta_2)}{\sin\theta_1 \sin\theta_2} > 0.$$

□

Remark 3.3. Let us consider an arbitrary symmetric and positive semidefinite 3×3 matrix C_e such that $\ker(C_e) = \operatorname{span}(\mathbf{1})$, where $\mathbf{1} = (1, 1, 1)^T$. As will be shown in the next chapter, subject to a scaling factor, C_e can be viewed as an element stiffness matrix for some elliptic problem (coefficient matrix) and/or properly set triangular finite element.

The next representation of the element stiffness matrix A_e simply follows from Lemma 3.2:

$$A_e = \frac{c}{2} \begin{bmatrix} \beta + 1 & -1 & -\beta \\ -1 & \alpha + 1 & -\alpha \\ -\beta & -\alpha & \alpha + \beta \end{bmatrix}, \tag{3.4}$$

$\alpha = a/c$, $\beta = b/c$, and $(\alpha, \beta) \in D$, where

$$D = \Big\{(\alpha, \beta) \in \mathbb{R}^2 : -\frac{1}{2} < \alpha \leq 1, \max\{-\frac{\alpha}{\alpha + 1}, |\alpha|\} \leq \beta \leq 1\Big\}. \tag{3.5}$$

The next purely algebraic inequality plays an important role in the following considerations of robust AMLI algorithms for linear conforming and nonconforming finite elements in the 2D case.

Lemma 3.4 ([14]). *For all $(\alpha, \beta) \in D$, there holds the inequality*

$$\frac{\alpha\beta + \alpha + \beta + 1}{(\alpha + \beta + 1)(\alpha + \beta + 2)} > \frac{4}{15}. \tag{3.6}$$

Proof. The inequality is equivalent to

$$4\alpha^2 + 4\beta^2 - 3(\alpha + \beta) - 7\alpha\beta < 7. \tag{3.7}$$

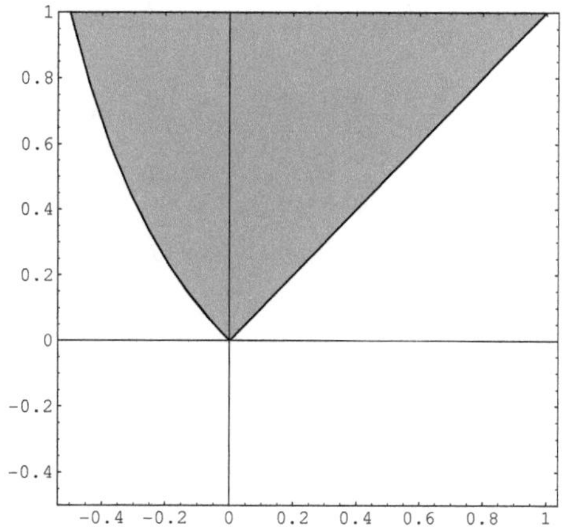

Figure 3.2: Domain of the parameters (α, β)

Let us introduce the auxiliary function $\psi(\alpha, \beta) = 4\alpha^2 + 4\beta^2 - 3(\alpha + \beta) - 7\alpha\beta$ defined in D (see Figure 3.2). From

$$\frac{\partial \psi}{\partial \alpha} = 8\alpha - 7\beta - 3,$$

it follows that if ψ has an extremum in some interior point $(\tilde{\alpha}, \tilde{\beta}) \in D$ then $\tilde{\alpha} = (7\tilde{\beta} + 3)/8$. Now we consider

$$\psi\Big(\frac{7\beta + 3}{8}, \beta\Big) = \tilde{\psi}(\beta) = \frac{1}{16}(15\beta^2 - 90\beta - 9),$$

which is strictly decreasing if $0 \leq \beta \leq 1$. This means that $\psi(\alpha, \beta)$ achieves its extremum on the boundary of D. From the expression (3.6) it follows that the extremal values must be taken either for $\alpha < 0$ and $|\alpha|\beta$ maximum, or for $\alpha = \beta = 1$. This simply leads to

$$\psi_{\max} = \psi\Big(-\frac{1}{2}, 1\Big) = 7,$$

which completes the proof. □

Remark 3.5. Let us consider the case of linear tetrahedral finite elements, and let A_e be the element stiffness matrix corresponding to the Laplace operator. Then the following 3D analog of formula (3.3) holds true (see, e.g., in [105]):

$$A_e = \frac{1}{6}\begin{bmatrix} \sum\limits_{1\neq i<j} l_{ij}\cot\theta_{ij} & -l_{34}\cot\theta_{34} & -l_{24}\cot\theta_{24} & -l_{23}\cot\theta_{23} \\ -l_{34}\cot\theta_{34} & \sum\limits_{2\neq i<j\neq 2} l_{ij}\cot\theta_{ij} & -l_{14}\cot\theta_{i4} & -l_{13}\cot\theta_{13} \\ -l_{24}\cot\theta_{24} & l_{14}\cot\theta_{14} & \sum\limits_{3\neq i<j\neq 3} l_{ij}\cot\theta_{ij} & -l_{12}\cot\theta_{12} \\ -l_{23}\cot\theta_{23} & l_{13}\cot\theta_{13} & -l_{12}\cot\theta_{12} & \sum\limits_{i<j\neq 4} l_{ij}\cot\theta_{ij} \end{bmatrix},$$

where l_{ij} denotes the length of the edge connecting the vertices v_i and v_j of the tetrahedron, and θ_{ij} stands for the dihedral angle at that edge.

3.2 Uniform estimates of the constant in the strengthened CBS inequality

Consider two consecutive meshes $\mathcal{T}_k \subset \mathcal{T}_{k+1}$. The following uniform refinement procedure will be considered as a default setting. The current coarse triangle $e \in \mathcal{T}_k$ is subdivided into four congruent triangles by joining the mid-edge nodes to get the macroelement $E \in \mathcal{T}_{k+1}$. The related macroelement stiffness matrix A_E (see (2.11)) consists of blocks which are 3×3 matrices and the local eigenproblem (2.36) to compute γ_E has a reduced dimension of 2×2.

In the so arising six node-points of the macroelement we can also use hierarchical basis functions, where we keep the linear basis functions in the vertex nodes and add piecewise quadratic basis functions in the mid-edge nodes with support on the whole triangle. The first (default assumption) refinement is referred to as h-version while the second alternative approach is called p-version ($p = 2$) (see Figure 3.3). Let us denote by $\widehat{A}_E$ and $\widehat{\gamma}_E$ the corresponding macroelement stiffness matrix and CBS constant.

The following relation between γ_E and $\widehat{\gamma}_E$ holds.

Theorem 3.6 ([85]). *Let us consider a piecewise Laplacian elliptic problem on an arbitrary finite element triangular mesh $\mathcal{T}_k$, and let each element from $\mathcal{T}_k$ be uniformly refined into four congruent elements to get $\mathcal{T}_{k+1}$. Then*

$$\widehat{\gamma}_E^2 = \frac{4}{3}\gamma_E^2, \tag{3.8}$$

where $\widehat{\gamma}_E$, γ_E are the local CBS constants for the hierarchical piecewise quadratic and the piecewise linear finite elements, respectively.

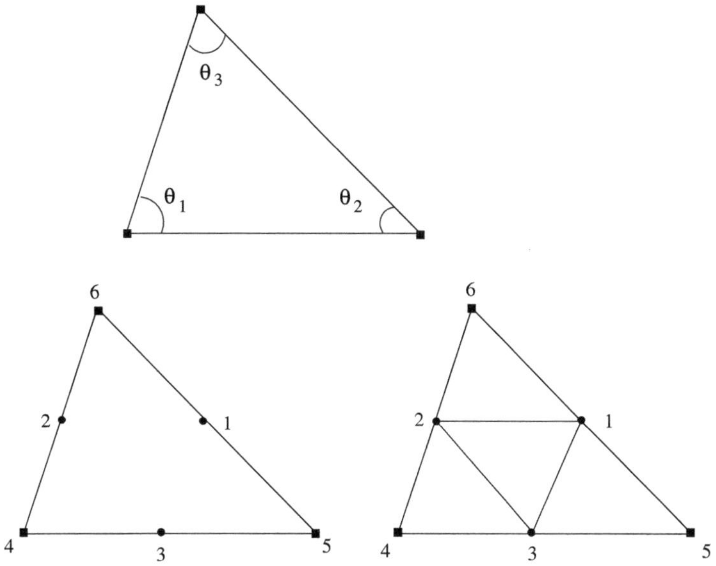

Figure 3.3: Uniform refinement of the coarse linear triangle: p-version (left) and h-version (right)

Proof. Let us consider the standard (linear FEM) hierarchical basis macroelement stiffness matrices A_E and the related quadratic FEM element stiffness matrix $\widehat{A}_E$ written in a two-by-two block form:

$$A_E = \begin{bmatrix} A_{E:11} & A_{E:12} \\ A_{E:21} & A_{E:22} \end{bmatrix}, \quad \widehat{A}_E = \begin{bmatrix} \widehat{A}_{E:11} & \widehat{A}_{E:12} \\ \widehat{A}_{E:21} & \widehat{A}_{E:22} \end{bmatrix}.$$

For the h-version refinement, we can apply formula (3.3) to assemble the macroelement stiffness matrix, and then use the simple hierarchical basis transformation. The corresponding blocks are:

$$A_{E:11} = \begin{bmatrix} a+b+c & -c & -b \\ -c & a+b+c & -a \\ -b & -a & a+b+c \end{bmatrix},$$

$$A_{E:22} = \frac{1}{2}\begin{bmatrix} b+c & -c & -b \\ -c & a+c & -c \\ -b & -c & a+b \end{bmatrix},$$

$$A_{E:12} = A_{E:21} = -A_{E:22}.$$

For the p-version refinement ($p = 2$), the (macro)element stiffness matrix can be derived in a similar way as in (3.3). The block $\widehat{A}_{E:11}$ has the form

$$\widehat{A}_{E:11} = \frac{4}{3}\begin{bmatrix} a+b+c & -c & -b \\ -c & a+b+c & -a \\ -b & -a & a+b+c \end{bmatrix},$$

i.e., $\widehat{A}_{E:11} = \frac{4}{3}A_{E:11}$. It follows by the construction of the hierarchical basis that

$$\widehat{A}_{E:22} = A_{E:22},$$

and finally

$$\widehat{A}_{E:12} = \widehat{A}_{E:21} = -\frac{4}{3}\widehat{A}_{E:22} = -\frac{4}{3}A_{E:22}.$$

It simply follows from (2.35)–(2.36) that γ_E^2 and $\widehat{\gamma}_E^2$ are the corresponding largest eigenvalues $\lambda_{\max}$ and $\widehat{\lambda}_{\max}$ of the local eigenproblems:

$$A_{E:22}A_{E:11}^{-1}A_{E:22}\mathbf{v}_{E:2} = \lambda A_{E:22}\mathbf{v}_{E:2}, \qquad \mathbf{v}_{E:2} \neq \text{const}, \tag{3.9}$$

$$\widehat{A}_{E:22}\widehat{A}_{E:11}^{-1}\widehat{A}_{E:22}\mathbf{v}_{E:2} = \widehat{\lambda}\widehat{A}_{E:22}\mathbf{v}_{E:2}, \qquad \mathbf{v}_{E:2} \neq \text{const}. \tag{3.10}$$

Using the relations between the blocks of A_E and $\widehat{A}_E$ we get the following equivalent representation of (3.9),

$$\frac{4}{3}A_{E:22}A_{E:11}^{-1}A_{E:22}\mathbf{v}_{E:2} = \widehat{\lambda}A_{E:22}\mathbf{v}_{E:2}, \qquad \mathbf{v}_{E:2} \neq \text{const}$$

which completes the proof. □

Corollary 3.7. *The local estimate*

$$\gamma_E^2 < \frac{3}{4} \tag{3.11}$$

holds uniformly with respect to the mesh anisotropy.

The next fundamental result follows directly from the local estimate (3.11), the equivalence relation (3.1) and Lemma 2.5.

Theorem 3.8. *Consider the problem (1.1) discretized by conforming linear finite elements, where the coarsest grid $\mathcal{T}_0$ is aligned with the discontinuities of the coefficient $\mathbf{a}(e)$, $e \in \mathcal{T}_0$. Let us assume also that $\mathcal{T}_k \subset \mathcal{T}_{k+1}$ are two consecutive* *meshes where each element from $\mathcal{T}_k$ is refined into four congruent elements to get $\mathcal{T}_{k+1}$. Then, the estimate*

$$\gamma^2 < \frac{3}{4} \tag{3.12}$$

of the CBS constant holds uniformly with respect to the coefficient jumps, mesh or/and coefficient anisotropy, and the refinement level k.

Remark 3.9. Let us consider the local problem under the assumptions of Theorem 3.6. In this setting the following explicit formula for the local CBS constant can be derived (see [85]):

$$\gamma_E^2 = \frac{3}{8} + \frac{1}{4}\sqrt{d - \frac{3}{4}}, \tag{3.13}$$

where

$$d = \sum_{i=1}^{3} \cos^2 \theta_i .$$

It is easy to observe that the optimal case corresponds to the equilateral triangle ($d = 3/4$) for which $\gamma_E^2 = 3/8$. The worst case of $\gamma_E^2 = 3/4$ is approached when the maximal angle tends to the limit case of π. In a more general setting, formula (3.13) could be used to get improved estimates of the global CBS constant if the minimal angle of the triangulation is known, which will be discussed briefly in Section 3.5, or if some other measure of the level of mesh and/or coefficient anisotropy is available.

We present here some further achievements in the robust estimates of the CBS constant. The estimate (3.12) is generalized in [8] to the case when each element from the current coarser mesh is subdivided into m^2, $m \geq 2$, congruent elements using a uniform refinement (with a multiplicity of m). Then, the following universal estimate holds:

$$\gamma^2 < \frac{m^2 - 1}{m^2}, \qquad m = 2, 3, \dots. \tag{3.14}$$

A similar result is obtained for the case of 3D linear finite elements. Let us assume that the current element from the coarser mesh is uniformly divided into m^3 tetrahedra. Then, the uniform estimate for $m = 2$ reads as

$$\gamma^2 < \frac{9}{10},$$

which is extended to the universal estimate

$$\gamma^2 < \frac{(m^2 - 1)(m^2 + 2)}{m^2(m^2 + 1)}, \qquad m = 2, 3, \dots, \tag{3.15}$$

see [8, 28] and the references therein.

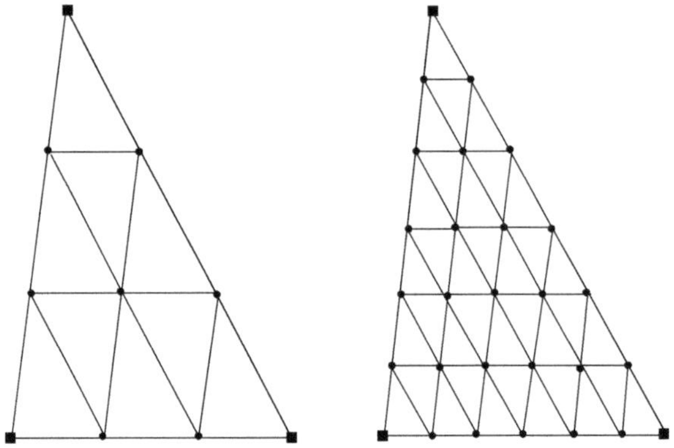

Figure 3.4: Uniform refinement of the current element into m^2 congruent triangles: left $m = 3$ and right $m = 6$

Another generalization of the AMLI theory concerns the Lamé system of linear elasticity. Here, the coefficient anisotropy is introduced by the Poisson ratio. The general scheme of analyzing the CBS constant is applied in the case of conforming linear and bilinear finite elements. Let us note, that the obtained estimates for linear elements are the same as for the scalar elliptic problems, being robust with respect to the mesh anisotropy and the Poisson ratio, see e.g., [2, 20, 87, 88].

3.3 Additive preconditioning of the pivot block

When applicable, we will skip the superscripts of the pivot block and its approximation. Here, we will write A_{11}, C_{11}, instead of $A_{11}^{(k)}$, $C_{11}^{(k)}$. The construction and the analysis of the preconditioners C_{11} are based on a macroelement-by-macroelement assembling procedure.

Following (3.2), we write A_{11} in the form

$$A_{11} = \sum_{E \in \mathcal{T}_{k+1}} {R_E}^T A_{E:11} R_E. \tag{3.16}$$

Now, we use the representation (3.3) of the element stiffness matrix to get

$$A_{E:11} = r_T \begin{bmatrix} a_T + b_T + c_T & -c_T & -b_T \\ -c_T & a_T + b_T + c_T & -a_T \\ -b_T & -a_T & a_T + b_T + c_T \end{bmatrix},$$

where: a) the factor r_T depends on the shape of $T \in \mathcal{T}_0$ and on the related coefficient matrix $\mathbf{a}(e)$ which is one and the same for all elements $e \in \mathcal{T}_k \cap T$, and b)

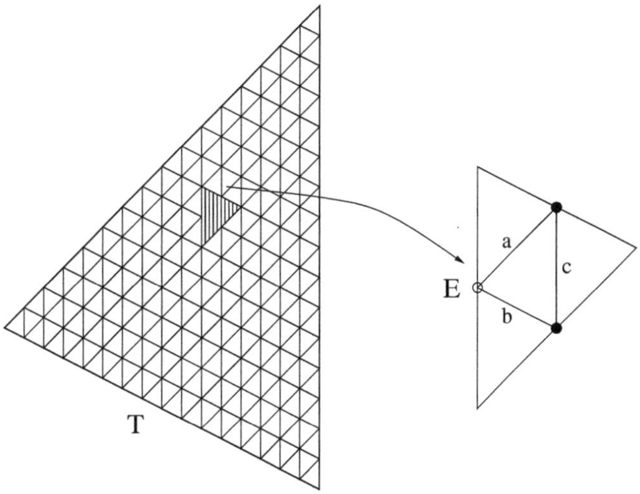

Figure 3.5: Four levels of uniform refinement of $T \in \mathcal{T}_0$ and a current macro-element $E \in \mathcal{T}_4$

$|a_T| \leq b_T \leq c_T$ equal the cotans of the angles in some triangle $\tilde{T}$ associated with T. Using the scaled representation (3.4) we get

$$A_{E:11} = r_T c_T \begin{bmatrix} \alpha+\beta+1 & -1 & -\beta \\ -1 & \alpha+\beta+1 & -\alpha \\ -\beta & -\alpha & \alpha+\beta+1 \end{bmatrix}, \tag{3.17}$$

where $\alpha = a_T/c_T$, $\beta = b_T/c_T$, and $(\alpha, \beta) \in D$ as introduced in (3.5). Then, the additive preconditioner of A_{11} is defined by

$$C_{11} = \sum_{E \in \mathcal{T}_{k+1}} R_E{}^T C_{E:11} R_E, \tag{3.18}$$

where

$$C_{E:11} = r_T c_T \begin{bmatrix} \alpha+\beta+1 & -1 & 0 \\ -1 & \alpha+\beta+1 & 0 \\ 0 & 0 & \alpha+\beta+1 \end{bmatrix}. \tag{3.19}$$

As one can see, the local matrix $C_{E:11}$ is obtained by preserving only the *strongest* off-diagonal entries of $A_{E:11}$.

To estimate the relative condition number of the preconditioner (3.18) with respect to A_{11} we consider the local generalized eigenproblem

$$A_{E:11}\mathbf{v}_E = \lambda_E C_{E:11}\mathbf{v}_E. \tag{3.20}$$

The characteristic equation for λ_E, $\det(A_{E:11} - \lambda_E C_{E:11}) = 0$ can be written in the form

$$\begin{vmatrix} (\alpha+\beta+1)\mu_E & -\mu_E & -\beta \\ -\mu_E & (\alpha+\beta+1)\mu_E & -\alpha \\ -\beta & -\alpha & (\alpha+\beta+1)\mu_E \end{vmatrix} = 0, \tag{3.21}$$

where $\mu_E = 1 - \lambda_E$. For the solutions of (3.21) we get

$$\mu_E^{(1)} = 0, \text{ and } \left(\mu_E^{(2,3)}\right)^2 = \frac{(\alpha+\beta+1)(\alpha^2+\beta^2)+2\alpha\beta}{(\alpha+\beta+1)[(\alpha+\beta+1)^2-1]},$$

or, after simplification,

$$\left(\mu_E^{(2,3)}\right)^2 = \frac{\alpha^2+\beta^2+\alpha+\beta}{(\alpha+\beta+1)(\alpha+\beta+2)} = 1 - 2\frac{\alpha+\beta+1+\alpha\beta}{(\alpha+\beta+1)(\alpha+\beta+2)}.$$

Hence, applying the inequality (3.6), it follows that $(\mu_E^{(2,3)})^2 < 7/15$, and thus the local eigenvalue estimate

$$1 - \sqrt{7/15} < \lambda_E < 1 + \sqrt{7/15} \tag{3.22}$$

holds. Now we are ready to prove the next theorem.

Theorem 3.10 ([14, 15]). *The additive preconditioner of A_{11} has a relative condition number uniformly bounded by*

$$\kappa\left(C_{11}{}^{-1}A_{11}\right) < \frac{1}{4}(11+\sqrt{105}) \approx 5.31. \tag{3.23}$$

This condition number estimate holds independently of the shape, the size of each element and of the coefficient matrix $\mathbf{a}(e)$ of the FEM problem.

Proof. Applying (3.22) we get

$$\begin{aligned} v^T A_{11} v &= \sum_{E\in\mathcal{T}_h} \mathbf{v}_E^T R_E{}^T A_{11:E} R_E \mathbf{v}_E < \sum_{E\in\mathcal{T}_h} \lambda_E^{\max} \mathbf{v}_E^T R_E{}^T C_{11:E} R_E \mathbf{v}_E \\ &< \left(1+\sqrt{7/15}\right) \sum_{E\in\mathcal{T}_h} \mathbf{v}_E^T R_E{}^T C_{11:E} R_E \mathbf{v}_E \\ &= \left(1+\sqrt{7/15}\right) \mathbf{v}^T C_{11} \mathbf{v} \end{aligned}$$

and, similarly,

$$\mathbf{v}^T A_{11} \mathbf{v} > \left(1-\sqrt{7/15}\right) \mathbf{v}^T C_{11} \mathbf{v}.$$

Combining the last two inequalities completes the proof, i.e.,

$$\kappa\left(C_{11}{}^{-1}A_{11}\right) < \frac{\lambda_{\max}\left(C_{11}{}^{-1}A_{11}\right)}{\lambda_{\min}\left(C_{11}{}^{-1}A_{11}\right)} < \frac{1+\sqrt{7/15}}{1-\sqrt{7/15}}.$$ □

This additive preconditioner was first introduced and analyzed in [15]. The above new proof, based on the algebraic inequality from Lemma 3.4, has been presented in [14]. As we will see later in this chapter, the inequality from Lemma 3.4 plays a key role in the analysis of various robust AMLI preconditioners for conforming linear finite elements.

3.4 Multiplicative preconditioning of the pivot block

Let us partition the nodes corresponding to the block A_{11} into two groups where the first one contains the centers of the parallelogram superelements Q (see Figure 3.6), which are weakly connected in the sense of the relations between the coefficients $|a_T| \le b_T \le c_T$. It is important to note that the parallelograms $Q \subset T \in \mathcal{T}_0$, i.e., it is not allowed to be composed of triangles of neighbor elements from the coarsest triangulation $\mathcal{T}_0$. With respect to this partitioning A_{11} admits the following two-by-two block factorization

$$A_{11} = \begin{bmatrix} D_{11} & F_{11} \\ F_{11}{}^T & E_{11} \end{bmatrix} = \begin{bmatrix} D_{11} & 0 \\ F_{11}{}^T & S_{11} \end{bmatrix} \begin{bmatrix} I & D_{11}{}^{-1} F_{11} \\ 0 & I \end{bmatrix} \tag{3.24}$$

where S_{11} stands for the related Schur complement. We define now C_{11} as the symmetric block Gauss–Seidel preconditioner of A_{11}, i.e.,

$$C_{11} = \begin{bmatrix} D_{11} & 0 \\ F_{11}{}^T & E_{11} \end{bmatrix} \begin{bmatrix} I & D_{11}{}^{-1} F_{11} \\ 0 & I \end{bmatrix}. \tag{3.25}$$

Since D_{11} is a diagonal matrix it follows that the Schur complement S_{11} can be

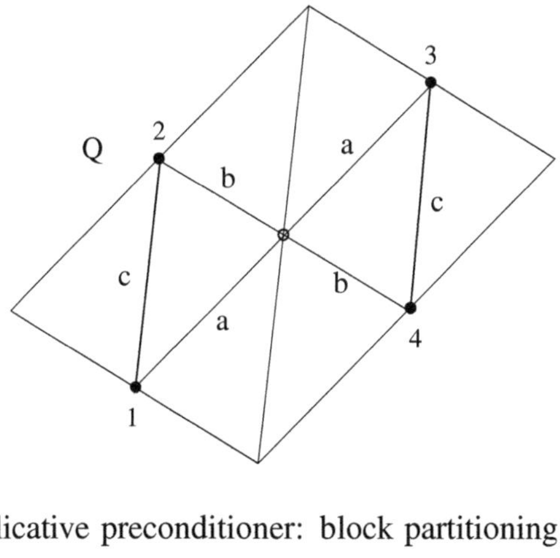

Figure 3.6: Multiplicative preconditioner: block partitioning of the nodes of the superelement Q

assembled from the corresponding superelement Schur complements

$$S_{Q:11} = E_{Q:11} - {F_{Q:11}}^T {D_{Q:11}}^{-1} F_{Q:11}.$$

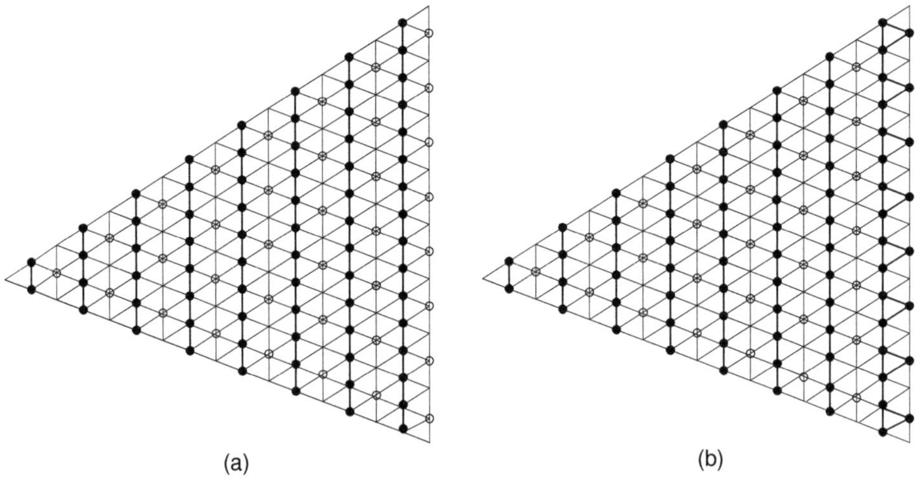

Figure 3.7: Connectivity pattern of $T \in \mathcal{T}_0$: (a) additive preconditioner of $A^{(4)}_{T:11}$ i.e., the matrix block arising after four refinement steps; (b) multiplicative preconditioner of $A^{(4)}_{T:11}$

Such a procedure is sometimes called *static condensation*. The obtained sparsity structure is such that solving systems with E_{11} requires: first, local elimination steps along lines of dominating anisotropy; and second, solving a sparse system the order and structure of which are similar to that of $A^{(0)}$ corresponding to $T \in \mathcal{T}_0$. It will be shown in the last section of this chapter that the computational cost to solve a system with the current matrix C_{11} is proportional to the size of this matrix. The connectivity pattern of the block E_{11} related to a given triangle $T \in \mathcal{T}_0$ is illustrated in Figure 3.7 (b). The only difference between the decoupled structure of the additive and the multiplicative preconditioners is in the boundary layer which is parallel to the direction of dominating anisotropy in the coarsest grid triangle $T \in \mathcal{T}_0$.

A similar construction was first introduced in [91] for the particular case of triangulations $\mathcal{T}_0$ consisting of right triangles with legs parallel to the coordinate axes and a diagonal coefficient matrix $\mathbf{a}(e)$. The general case is studied in [14]. As for the additive algorithm, the local analysis is based on the algebraic inequality from Lemma 3.4.

Lemma 3.11. *Consider the generalized eigenproblem*

$$S_{Q:11}\mathbf{v}_Q = \lambda_Q E_{Q:11}\mathbf{v}_Q. \tag{3.26}$$

Then the minimal eigenvalue $\lambda_Q^{\min}$ *is uniformly bounded by*

$$\lambda_Q^{\min} > \frac{8}{15} \tag{3.27}$$

and the remaining eigenvalues are equal to 1.

Proof. The required superelement matrices read as follows:

$$A_{Q:11} = \frac{r_T}{c_T}\begin{bmatrix} 2\delta & -\alpha & -\beta & -\alpha & -\beta \\ -\alpha & \delta & -1 & 0 & 0 \\ -\beta & -1 & \delta & 0 & 0 \\ -\alpha & 0 & 0 & \delta & -1 \\ -\beta & 0 & 0 & -1 & \delta \end{bmatrix},$$

$$E_{Q:11} = \frac{r_T}{c_T}\begin{bmatrix} \delta & -1 & 0 & 0 \\ -1 & \delta & 0 & 0 \\ 0 & 0 & \delta & -1 \\ 0 & 0 & -1 & \delta \end{bmatrix},$$

$$S_{Q:11} = \frac{r_T}{c_T}\begin{bmatrix} \delta-\alpha^2\omega & -1-\alpha\beta\omega & -\alpha^2\omega & -\alpha\beta\omega \\ -1-\alpha\beta\omega & \delta-\beta^2\omega & -\alpha\beta\omega & -\beta^2\omega \\ -\alpha^2\omega & -\alpha\beta\omega & \delta-\alpha^2\omega & -1-\alpha\beta\omega \\ -\alpha\beta\omega & -\beta^2\omega & -1-\alpha\beta\omega & \delta-\beta^2\omega \end{bmatrix},$$

where $\delta = \alpha + \beta + 1$ and $\omega = \frac{1}{2\delta}$. Then, for the solution of the generalized eigenproblem (3.26) we obtain

$$\lambda_Q^{(1)} = 2\frac{\alpha\beta + \alpha + \beta + 1}{(\alpha + \beta + 1)(\alpha + \beta + 2)}$$

and $\lambda_Q^{(2)} = \lambda_Q^{(3)} = \lambda_Q^{(4)} = 1$. It is easy to see that $\lambda_Q^{(1)}$ is indeed the minimal eigenvalue because the inequality $\lambda_Q^{(1)} < 1$ is equivalent to the obviously satisfied inequality $\alpha^2 + \beta^2 + \alpha + \beta > 0$. Finally we have to show that $\lambda_Q^{(1)} > \frac{8}{15}$ which follows immediately from (3.6). □

The main result of this section is given in the theorem below.

Theorem 3.12 ([14]). *The multiplicative preconditioner of A_{11} has a relative condition number uniformly bounded by*

$$\kappa\left(C_{11}{}^{-1}A_{11}\right) < \frac{15}{8} = 1.875. \tag{3.28}$$

This result is proved in the same way as Theorem 3.10 applying the estimate from Lemma 3.11. As one can observe, the new estimate considerably improves the result (3.23) for the additive preconditioner.

3.5 Locally improved estimates of the AMLI parameters

Let us consider the isotropic elliptic problem associated with the bilinear form

$$\mathcal{A}(u,v) = \sum_{e\in\mathcal{T}_1}\int_e a(e)\left(\frac{\partial u}{\partial x}\frac{\partial v}{\partial x} + \frac{\partial u}{\partial y}\frac{\partial v}{\partial y}\right) de.$$

We will study in this short section the case of pure mesh anisotropy where the minimal angle of the initial triangulation $\mathcal{T}_0$ is bounded by a given parameter $\pi/3 \geq \tau > 0$. Let us note that the minimal angle condition is explicitly set in some of the mesh generators. We will show here how the general estimates of the CBS constant γ, and the condition number estimates of the additive (κ_A) and the multiplicative (κ_M) preconditioners for the pivot block A_{11}, can be improved in terms of τ.

The locally improved estimate of the CBS constant follows directly from the formula (3.13), i.e.,

$$\gamma^2 \leq \max_{\theta_1\geq\theta_2\geq\theta_3\geq\tau}\left(\frac{3}{8} + \frac{1}{4}\sqrt{\sum_{i=1}^{3}\cos^2\theta_i - \frac{3}{4}}\right). \tag{3.29}$$

Let T be an arbitrary triangle from $\mathcal{T}_0$. For this element

$$\theta_1 \geq \theta_2 \geq \theta_3 \geq \tau$$

which is equivalent to

$$|a| \leq b \leq c \leq t = \cot\tau.$$

The related scaled parameters are $(\alpha,\beta) \in \tilde{D} \subset D$ (see (3.5)), where

$$\tilde{D} = \left\{(\alpha,\beta)\in\mathbb{R}^2 : \frac{1-t^2}{2t^2} < \alpha \leq 1, \max\left\{-\frac{1-\alpha t^2}{2t^2(\alpha+1)}, |\alpha|\right\} \leq \beta \leq 1\right\}.$$

What we need to get better estimates of κ_A and κ_M is to improve the inequality of Lemma 3.4 which can be done, e.g., using

$$\max_{(\alpha,\beta)\in\tilde{D}} \frac{\alpha\beta+\alpha+\beta+1}{(\alpha+\beta+1)(\alpha+\beta+2)} = \mathcal{R}(\tau).$$

This approach is applied to get the improved estimates presented in Table 3.1.

Table 3.1: Locally improved estimates for the AMLI parameters, applying a minimal angle condition

τ	0	15°	30°	45°
γ^2	0.75	0.667	0.525	0.5
κ_A	5.212	5.071	4.442	3.732
κ_M	1.875	1.820	1.667	1.5

The limit case shown in the first column ($\tau = 0°$) corresponds to arbitrary mesh anisotropy. As is expected, all the parameters are improved with the raise of the minimal angle τ, tending to the optimal case of $\tau = 60°$.

3.6 Optimal complexity solution algorithms for systems with C_{11}

The ability for efficient solution of systems with the introduced preconditioning matrices C_{11} is determined by their connectivity pattern, assuming that rapid solution methods are used at this step of the algorithm.

3.6.1 A model problem

Let us begin with the model problem in $\Omega = (0,1) \times (0,1)$ where the mesh is rectangular and uniform and the bilinear functional is as follows:

$$a(u,v) = \int_\Omega a_1 \frac{\partial u}{\partial x_1}\frac{\partial v}{\partial x_1} + a_2 \frac{\partial u}{\partial x_2}\frac{\partial v}{\partial x_2} d\Omega,$$

where the coefficients (a_1, a_2) are piecewise constant in the subdomains Ω_i, $i \in \{1,2,3,4\}$ varying the anisotropy ratio as follows: in $\Omega_1 = (0,1/2) \times (0,1/2)$ $a_1 > a_2$; in $\Omega_2 = (1/2,1)\times(0,1/2)$ $a_1 < a_2$; in $\Omega_3 = (1/2,1)\times(1/2,1)$ $a_1 > a_2$; and in $\Omega_4 = (0,1/2) \times (1/2,1)$ $a_1 < a_2$. Figure 3.8 illustrates the connectivity pattern of C_{11} where the dense circles and bold lines show the remaining links after the local modification in the additive algorithm, and after the static condensation in the multiplicative variant.

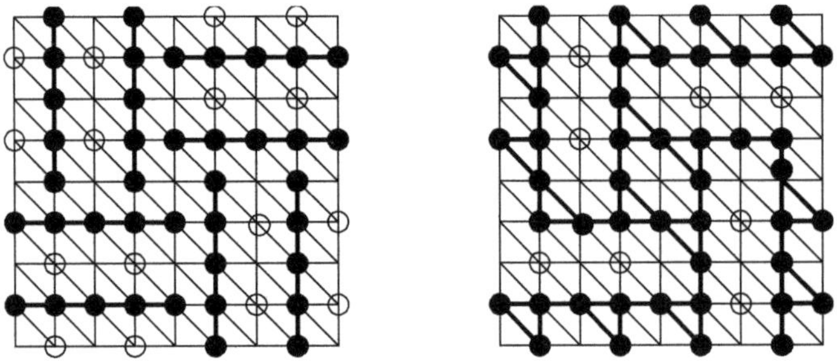

(a) (b)

Figure 3.8: Connectivity pattern of C_{11} for the model problem of varying ratio of anisotropy in $(0,1)^2$: (a) additive preconditioner; (b) multiplicative preconditioner

What one can see for this example is that the solution of systems with the preconditioning matrices C_{11} is either split into a number of decoupled tridiagonal systems (additive algorithm) or after solving such tridiagonal blocks, the reduced system is sparse and relatively small (multiplicative algorithm). Our final goal is to generalize these observations.

3.6.2 Additive algorithm

Consider now a general irregular mesh, as shown in Figure 3.9. It is readily seen, that in this case, the matrix C_{11} has a generalized tridiagonal structure (see [15] and also [95]), that is, the solution of linear systems with C_{11} has a computational cost which is proportional to the related problem size. In some more details, due to the form of the corresponding element matrices $C_{11:E}$, the related connectivity pattern of the preconditioner C_{11} is such as shown in Figure 3.9. This means that the coupled nodes form either a single point, a polyline or a polygon. Therefore, there are no cross-points. Finally, we summarize the major result of this subsection in the next statement.

Theorem 3.13. *The additive preconditioner of A_{11} has an optimal computational complexity with respect to both problem and discretization parameters.*

3.6.3 Multiplicative algorithm

For this algorithm, the passing from the model problem to the general case, turns out to be also almost trivial. For this purpose we will use the *nested dissection* (ND) algorithm which is known as a fast rapid solution method for sparse linear systems. If the graph representing the connectivity of the matrix is planar, i.e., it

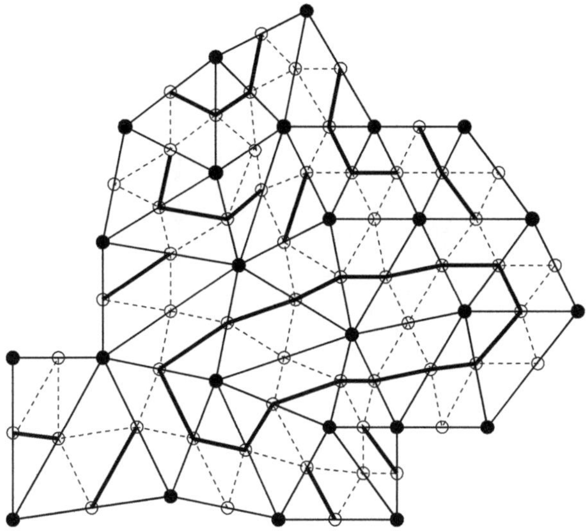

Figure 3.9: An example of the connectivity pattern for the additive preconditioner for a mesh after one refinement step of $\mathcal{T}_0$, i.e., for $\mathcal{T}_1$

can be drawn in a plane such that no edges cross each other, then the computational complexity of the ND method is

$$\mathcal{N}_{ND} = \mathcal{O}(n^{3/2}),$$

where n stands for the size of the problem (see [59]).

Now, let us denote by $\overline{N}$ the size of C_{11}. At the first step of our solution algorithm we eliminate the unknowns corresponding to the nodes from the interior of the triangles from the coarsest mesh $\mathcal{T}_0$. This is performed by solving a number of tridiagonal systems (see Figure 3.8) and therefore requires a number of arithmetic operations which is proportional to $\overline{N}$, i.e.,

$$\mathcal{N}_1 = \mathcal{O}\left(\overline{N}\right).$$

The obtained reduced problem has a planar graph of connectivity (see Figure 3.10). Its size is $\mathcal{O}(\overline{N}^{1/2})$, and therefore, the solution of the reduced problem by the ND algorithm has a computational cost

$$\mathcal{N}_2 = \mathcal{O}(\overline{N}^{3/4}),$$

and the total complexity is

$$\mathcal{N} = \mathcal{N}_1 + \mathcal{N}_2 = \mathcal{O}\left(\overline{N}\right).$$

As for the additive algorithm we get the final theorem.

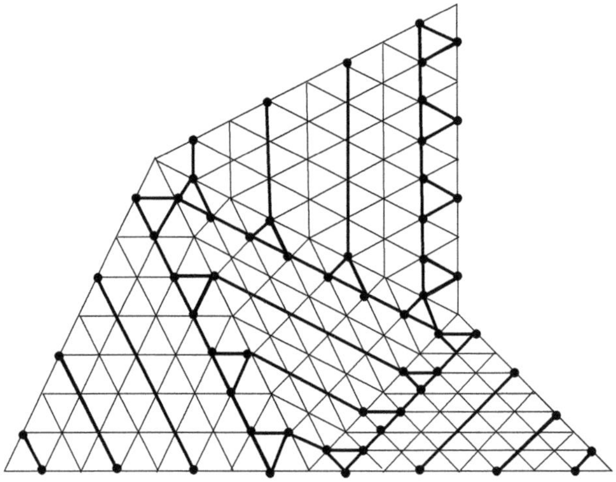

Figure 3.10: Connectivity pattern of the reduced problem for the multiplicative algorithm: aggregate of four elements from $\mathcal{T}_0$ with varying directions of dominating anisotropy and three steps of refinement

Theorem 3.14. *The multiplicative preconditioner of A_{11} has an optimal computational complexity with respect to both problem and discretization parameters.*

We summarize here some of the benefits from using the aforementioned preconditioning methods. The additive preconditioner has a block diagonal structure which makes the implementation of the algorithm easier. Furthermore it has an optimal computational complexity with respect to the size of the problem. The multiplicative preconditioner results in a better condition number. However, its optimal complexity heavily depends on the implementation of the nested dissection algorithm.

Remark 3.15. Consider the problem (1.1) discretized by conforming linear finite elements, where the coarsest grid $\mathcal{T}_0$ is aligned with the discontinuities of the coefficient $\mathbf{a}(e)$, $e \in \mathcal{T}_0$. Let us assume also that the multiplicative preconditioner C_{11} of the pivot block A_{11} is used. Then, the estimates (2.81) and (2.85) lead to the efficient hierarchical error estimate

$$\|\overline{\mathbf{w}}_1\|_{C_{11}} \leq \|u - u_H\|_{\mathcal{A}} \leq \frac{5}{2(1-\zeta^2)} \|\overline{\mathbf{w}}_1\|_{C_{11}}, \tag{3.30}$$

which is robust with respect to the mesh and coefficient anisotropy as well as to the coefficient jumps. Let us note that the coefficient in (3.30) is significantly better than the related coefficient in (2.96) which was derived in the case of a simplified AMLI setting for the elliptic model problem (2.80).

4 Robust AMLI algorithms: Nonconforming linear finite elements

For the nonconforming Crouzeix–Raviart finite element, where the nodal basis functions are defined at the midpoints along the edges of the triangle rather than at its vertices (cf. Figure 4.1), the natural vector spaces $\mathcal{V}_H(E) := \text{span}\,\{\phi_I, \phi_{II}, \phi_{III}\}$ and $\mathcal{V}_h(E) := \text{span}\,\{\phi_i\}_{i=1}^9$ (cf. the macroelement in Figure 4.1(b)) are no longer nested, i.e. $\mathcal{V}_H(E) \not\subseteq \mathcal{V}_h(E)$. This makes the direct construction with $\mathcal{V}_2(E) := \mathcal{V}_H(E)$, as used for conforming elements, impossible. Consequently, the hierarchical basis functions have to be chosen in a way that for the resulting subspaces $\mathcal{V}_1(E)$ and $\mathcal{V}_2(E)$, and hence, for the global finite element subspaces $\mathcal{V}_1$ and $\mathcal{V}_2$, the direct sum condition

$$\mathcal{V} = \mathcal{V}_1 \oplus \mathcal{V}_2 \tag{4.1}$$

is satisfied again.

4.1 Crouzeix–Raviart finite elements

A simple computation shows that the standard nodal basis element stiffness matrix for a nonconforming Crouzeix–Raviart (CR) linear finite element A_e^{CR} coincides with that of the corresponding conforming linear element up to a factor four, i.e.,

$$A_e^{\text{CR}} = 2 \begin{bmatrix} b+c & -c & -b \\ -c & a+c & -a \\ -b & -a & a+b \end{bmatrix}, \tag{4.2}$$

cf. (3.3). The construction of the hierarchical stiffness matrix at macroelement level starts with the assembly of four such matrices according to the numbering of the nodal points as shown in Figure 4.1(b). It further utilizes a transformation, which is based on a proper decomposition of the vector space $\mathcal{V}(E) = \mathcal{V}_h(E)$, which is associated with the fine-grid basis functions related to this macroelement E. This will be addressed in the following.

We consider three different splittings, each of which makes use of half-difference and half-sum basis functions in order to guarantee the condition

$$\mathcal{V}(E) := \text{span}\,\{\Phi_E\} = \mathcal{V}_1(E) \oplus \mathcal{V}_2(E). \tag{4.3}$$

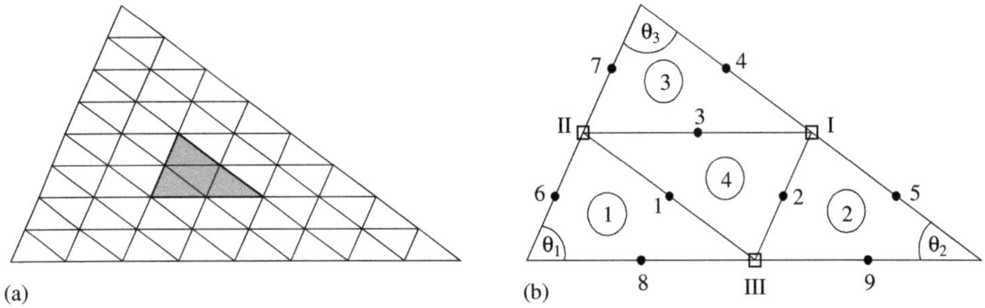

Figure 4.1: Crouzeix–Raviart finite element. (a) Discretization. (b) Macro-element.

Here $\Phi_E := \{\phi^{(i)}\}_{i=1}^9$ denotes the set of the "midpoint" basis functions of the four congruent elements in the macro-element E, as depicted in Figure 4.1(b). The splitting of $\mathcal{V}(E)$ can be defined in the general form

$$\begin{aligned}\mathcal{V}_1(E) &:= \text{span}\,\{\phi_1, \phi_2, \phi_3, \phi_1^D + \phi_4 - \phi_5, \phi_2^D + \phi_6 - \phi_7, \phi_3^D + \phi_8 - \phi_9\},\\ \mathcal{V}_2(E) &:= \text{span}\,\{\phi_1^C + \phi_4 + \phi_5, \phi_2^C + \phi_6 + \phi_7, \phi_3^C + \phi_8 + \phi_9\},\end{aligned} \tag{4.4}$$

where $\phi_i^D := \sum_k d_{ik}\phi_k$ and $\phi_i^C := \sum_k c_{ik}\phi_k$ with $i, k \in \{1, 2, 3\}$. The transformation matrix corresponding to this general splitting is given by

$$J_E = J_E(C_E, D_E) = \begin{bmatrix} I & D_E & C_E \\ 0 & J_{E-} & J_{E+} \end{bmatrix} \quad (\in \mathbb{R}^{9\times 9}), \tag{4.5}$$

where I denotes the 3×3 identity matrix, and C_E and D_E are 3×3 matrices whose entries c_{ij} respectively d_{ij} will be specified later. The 3×6 matrices

$$J_{E-} = \frac{1}{2}\begin{bmatrix} 1 & -1 & & & & \\ & & 1 & -1 & & \\ & & & & 1 & -1 \end{bmatrix}^T, \quad J_{E+} = \frac{1}{2}\begin{bmatrix} 1 & 1 & & & & \\ & & 1 & 1 & & \\ & & & & 1 & 1 \end{bmatrix}^T \tag{4.6}$$

introduce the so-called half-difference and half-sum basis functions associated with the sides of the macro-element triangle. The matrix J_E transforms the vector of the macro-element basis functions $\Phi_E := (\phi^{(i)})_{i=1}^9$ to the hierarchical basis vector

$$\tilde{\Phi}_E := (\tilde{\phi}^{(i)})_{i=1}^9 = J_E^T\, \Phi_E \tag{4.7}$$

and the hierarchical stiffness matrix at macro-element level is obtained as

$$\tilde{A}_E = J_E^T A_E J_E = \begin{bmatrix} \tilde{A}_{E:11} & \tilde{A}_{E:12} \\ \tilde{A}_{E:12}^T & \tilde{A}_{E:22} \end{bmatrix} \begin{matrix} \} \in \mathcal{V}_1(E) \\ \} \in \mathcal{V}_2(E) \end{matrix}. \tag{4.8}$$

The related global stiffness matrix is obtained as $\tilde{A}_h := \sum_{E \in \mathcal{T}_H} R_E^T \tilde{A}_E R_E$. Here, and in what follows, R_E^T is the natural inclusion (canonical injection), i.e., R_E^T transforms a macroelement vector to the corresponding global vector by extending it with zeros (outside the macroelement).

The transformation matrix $J = J(C_E, D_E : E \in \mathcal{T}_H)$ such that $\tilde{\Phi} = J^T \Phi$ is then used for the transformation of the global matrix A_h to its hierarchical form $\tilde{A}_h = J^T A_h J$, and (by a proper permutation of rows and columns) the latter admits the 3×3-block representation

$$\tilde{A}_h = \begin{bmatrix} \tilde{A}_{11} & \tilde{A}_{12} & \tilde{A}_{13} \\ \tilde{A}_{12}^T & \tilde{A}_{22} & \tilde{A}_{23} \\ \tilde{A}_{13}^T & \tilde{A}_{23}^T & \tilde{A}_{33} \end{bmatrix} \begin{matrix} \left.\vphantom{\begin{matrix} a \\ a \end{matrix}}\right\} \in \mathcal{V}_1 \\ \} \in \mathcal{V}_2 \end{matrix} \tag{4.9}$$

according to the interior, half-difference, and half-sum basis functions, which are associated with the locally introduced splitting (4.4). The upper-left 2×2 block is thus related to the vector space $\mathcal{V}_1$, while the lower-right block $\tilde{A}_{33}$ relates to $\mathcal{V}_2$. Note that due to the structure of J_E, the relation $\tilde{A}_{11} = A_{11}$ still holds.

4.2 Two-level splittings: "First Reduce" and "Differences and Aggregates"

The so-called First Reduce (FR) splitting [31, 77, 92] can be described as a combination of a basis transformation involving the matrix

$$J_{\pm} := \begin{bmatrix} I & 0 & 0 \\ 0 & J_{-} & J_{+} \end{bmatrix}, \tag{4.10}$$

where the global matrices J_{-} and J_{+} correspond to the macroelement terms as introduced in (4.6), and a reduction step. In the latter the degrees of freedom corresponding to the so-called interior basis functions are eliminated. If

$$\overline{A}_h := J_{\pm}^T A_h J_{\pm} = \begin{bmatrix} \overline{A}_{11} & \overline{A}_{12} & \overline{A}_{13} \\ \overline{A}_{12}^T & \overline{A}_{22} & \overline{A}_{23} \\ \overline{A}_{13}^T & \overline{A}_{23}^T & \overline{A}_{33} \end{bmatrix} \begin{matrix} \left.\vphantom{\begin{matrix} a \\ a \end{matrix}}\right\} \in \mathcal{V}_1 \\ \} \in \mathcal{V}_2 \end{matrix} \tag{4.11}$$

denotes the matrix after the transformation step, then the unknowns related to the block $\overline{A}_{11}$ $(= A_{11})$ are first eliminated and the system with $\overline{A}_h$ is reduced to a system with its Schur complement

$$B = \begin{bmatrix} \overline{A}_{22} & \overline{A}_{23} \\ \overline{A}_{23}^T & \overline{A}_{33} \end{bmatrix} - \begin{bmatrix} \overline{A}_{12}^T \\ \overline{A}_{13}^T \end{bmatrix} A_{11}^{-1} [\, \overline{A}_{12}, \; \overline{A}_{13} \,]. \tag{4.12}$$

Note that the exact computation of B is cheap since $\overline{A}_{11} = A_{11}$ is block-diagonal with blocks of size 3×3. We consider then the partitioning of the matrix B into a 2×2 block form

$$B = \begin{bmatrix} B_{11} & B_{12} \\ B_{12}^T & B_{22} \end{bmatrix}, \tag{4.13}$$

where B_{11} corresponds to the half-difference basis functions, while the block B_{22} is related to the subspace $\mathcal{V}_2$ and is thus associated with the coarse grid. This partitioning of B is now used for the construction of two-level preconditioners. As shown in the following remark, the FR splitting also fits into the framework of our general splitting.

Remark 4.1 (FR splitting). The reduction step in the FR approach can (due to the symmetry of $\overline{A}_h$ and the fact that $\overline{A}_{11} = A_{11}$) be written in multiplicative form

$$J_B^T \overline{A}_h J_B = \begin{bmatrix} A_{11} & 0 \\ 0 & B \end{bmatrix}, \quad J_B = \begin{bmatrix} I & -A_{11}^{-1}\overline{A}_{12} & -A_{11}^{-1}\overline{A}_{13} \\ 0 & I & 0 \\ 0 & 0 & I \end{bmatrix}, \tag{4.14}$$

where J_B denotes the matrix corresponding to the reduction step. Combining the transformation (4.11) with this reduction step yields the FR transformation matrix

$$J_{\mathrm{FR}} := J_{\pm} J_B = \begin{bmatrix} I & -A_{11}^{-1}\overline{A}_{12} & -A_{11}^{-1}\overline{A}_{13} \\ 0 & J_- & J_+ \end{bmatrix}. \tag{4.15}$$

Note that J_{FR}, as defined in (4.15), exhibits the structure of the transformation matrix of the general splitting (4.5), which is obtained when choosing $D = -A_{11}^{-1}\overline{A}_{12}$ and $C = -A_{11}^{-1}\overline{A}_{13}$. The FR hierarchical basis matrix thus has the form

$$\tilde{A}_h = J_{\mathrm{FR}}^T A_h J_{\mathrm{FR}} = \begin{bmatrix} A_{11} & 0 \\ 0 & B \end{bmatrix} \tag{4.16}$$

where B is given by (4.13). That means, in particular, we have $\tilde{A}_{33} = B_{22}$.

On the other hand, any splitting based on differences and aggregates can be written in the product representation (4.15) with a properly defined matrix J_B (actually obtained by replacing the first row of J_B in (4.14) by $[I \quad D \quad C]$ and using appropriate matrices C and D).

Since the constant in the strengthened CBS inequality does not depend on the choice of the matrix D, cf. Section 4.3, we make use of this fact in the following definition of the transformation related to the FR splitting, which is formulated in the framework of differences and aggregates.

Definition 4.2 (First Reduce (FR) splitting). The splitting based on differences and aggregates incorporating the "first reduce" step (in short FR splitting), cf. [92] and [31], is characterized by using $D_E = 0$ and $C_E = -A_{E:11}^{-1}\overline{A}_{E:13}$ in the general transformation matrix (4.5). The matrices $\overline{A}_{E:11} = A_{E:11}$ and $\overline{A}_{E:13}$ are to be taken according to (4.11).

Definition 4.3 (Differences and Aggregates (DA)). The basis transformation for the standard splitting based on differences and aggregates (DA), cf. [31], follows from the general transformation (4.5) for the choice $D_E = 0$ and $C_E = \frac{1}{2}\operatorname{diag}(1, 1, 1)$.

Definition 4.4 (Generalized DA (GDA)). A generalized splitting based on differences and aggregates (GDA), as considered in [90] and analyzed in [77], is retrieved by substituting in (4.5) the matrices $D_E = 0$ and

$$C_E = \frac{1}{2}I + \mu(\mathbb{1} - 3I) \tag{4.17}$$

whereby the locally optimal choice of $\mu \in [0, \frac{1}{4}]$ depends on the minimum angle condition used in the triangular mesh.[1]

4.3 Uniform estimates of the CBS constant in case of non-nested spaces

Similar to the conforming finite element spaces the constant γ in the CBS inequality, which measures the cosine of the abstract angle between the two subspaces $\mathcal{V}_1$ and $\mathcal{V}_2$, i.e.,

$$\gamma = \cos(\mathcal{V}_1, \mathcal{V}_2),$$

is an important quality criterion for the splitting. In this section we will therefore deal again with upper bounds for γ but this time for discretizations using nonconforming Crouzeix–Raviart type finite elements.

Let us consider the hierarchical basis matrix $\tilde{A}_h = J^T A_h J$, partitioned as in (2.33), where J is defined according to (4.5). Then the CBS constant is given by

$$\gamma^2 = \max_{\mathbf{v}_3 \neq \mathbf{c}} \left[1 - \frac{\mathbf{v}_3^T \tilde{S} \mathbf{v}_3}{\mathbf{v}_3^T \tilde{A}_{33} \mathbf{v}_3} \right] = 1 - \min_{\mathbf{v}_3 \neq \mathbf{c}} \frac{\mathbf{v}_3^T \tilde{S} \mathbf{v}_3}{\mathbf{v}_3^T \tilde{A}_{33} \mathbf{v}_3} \tag{4.18}$$

where $\tilde{S}$ is the Schur complement obtained from $\tilde{A}_h$ by eliminating the degrees of freedom corresponding to basis functions in $\mathcal{V}_1$, i.e., by reducing the system to the

[1]Here $\mathbb{1}$ denotes the 3×3 matrix of all ones.

lower-right block. Let S be the corresponding Schur complement obtained in the same way from A_h. Then the following lemma provides the starting point for our discussion.

Lemma 4.5. *Under the above assumptions we have*

$$\tilde{S} = S$$

and thus $\tilde{S}$ is invariant with respect to the matrices C and D, which appear in the transformation matrix J.

Proof. Follows from direct calculation. □

Lemma 4.6. *The CBS constant γ, given by (4.18), for the splitting (4.3) associated with the hierarchical basis transformation matrix that is induced by the local transformation (4.5) is invariant to the matrix D.*

Proof. Direct calculations are applied again. Following [77] we get the following six blocks of the symmetric matrix $\tilde{A}_h = J^T A_h J$ (see (4.9)):

$$\begin{aligned}
\tilde{A}_{11} &= A_{11} \\
\tilde{A}_{12} &= A_{11} D + A_{12} J_{-22} + A_{13} J_{-23} \\
\tilde{A}_{13} &= A_{11} C + A_{12} J_{+23} + A_{13} J_{+33} \\
\tilde{A}_{22} &= D^T \left(A_{11} D + A_{12} J_{-22} + A_{13} J_{-23}\right) + J_{-22}^T A_{12}^T D + J_{-32}^T A_{13}^T D \\
&\quad + J_{-22}^T A_{22}^T J_{-22} + J_{-32}^T A_{23}^T J_{-22} + J_{-22}^T A_{23} J_{-32} + J_{-32}^T A_{33} J_{-32} \\
\tilde{A}_{23} &= D^T \left(A_{11} C + A_{12} J_{+23} + A_{13} J_{+33}\right) + J_{-22}^T A_{12}^T C + J_{-32}^T A_{13}^T C \qquad (4.19) \\
&\quad + J_{-22}^T A_{22} J_{+23} + J_{-32}^T A_{23}^T J_{+23} + J_{-22}^T A_{23} J_{+33} + J_{-32}^T A_{33} J_{+33} \\
\tilde{A}_{33} &= C^T \left(A_{11} C + A_{12} J_{+23} + A_{13} J_{+33}\right) + J_{+23}^T A_{12}^T C + J_{+33}^T A_{13}^T C \\
&\quad + J_{+23}^T A_{22} J_{+23} + J_{+33}^T A_{23}^T J_{+23} + J_{+23}^T A_{23} J_{+33} + J_{+33}^T A_{33} J_{+33} \,.
\end{aligned}$$

The notations $J_- := [J_{-22}, J_{-32}]^T$ and $J_+ := [J_{+23}, J_{+33}]^T$ are used to indicate the position of the respective blocks in the transformation matrix J. From the last equation in (4.19) we see that $\tilde{A}_{33} = \tilde{A}_{33}(C)$ depends on the matrix C but not on D. By Lemma 4.5 we have $\tilde{S} = S$, which does neither depend on C nor on D. Hence, in view of (4.18) the assertion is proved true. □

We end the discussion of the general splitting with an important observation that is relevant in the context of the multilevel hierarchical splitting, i.e., for the multilevel extension of the two-level preconditioners. There, for both variants, the multiplicative and the additive AMLI method, the construction is such that the matrix block $\tilde{A}_{33}^{(k)}$ for $k = \ell, \ldots, 1$ defines the coarse-level matrix $A^{(k-1)}$. Then, as is well known, the condition

$$\ker\left(\tilde{A}_{E:33}^{(k)}\right) = \ker\left(A_e\right) = \operatorname{span}\left(\mathbf{1}\right) \quad \forall k \in \{\ell, \ldots, 1\} \tag{4.20}$$

($\mathbf{1}$ stands for the vector of all ones) is necessary for any splitting to result in a local CBS constant γ_E that is strictly less than one.

The next lemma gives a general characterization of the sets $\mathcal{A}^{\mathrm{C}} = \mathcal{A}^{\mathrm{CR}}$ of local element stiffness matrices for Courant (C) linear conforming and Crouzeix–Raviart (CR) linear nonconforming finite elements.

Lemma 4.7. *Let us denote (see (4.2)) by*

$$\mathcal{A}^{\mathrm{C}} = \mathcal{A}^{\mathrm{CR}} := \{d\, A_e : (a, b, c) = (\cot\theta_1, \cot\theta_2, \cot\theta_3),\ d > 0\} \tag{4.21}$$

where

$$A_e = \begin{bmatrix} b+c & -c & -b \\ -c & a+c & -a \\ b & a & a+b \end{bmatrix}.$$

Further let $\overline{A}_e$ *be an arbitrary* 3×3 *symmetric and positive semidefinite matrix such that* $\ker\left(\overline{A}_e\right) = \operatorname{span}\left(\mathbf{1}\right)$. *Then* $\overline{A}_e \in \mathcal{A}^{\mathrm{C}} = \mathcal{A}^{\mathrm{CR}}$.

Proof. By the assumption $\overline{A}_e$ is symmetric and has the property (4.20), i.e., the row sums equal zero. Therefore

$$\overline{A}_e = \begin{bmatrix} \overline{b}+\overline{c} & -\overline{c} & -\overline{b} \\ -\overline{c} & \overline{a}+\overline{c} & -\overline{a} \\ \overline{b} & -\overline{a} & \overline{a}+\overline{b} \end{bmatrix}, \tag{4.22}$$

where (eventually after a proper permutation of rows and columns) $\overline{a} \le \overline{b} \le \overline{c}$. Further, since $\overline{A}_e$ is SPSD, we have $\ker(\overline{A}_e) = \operatorname{span}(\mathbf{1})$ if and only if both the conditions

$$\overline{b} > 0 \quad \text{and} \quad -\frac{\overline{b}\,\overline{c}}{\overline{b}+\overline{c}} < \overline{a} \tag{4.23}$$

are satisfied. Note also that in this case, $-\frac{\overline{b}\,\overline{c}}{\overline{b}+\overline{c}} \leq \overline{a}$ implies $|\overline{a}| \leq \overline{b}$, and therefore

$$|\overline{a}| \leq \overline{b} \leq \overline{c}. \tag{4.24}$$

The latter inequality from (4.23) equivalently can be written as

$$\overline{d}^2 = \overline{a}\,\overline{b} + \overline{a}\,\overline{c} + \overline{b}\,\overline{c} > 0. \tag{4.25}$$

We rewrite (4.25), i.e.,

$$a\,b + a\,c + b\,c = 1 \tag{4.26}$$

where $a := \overline{a}/\overline{d}$, $b := \overline{b}/\overline{d}$ and $c := \overline{c}/\overline{d}$. Now let $\cot\theta_2 := b$ and $\cot\theta_3 := c$ where $b > 0$ and $c > 0$ implies $\theta_i \in (0, \pi/2)$, $i = 1, 2$. Then, cf. (4.26),

$$a = \frac{1 - b\,c}{b + c} = \frac{1 - \cot\theta_2\,\cot\theta_3}{\cot\theta_2 + \cot\theta_3} = \cot(\pi - \theta_2 - \theta_3) := \cot\theta_1,$$

which completes the proof. □

In order to ensure that the local estimates presented in the rest of this chapter can repeatedly be applied on the coarser levels, it will be shown in the next lemma that starting with a matrix (at some level ℓ), which has been obtained from a global assembly of nonconforming Crouzeix–Raviart (CR) linear finite elements A_e^{CR}, as defined in (4.2), the matrices $\tilde{A}_{E:33}^{(k)}$ are of the same type on all levels $k = \ell, \ell-1, \ldots, 1$.

Lemma 4.8. *Let $A_e^{(k)} \in \mathcal{A}^{\mathrm{CR}}$ be an element matrix at some level k where $k = \ell, \ldots, 1$, and let $A_e^{(k-1)} := \tilde{A}_{E:33}^{(k)}$ be the element matrix retrieved from the lower* 3×3 *block of the macroelement matrix $\tilde{A}_E^{(k)}$, which satisfies (4.20). Then*

$$A_e^{(k-1)} \in \mathcal{A}^{\mathrm{CR}},$$

i.e., it is a Crouzeix–Raviart element stiffness matrix.

Proof. First we note that by assumption $\ker(A_e^{(k-1)}) = \operatorname{span}(\mathbf{1})$. Moreover, since $\tilde{A}_E^{(k)}$ is symmetric and positive semidefinite (SPSD) and $A_e^{(k-1)}$ is a principal submatrix of $\tilde{A}_E^{(k)}$, it follows that $A_e^{(k-1)}$ is SPSD as well. Then, the statement follows directly from Lemma 4.7. □

In the following we shall provide a summary of the main results from references [31, 32, 77] related to a local estimation of the CBS constant and the resulting AMLI methods.

DA splitting:

Now, we are in a position to analyze the constant

$$\gamma = \cos(\mathcal{V}_1, \mathcal{V}_2)$$

for the DA splitting for which, cf. Definition 4.3,

$$\begin{aligned}\mathcal{V}_1(E) &:= \operatorname{span}\{\phi_1, \phi_2, \phi_3, \phi_4 - \phi_5, \phi_6 - \phi_7, \phi_8 - \phi_9\},\\ \mathcal{V}_2(E) &:= \operatorname{span}\Big\{\frac{1}{2}(\phi_1 + \phi_2 + \phi_3) + \phi_4 + \phi_5,\\ &\quad \frac{1}{2}(\phi_1 + \phi_2 + \phi_3) + \phi_6 + \phi_7, \frac{1}{2}(\phi_1 + \phi_2 + \phi_3) + \phi_8 + \phi_9\Big\}.\end{aligned} \tag{4.27}$$

Again, this analysis is performed locally, by considering the corresponding problems on macroelements.[2]

Here we follow the procedure from [31, 32], which slightly differs from the one which was used in the case of conforming FEs in the sense that the reference right angle macroelement is considered first, see Figure 4.2.

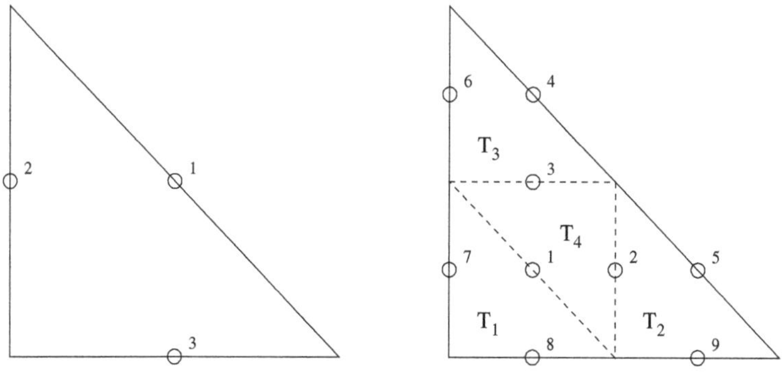

Figure 4.2: The reference coarse-grid triangle and the macroelement $\hat{E}$.

Let $\mathcal{V}_1(\hat{E})$, $\mathcal{V}_2(\hat{E})$ define the two-level DA splitting for the reference macroelement $\hat{E}$, and let us denote by $\boldsymbol{d}^{(k)} = \boldsymbol{d}^{(k)}(u) = \nabla u\,|_{T_k}$, $\boldsymbol{\delta}^{(k)} = \boldsymbol{\delta}^{(k)}(v) = \nabla v\,|_{T_k}$ for $u \in \mathcal{V}_1(\hat{E})$, $v \in \mathcal{V}_2(\hat{E})$. Then the relations between the function values in some nodal points, namely $u(P_4) = -u(P_5)$, $u(P_6) = -u(P_7)$, $u(P_8) = -u(P_9)$ and $v(P_1) = v(P_4) = v(P_5)$, $v(P_2) = v(P_6) = v(P_7)$, $v(P_3) = v(P_8) = v(P_9)$, imply that

[2] Note that (4.20) holds true for all the splittings DA, GDA and FR.

$$\boldsymbol{d}^{(1)} + \boldsymbol{d}^{(2)} + \boldsymbol{d}^{(3)} + \boldsymbol{d}^{(4)} = \mathbf{0}, \tag{4.28}$$

$$\boldsymbol{\delta}^{(1)} = \boldsymbol{\delta}^{(2)} = \boldsymbol{\delta}^{(3)} = -\boldsymbol{\delta}^{(4)} = \boldsymbol{\delta}. \tag{4.29}$$

Hence,

$$\begin{aligned} \mathcal{A}_{\hat{E}}(u,v) &= \sum_{k=1}^{4} \int_{T_k} \mathbf{a}\nabla u \cdot \nabla v dx = \sum_{k=1}^{4} \Delta \langle \mathbf{a}\boldsymbol{\delta}^{(k)}, \boldsymbol{d}^{(k)} \rangle \\ &= \Delta \langle \mathbf{a}\boldsymbol{\delta}, \boldsymbol{d}^{(1)} + \boldsymbol{d}^{(2)} + \boldsymbol{d}^{(3)} - \boldsymbol{d}^{(4)} \rangle \\ &= -2\Delta \langle \mathbf{a}\boldsymbol{\delta}, \boldsymbol{d}^{(4)} \rangle \le 2\Delta \|\boldsymbol{\delta}\|_{\mathbf{a}} \|\boldsymbol{d}^{(4)}\|_{\mathbf{a}} \end{aligned} \tag{4.30}$$

where $\Delta = \text{area}(T_k)$, $\langle \boldsymbol{x}, \mathbf{y} \rangle = \boldsymbol{x}^T \mathbf{y}$ denotes the inner product in $\mathbb{R}^2$, and $\|\boldsymbol{x}\|_{\mathbf{a}} = \sqrt{\langle \mathbf{a}\boldsymbol{x}, \boldsymbol{x} \rangle}$.

Further,

$$\|\boldsymbol{d}^{(4)}\|_{\mathbf{a}}^2 = \|\boldsymbol{d}^{(1)} + \boldsymbol{d}^{(2)} + \boldsymbol{d}^{(3)}\|_{\mathbf{a}}^2 \le 3 \sum_{k=1}^{3} \|\boldsymbol{d}^{(k)}\|_{\mathbf{a}}^2$$

leads to

$$\mathcal{A}_{\hat{E}}(u,u) = \sum_{k=1}^{4} \|\boldsymbol{d}^{(k)}\|_{\mathbf{a}}^2 \Delta \ge \left(1 + \frac{1}{3}\right) \Delta \|\boldsymbol{d}^{(4)}\|_{\mathbf{a}}^2 \tag{4.31}$$

and

$$\mathcal{A}_{\hat{E}}(v,v) = 4\Delta \|\boldsymbol{\delta}\|_{\mathbf{a}}^2. \tag{4.32}$$

Thus,

$$\begin{aligned} \mathcal{A}_{\hat{E}}(u,v) &\le 2\Delta \sqrt{\frac{3}{4\Delta} \mathcal{A}_{\hat{E}}(u,u)} \sqrt{\frac{1}{4\Delta} \mathcal{A}_{\hat{E}}(v,v)} \\ &= \sqrt{\frac{3}{4}} \sqrt{\mathcal{A}_{\hat{E}}(u,u)} \sqrt{\mathcal{A}_{\hat{E}}(v,v)}. \end{aligned} \tag{4.33}$$

In the case of an arbitrarily shaped macroelement E we can use the affine mapping $F : \hat{E} \to E$ which transforms the problem to the reference macroelement (for more details see, e.g., [8]). This transformation changes the anisotropy of the problem, cf. (3.1), but the estimate $\gamma_{DA,E}^2 \le \frac{3}{4}$ will still hold since the bound (4.33) for the reference macroelement does not depend on anisotropy.

The obtained result is summarized in the following theorem.

Theorem 4.9. *Let us consider the DA splitting. Then the corresponding strengthened CBS inequality constant γ is uniformly bounded with respect to both, coefficient and mesh anisotropy,*

$$\gamma_{DA}^2 \leq \frac{3}{4}. \tag{4.34}$$

The latter estimate is independent of the discretization (mesh) parameter h and possible coefficient jumps aligned with the finite element partitioning $\mathcal{T}_H$.

Let us consider the DA splitting with two particular test functions $u \in \mathcal{V}_1(\widehat{E})$, $v \in \mathcal{V}_2(\widehat{E})$ determined by the values in the nodes P_i, i.e., $u(P_i) = u_i$, $v(P_i) = v_i$, $i = 1, \dots, 9$, see Figure 4.2. Now let

$$u_1 = 2/3,\ u_2 = u_3 = u_6 = u_9 = -1/3,\ u_4 = u_5 = 0,\ u_7 = u_8 = 1/3,$$

$$v_1 = v_4 = v_5 = 1,\ v_2 = v_6 = v_7 = 0,\ v_3 = v_8 = v_9 = 0.$$

Using the above introduced notations we find

$$\boldsymbol{d}^{(1)} = \boldsymbol{d}^{(2)} = \boldsymbol{d}^{(3)} = \frac{1}{3}(1,1)^T = -\frac{1}{3}\boldsymbol{d}^{(4)}$$

and

$$\boldsymbol{\delta}^{(1)} = \boldsymbol{\delta}^{(2)} = \boldsymbol{\delta}^{(3)} = (1,1)^T = -\boldsymbol{\delta}^{(4)}.$$

Thus

$$\begin{aligned}
\mathcal{A}_{\widehat{E}}(u,v) &= 2\Delta\langle \mathbf{a}\boldsymbol{d}^{(4)}, \boldsymbol{\delta}^{(4)}\rangle = 6\Delta\langle \mathbf{a}\boldsymbol{d}^{(4)}, \boldsymbol{d}^{(4)}\rangle \\
\mathcal{A}_{\widehat{E}}(u,u) &= \frac{4}{3}\Delta\langle \mathbf{a}\boldsymbol{d}^{(4)}, \boldsymbol{d}^{(4)}\rangle \\
\mathcal{A}_{\widehat{E}}(v,v) &= 4\Delta\,\langle \mathbf{a}\boldsymbol{\delta}^{(4)}, \boldsymbol{\delta}^{(4)}\rangle = 36\Delta\,\langle \mathbf{a}\boldsymbol{d}^{(4)}, \boldsymbol{d}^{(4)}\rangle
\end{aligned}$$

i.e.

$$\mathcal{A}_{\widehat{E}}(u,v) = \sqrt{\frac{3}{4}}\sqrt{\mathcal{A}_{\widehat{E}}(u,u)}\sqrt{\mathcal{A}_{\widehat{E}}(v,v)}. \tag{4.35}$$

Remark 4.10. As (4.35) holds independently of the coefficient matrix **a**, the macroelement CBS constant can be estimated by $\sqrt{3/4}$ for both, mesh and coefficient anisotropy.

The following theorem is useful to extend the two-level constructions and estimates to the multilevel case.

Theorem 4.11. *Let $\tilde{A}_{33}$ be the stiffness matrix corresponding to the space $\mathcal{V}_2$ induced by the DA splitting (4.27), and let A_H be the stiffness matrix corresponding to the finite element space $\mathcal{V}_H$ corresponding to the coarse discretization $\mathcal{T}_H$, equipped with the standard nodal finite element basis Φ_H. Then*

$$\tilde{A}_{33} = 4\, A_H \,. \tag{4.36}$$

Proof. The result follows directly from $\tilde{A}_E = J_E^T A_E J_E$ where J_E is given by (4.5), and by choosing the local matrices $D_E := 0$ and $C_E := \frac{1}{2}\operatorname{diag}(1,1,1)$, which is in accordance to Definition 4.3. □

The DA algorithm allows for a direct recursive extension of the estimate (4.34) to the multi-level case, which follows from Theorem 4.11. The same does not hold automatically for the FR and GDA algorithms since the related blocks $\tilde{A}_{33}$ are only associated with the coarse grid. The needed extra theoretical analysis for a robust multilevel extension of the latter splittings is supported by Lemma 4.8.

GDA splitting:

This generalization of the standard DA splitting is based on the assumption of a minimum angle condition, which is commonly used in commercial mesh generators, and on symmetry (or anisotropy) assumptions for the matrices C_E (in the transformation J) which lead to a local dependence of a single parameter $\mu \in [0,\ 1/4]$, cf. [90].

The CBS constant, as defined in (4.18), clearly depends on C_E. We now seek a matrix C_E that provides the minimal (local) $\gamma_E = \gamma_{E:\min}$ where

$$\begin{aligned}\gamma^2_{E:\min} &:= \min_{C_E}\,[\gamma^2_{E:\min}] = 1 - \max_{C_E}\left[\min_{\mathbf{v}_{E:3}\neq\mathbf{c}} \frac{\mathbf{v}^T_{E:3} S_E \mathbf{v}_{E:3}}{\mathbf{v}^T_{E:3}\tilde{A}_{E:33}(C_E)\mathbf{v}_{E:3}}\right] \\ &= 1 - \min_{\mathbf{v}_{E:3}\neq\mathbf{c}}\left[\max_{C_E} \frac{\mathbf{v}^T_{E:3} S_E \mathbf{v}_{E:3}}{\mathbf{v}^T_{E:3}\tilde{A}_{E:33}(C_E)\mathbf{v}_{E:3}}\right],\end{aligned} \tag{4.37}$$

$\mathbf{c} = (c, c, c)^T$, subject to the constraint that C_E is of the form (4.17) for which we use the short notation $C_E \in \mathcal{C}_{\mathrm{GDA}}$. The following theorem is published in [77].

Theorem 4.12. *Consider a macro element matrix A_E, assembled from four element matrices of the form (4.2) corresponding to the four similar triangles obtained by a regular refinement step, cf. Figure 4.1. Then the local CBS constant $\gamma_E = \gamma_{\mathrm{GDA},E}$ for the GDA splitting, cf. Definition 4.4, is minimal for*

$$\mu = \mu_{\mathrm{opt}} = \frac{s - r}{5s - 4r} \tag{4.38}$$

where $s = a + b + c$ and $r = \sqrt{s^2 - 3}$. The minimal local CBS constant then is given by

$$\gamma^2_{\mathrm{GDA},E:\min} = \frac{s\,(2s + r)}{4\,(1 + s^2)}. \tag{4.39}$$

Proof. First we note that the Schur complement S_E in (4.37) is invariant to the matrix $C_E = C_{\mathrm{GDA}}$ (by Lemma 4.5). Further the local transformation matrix related to the GDA splitting can be decomposed in the form

$$J_{\mathrm{GDA}} = J_{E\pm} \begin{bmatrix} I & 0 & C_{\mathrm{GDA}} \\ 0 & I & 0 \\ 0 & 0 & I \end{bmatrix}$$

with $J_{E\pm}$ as given in (4.10). This implies that the matrix $\tilde{A}_{E:33} = \tilde{A}_{E:33}(C_{\mathrm{GDA}})$ has the representation

$$\tilde{A}_{E:33}(C_{\mathrm{GDA}}) = \begin{bmatrix} C^T_{\mathrm{GDA}} & I \end{bmatrix} \begin{bmatrix} \overline{A}_{E:11} & \overline{A}_{E:13} \\ \overline{A}^T_{E:13} & \overline{A}_{E:33} \end{bmatrix} \begin{bmatrix} C_{\mathrm{GDA}} \\ I \end{bmatrix}, \tag{4.40}$$

with

$$\overline{A}_{E:13} = A_{E:12} J_{+23} + A_{E:13} J_{+33} \tag{4.41}$$

and

$$\begin{aligned} \overline{A}_{E:33} &= J^T_{+23} A_{E:22} J_{+23} + J^T_{+23} A_{E:23} J_{+33} \\ &\quad + J^T_{+33} A^T_{E:23} J_{+23} + J^T_{+33} A_{E:33} J_{+33}, \end{aligned} \tag{4.42}$$

cf. (4.19). Hence, we have the decomposition

$$\begin{aligned} \tilde{A}_{33,E}(C_{\mathrm{GDA}}) &= \begin{bmatrix} C^T_{\mathrm{GDA}} + \overline{A}^T_{E:13} A^{-1}_{E:11} & I \end{bmatrix} \\ &\quad \begin{bmatrix} A_{E:11} & 0 \\ 0 & \overline{S}_E \end{bmatrix} \begin{bmatrix} C_{\mathrm{GDA}} + A^{-1}_{E:11} \overline{A}_{E:13} \\ I \end{bmatrix} \end{aligned} \tag{4.43}$$

where $\overline{S}_E = \overline{A}_{E:33} - \overline{A}^T_{E:13} A^{-1}_{E:11} \overline{A}_{E:13}$. By using (4.17) in (4.43) one finds

$$\tilde{A}_{E:33} = \tilde{A}_{E:33}(\mu) = \mu^2 G_{11} + \mu(G_{12} + G^T_{12}) + G_{22} \tag{4.44}$$

where

$$\begin{aligned} G_{11} &= (\mathbb{1} - 3I)^T A_{E:11} (\mathbb{1} - 3I) \\ G_{12} &= \Big(\frac{1}{2} A_{E:11} + \overline{A}^T_{E:13}\Big)(\mathbb{1} - 3I) \\ G_{22} &= \frac{1}{4} A_{E:11} + \frac{1}{2}(\overline{A}_{E:13} + \overline{A}^T_{E:13}) + \overline{A}_{E:33}. \end{aligned} \tag{4.45}$$

Then, solving the generalized eigenproblem

$$S_E \mathbf{v} = \lambda_E \tilde{A}_{E:33} \mathbf{v}, \qquad \mathbf{v} \neq \mathbf{c},$$

(using a computer algebra software) the minimal eigenvalue that determines the local estimate on the CBS constant is found to be

$$\lambda_{E:\min} = \lambda_{\mathrm{GDA},E:\min} = \frac{1}{4\,[w(\mu) + 2\mu r s]}, \tag{4.46}$$

where $w(\mu) := 1-6\mu+12\mu^2+\mu^2 s(s-r)$ is positive. Moreover, one finds that the optimal choice of the parameter μ in the GDA splitting, which leads to the smallest possible local CBS constant, equivalently determines the matrix $C_{\mathrm{GDA}}(\mu_{\mathrm{opt}})$ that minimizes $\|\tilde{A}_{E:33}(\mu)\|_2$. Using (4.44) and (4.45), and following standard arguments, the formula (4.38) then is easily derived. Finally, inserting in (4.46) the optimal value (4.38) for μ and using $\gamma^2_{E:\min} = (1 - \lambda_{E:\min})$ one obtains (4.39). □

At the end of this section we want to collect (skipping the proofs) some results from reference [90] where it is shown that repeated application of the GDA splitting yields a sequence of element matrices with an *improving* shape of the associated triangles.

Lemma 4.13. *Let $B_e^{(0)} = A_e^{(\ell)} = A_e^{\mathrm{CR}}$ be defined by (4.2), and let $(B_e^{(n)})_{n\geq 0}$ denote the sequence of properly normalized element matrices arising during the coarsening process according to the GDA splitting. Then, under the assumption of convergence and for $\mu \neq 0$ we denote by $B_{eq} = \lim_{n\to\infty} B_e^{(n)}$. The matrix B_{eq} corresponds to the Laplace operator where the limiting element e is an equilateral triangle, i.e., $a = b = c = 1/\sqrt{3}$.*

For $\mu > 0$, where the limiting case is given by the equilateral triangle, the following convergence result has been shown.

Theorem 4.14. *Let $(B_e^{(n)})_{n\geq 0}$ be the sequence of normalized stiffness matrices as in Lemma 4.13. Further, let μ be bounded away from zero, i.e., $\mu \in [\mu_{\min}, 1/4]$ with $\mu_{\min} > 0$. Then,*

$$\|B_e^{(n+1)} - B_e^{eq}\|_F \;\leq\; q\, \|B_e^{(n)} - B_e^{eq}\|_F \tag{4.47}$$

is satisfied for some positive $q < 1$, where $\|.\|_F$ denotes the Frobenius norm of the given matrix.

Remark 4.15. The convergence factor

$$q = \frac{c\,g(\mu)}{c\,g(\mu) + p(\mu)}$$

in (4.47) is bounded from above by

$$\overline{q}(\mu) := \frac{g(\mu)}{g(\mu) + \mu^2},$$

where $g(\mu) = 1 - 6\mu + 12\mu^2$ and $p(\mu) = \mu^2\,(a + b + c)$. This upper bound $\overline{q}$, which is shown in Figure 4.3, attains its maximum at $\mu = 0$ with $\overline{q}(0) = 1$ and its minimum at $\mu = 1/3$ with $\overline{q}(1/3) = 3/4$. For $\mu \in [\mu_{\min},\ 1/4]$ with $\mu_{\min} > 0$, as used in Theorem 4.14, one obtains $\overline{q} \in [4/5,\ 1)$. That means, the larger the value of μ the smaller (better) the convergence factor in (4.47).

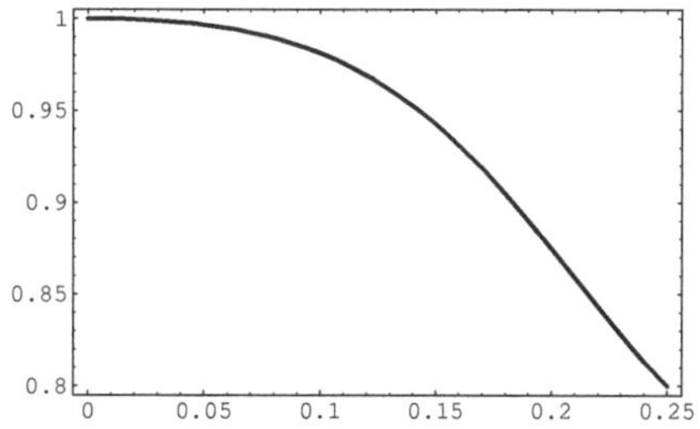

Figure 4.3: Upper bound $\overline{q}$ of the convergence factor q against μ

Remark 4.16. Note that if the minimal angle in the triangle tends to zero, then the optimized value of the parameter $\mu(a, b, c)$ which governs the generalized DA-splitting (GDA) also tends to zero, which renders the GDA to become closer and closer to the standard DA-algorithm.

FR splitting:

The GDA splitting with the optimal choice of the parameter μ as given by (4.38), however, in general is still not the best splitting, which is obtained for the FR basis transformation. The following theorem was originally published in [77].

Theorem 4.17. *Let $\tilde{A}_{E:33}(C_E)$ be given according to (4.19). Then, the minimum value for γ_E, as defined in (4.37), is attained for $C_{E:\mathrm{opt}} = -A_{E:11}^{-1}\overline{A}_{E:13}$ with $\overline{A}_{E:13} = A_{E:12}J_{+23} + A_{E:13}J_{+33}$.*

Proof. According to Lemma 4.6 the constant γ is invariant to the matrix D. Without loss of generality we can therefore set $D := 0$ in the transformation matrix J, which we can then write in the form

$$J = \begin{bmatrix} I & 0 & C \\ 0 & J_- & J_+ \end{bmatrix} = J_\pm \begin{bmatrix} I & 0 & C \\ 0 & I & 0 \\ 0 & 0 & I \end{bmatrix}, \tag{4.48}$$

where $J_\pm$ is defined by (4.10). In analogy to the proof of Theorem 4.12 (but now for a general matrix C) we decompose the matrix functional $\tilde{A}_{E:33}(C_E)$ in the form

$$\begin{aligned} \tilde{A}_{E:33}(C_E) &= \begin{bmatrix} C_E^T + \overline{A}_{E:13}^T A_{E:11}^{-1}, & I \end{bmatrix} \\ &\quad \begin{bmatrix} A_{E:11} & 0 \\ 0 & \overline{S}_E \end{bmatrix} \begin{bmatrix} C_E + A_{E:11}^{-1}\overline{A}_{E:13} \\ I \end{bmatrix}, \end{aligned} \tag{4.49}$$

where $\overline{S}_E = \overline{A}_{E:33} - \overline{A}_{E:13}^T A_{E:11}^{-1} \overline{A}_{E:13}$ and the matrices $\overline{A}_{E:13}$ and $\overline{A}_{E:33}$ are given by (4.41) and (4.42), respectively (cf. (4.40)–(4.43)). Since $\tilde{A}_{E:33}$ appears in the denominator in (4.37) we aim at minimizing the spectral norm of $\tilde{A}_{E:33}(C_E)$ with respect to the matrix C_E using the representation (4.49). But it is readily seen that the minimum is given by $\min_{C_E} \tilde{A}_{E:33}(C_E) = \overline{S}_E$, which is obtained for $C_{E:\mathrm{opt}} = -A_{E:11}^{-1}\overline{A}_{E:13}$. This completes the proof. □

Remark 4.18. Note that the optimal choice $C_{E:\mathrm{opt}}$, in Theorem 4.17, provides the harmonic extension obtained via the FR approach. Equivalently, $C_{E:\mathrm{opt}}$ can be shown to be the solution of the optimization problem

$$C_{E:\mathrm{opt}} = \min_{C_E} \|I - S_E^{-1}\tilde{A}_{E:33}(C_E)\|_2.$$

Remark 4.19. An explicit expression for the local CBS constant corresponding to the FR splitting was derived in reference [31]. Using the substitutions $s := a + b + c$ and $t := abc$ this formula reads

$$\gamma_{\mathrm{FR},E}^2 = \frac{3}{8} + \frac{1}{8}\sqrt{\frac{s-9t}{s-t}}. \tag{4.50}$$

As it can be seen from Figure 4.4 the estimates $\gamma_{\mathrm{GDA},E:\min}^2$ as given by (4.39) for the GDA splitting and $\gamma_{\mathrm{FR},E}^2$ as given by (4.50) for the FR splitting coincide for any isosceles triangle.

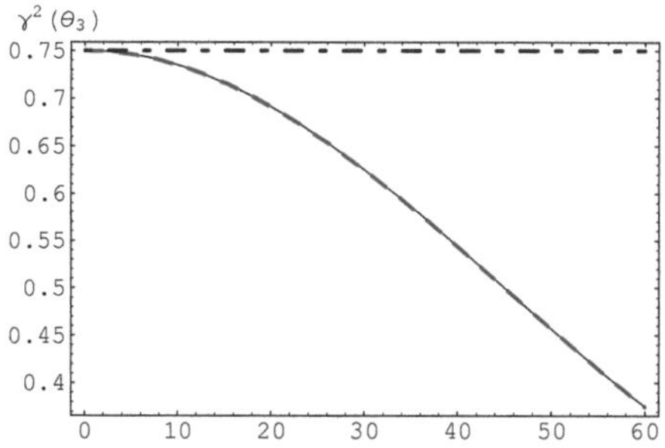

Figure 4.4: Local estimates of γ_E^2 for isosceles triangle depending on θ_3: FR (solid), GDA (dashed), and DA (dot-dashed) splitting

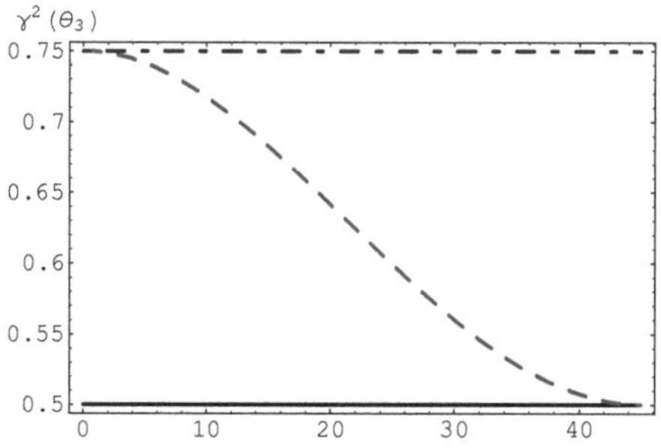

Figure 4.5: Local estimates of γ_E^2 for right triangle depending on θ_3: FR (solid), GDA (dashed), and DA (dot-dashed) splitting

For right triangles (4.50) yields $\gamma_{\mathrm{FR},E}^2 = 1/2$ independently of the other angles and the value of $\gamma_{\mathrm{GDA},E:\min}^2$ approaches the value of the squared local CBS constant for the DA splitting (which is always $3/4$) when the minimal angle tends to zero, cf. Figure 4.5.

Thus $\gamma_{\mathrm{GDA},E:\min}^2$ for an arbitrary but fixed macroelement E is always in between $\gamma_{\mathrm{FR},E}^2$ and $\gamma_{\mathrm{DA},E}^2$ (and since those estimates are sharp) we arrive at the following theorem, which characterizes the global CBS constants, cf. [77]:

Theorem 4.20. *Consider the discrete problem derived from the weak formulation (1.1) of the boundary value problem (1.4) using linear Crouzeix–Raviart finite elements for discretization, and assume that the coefficient matrix* $\mathbf{a}(e)$ *in (1.1) is piecewise constant and SPD on the coarsest mesh partitioning. Then, comparing the different decompositions of the finite element space according to the standard DA splitting (cf. Definition 4.3), the GDA splitting (cf. Definition 4.4), using locally the optimal parameter* $\mu = \mu_{\mathrm{opt}}$ *from (4.38), and the FR splitting (cf. Def-*

inition 4.2), we have the following relations between their respective global CBS constants:

$$\gamma_{\mathrm{FR}}^2 \le \gamma_{\mathrm{GDA}}^2 \le \gamma_{\mathrm{DA}}^2 \le 3/4. \tag{4.51}$$

Proof. As shown by Lemma 4.5 the global Schur complement related to the two-level splitting is invariant with respect to a change of basis using the transformation

$$J = J(C) = \begin{bmatrix} I & 0 & C \\ 0 & J_- & J_+ \end{bmatrix}, \qquad C \in (\mathcal{C}_{\mathrm{DA}} \cup \mathcal{C}_{\mathrm{GDA}} \cup \mathcal{C}_{\mathrm{FR}})$$

where the sets $\mathcal{C}_{\mathrm{DA}}$, $\mathcal{C}_{\mathrm{GDA}}$, $\mathcal{C}_{\mathrm{FR}}$ contain all the admissible matrices for the DA, GDA, and FR splitting, respectively. In this setting, we consider the particular local matrices C_E as a solution of the related minimization problems. Then, by construction we have

$$\mathcal{C}_{\mathrm{DA}} \subset \mathcal{C}_{\mathrm{GDA}} \subset \mathcal{C}_{\mathrm{FR}}. \tag{4.52}$$

Moreover, there is a triple of such sets containing the admissible matrices C_E for the respective local (macroelement level) transformation and one containing the corresponding (induced) matrices C for the respective global transformation, which, for convenience, we do not distinguish in the notations. Next we note that, due to (4.52) and the optimization in the related nested parameter spaces, we have the relations

$$\begin{aligned} \max_{C\in\mathcal{C}_{\mathrm{FR}}} \frac{\mathbf{v}_E^T S_E \mathbf{v}_E}{\mathbf{v}_E^T \tilde{A}_{33,E}(C)\mathbf{v}_E} &\ge \max_{C\in\mathcal{C}_{\mathrm{GDA}}} \frac{\mathbf{v}_E^T S_E \mathbf{v}_E}{\mathbf{v}_E^T \tilde{A}_{33,E}(C)\mathbf{v}_E} \\ &\ge \max_{C\in\mathcal{C}_{\mathrm{DA}}} \frac{\mathbf{v}_E^T S_E \mathbf{v}_E}{\mathbf{v}_E^T \tilde{A}_{33,E}(C)\mathbf{v}_E}, \end{aligned} \tag{4.53}$$

for all $\mathbf{v}_E \neq \mathbf{c}$, or, equivalently,

$$\mathbf{v}_E^T \tilde{A}_{33,E}(C_{\mathrm{FR}})\mathbf{v}_E \le \mathbf{v}_E^T \tilde{A}_{33,E}(C_{\mathrm{GDA}})\mathbf{v}_E \le \mathbf{v}_E^T \tilde{A}_{33,E}(C_{\mathrm{DA}})\mathbf{v}_E \quad \forall \mathbf{v}_E. \tag{4.54}$$

Then, since

$$\tilde{A}_{33}(X) = \sum_{E\in\mathcal{T}_H} \tilde{A}_{33,E}(X), \quad X \in \{C_{\mathrm{FR}}, C_{\mathrm{GDA}}, C_{\mathrm{DA}}\},$$

it follows that the relations (4.54) also hold for the global $\tilde{A}_{33}$-blocks, i.e.,

$$\mathbf{v}^T \tilde{A}_{33}(C_{\mathrm{FR}})\mathbf{v} \le \mathbf{v}^T \tilde{A}_{33}(C_{\mathrm{GDA}})\mathbf{v} \le \mathbf{v}^T \tilde{A}_{33}(C_{\mathrm{DA}})\mathbf{v} \quad \forall \mathbf{v}. \tag{4.55}$$

But this, due to the transformation invariance of the Schur complement already implies that

$$\frac{\mathbf{v}^T S \mathbf{v}}{\mathbf{v}^T \tilde{A}_{33}(C_{\mathrm{FR}})\mathbf{v}} \geq \frac{\mathbf{v}^T S \mathbf{v}}{\mathbf{v}^T \tilde{A}_{33}(C_{\mathrm{GDA}})\mathbf{v}} \geq \frac{\mathbf{v}^T S \mathbf{v}}{\mathbf{v}^T \tilde{A}_{33}(C_{\mathrm{DA}})\mathbf{v}} \quad \forall \mathbf{v} \neq \mathbf{c}. \tag{4.56}$$

Since every of the three quotients of quadratic forms in (4.56) defines the CBS constant related to the corresponding global splitting the inequalities in (4.51) are proven true. □

4.4 Preconditioning of the pivot block

We continue the study of the introduced FR, DA and GDA splittings. The presentation in this section includes results which are published in [32], see also [107].

Let us start with the FR splitting. After the exact local elimination of the block A_{11} we get the Schur complement B. The problem of optimal order preconditioning of the pivot block in the two-level splitting is now related to the block B_{11} which is associated with the half-differences of the hierarchical basis. Let us note, that the structure of B_{11}, see (4.13), is the same as of the pivot block A_{11} in the case of conforming elements.

The macroelement block $B_{E:11}$ is found explicitly, namely,

$$B_{E:11} = \frac{2p}{q} \begin{bmatrix} 3q + 2\beta' & q+2 & -q-2\beta^2 \\ q+2 & 3q+2\alpha' & -q-2\alpha^2 \\ -q-2\beta^2 & -q-2\alpha^2 & 3q+2(\alpha^2+\alpha\beta+\beta^2) \end{bmatrix},$$

where $\alpha' = 1+\alpha+\alpha^2$, $\beta' = 1+\beta+\beta^2$, and p, q are given by $q = \alpha+\alpha\beta+\beta$ and $p = 3(\alpha+\alpha\beta+\beta) + 3(\alpha^2+\alpha\beta+\beta^2) + \alpha\beta(3\alpha+3\beta+1)$. The above expressions are rather more complicated than the related representation (3.17) of $A_{E:11}$. Nevertheless, the additive and multiplicative preconditioners discussed in Sections 3.3–3.4 are directly applicable. Moreover, the condition number estimates are shown to be completely the same as for the Courant conforming finite elements and the following theorem holds.

Theorem 4.21. *The following statements hold for any element size and shape and any coefficient anisotropy:*

(a) *If C_{11} is the additive preconditioner to B_{11} then*

$$\kappa\left(C_{11}^{-1} B_{11}\right) < \frac{1}{4}(11 + \sqrt{105}). \tag{4.57}$$

(b) *If* C_{11} *is the multiplicative preconditioner to* B_{11} *then*

$$\kappa\left(C_{11}^{-1} B_{11}\right) < \frac{15}{8}. \tag{4.58}$$

(c) *The cost of the application of the preconditioner in both cases is proportional to the number of unknowns.*

Proof. We follow here the proof from [32]. Let us start with item (a) of additive preconditioning to B_{11}. As for the conforming elements, the estimate (4.57) is derived using the inequality from Lemma 3.4, where the local condition numbers are derived in terms of $(\alpha, \beta) \in D$. Consider the local (macroelement) generalized eigenproblem $B_{E:11}\mathbf{v} = \lambda C_{E:11}\mathbf{v}$ and the corresponding characteristic equation for λ,

$$|B_{E:11} - \lambda C_{E:11}| = 0. \tag{4.59}$$

The explicit form of the determinant of $B_{E:11} - \lambda C_{E:11}$ is found to be

$$\begin{vmatrix} s & (1-\lambda)(q+2) & -q-2\beta^2 \\ (1-\lambda)(q+2) & s & -q-2\alpha^2 \\ -q-2\beta^2 & -q-2\alpha^2 & s \end{vmatrix},$$

where

$$s = (1-\lambda)(3q + 2(1+\beta+\beta^2)).$$

Straightforward computation shows that $\mu_i = 1 - \lambda_i, i = 1, 2, 3$ satisfy

$$\begin{aligned} \mu_1 &= 0, \quad \text{i.e.,} \quad \lambda_1 = 1 \\ \mu_{2,3}^2 &= \frac{(\alpha+\beta)(\alpha+\alpha\beta+\beta+2(\alpha^2-\alpha\beta+\beta^2))}{(\alpha+\beta+2)[2(\alpha^2+\alpha\beta+\beta^2)+3(\alpha+\alpha\beta+\beta)]}. \end{aligned} \tag{4.60}$$

We show below that

$$\mu_{2,3}^2 \le \frac{7}{15}, \tag{4.61}$$

and thus

$$1 - \sqrt{\frac{7}{15}} \le \lambda_{2,3} \le 1 + \sqrt{\frac{7}{15}}. \tag{4.62}$$

From (4.62) the claimed result (4.57) follows immediately.

To show (4.61), we first observe that the denominator in the expression for $\mu_{2,3}^2$ in (4.60) is positive. The expanded form of inequality (4.61) becomes

$$\begin{aligned} \mathcal{F}(\alpha,\beta) \;\equiv\; & 16\alpha^3 + 16\beta^3 - 34\alpha^2 - 34\beta^2 - 34\alpha^2\beta - 34\alpha\beta^2 \\ & -82\alpha\beta - 42\alpha - 42\beta \le 0. \end{aligned} \tag{4.63}$$

We shall now prove that (4.63) holds for all $(\alpha, \beta) \in D$, where

$$D = \{(\alpha, \beta) \in \mathbb{R}^2 : -\frac{1}{2} < \alpha \leq 1, \max\Big\{-\frac{\alpha}{\alpha+1}, |\alpha|\} \leq \beta \leq 1\Big\}.$$

- Case 1: Let $\alpha = 0$. In this case

$$\mathcal{F}(0, \beta) \leq 16\beta^2 - 34\beta^2 - 42\beta = -18\beta^2 - 42\beta \leq 0.$$

- Case 2: Let $\alpha > 0$. Then

$$16\alpha^3 + 16\beta^3 - 34\beta^2 \leq 16\beta^2 + 16\beta^2 - 34\beta^2 \leq 0$$

  and the remaining terms in $\mathcal{F}$ are negative, so $\mathcal{F} \leq 0$.
- Case 3: Let $-\frac{1}{2} < \alpha < 0$. We use the fact that $\beta \leq 1$, i.e., $\beta^2 \leq \beta$ and that $-\alpha\beta \leq \alpha + \beta$. We have

$$\begin{aligned} \mathcal{F} &= 16\alpha^3 + 16\beta^3 - 34\alpha^2 - 34\beta^2 - 34\alpha\beta(\alpha+\beta) - 82\alpha\beta - 42(\alpha+\beta) \\ &\leq 16\alpha^3 + 16\beta^3 - 14\alpha\beta - 42(\alpha+\beta) \\ &\leq 16(\alpha^3 + \beta^3) + 14(\alpha+\beta) - 42(\alpha+\beta) \\ &= 16(\alpha+\beta)(\alpha^2 - \alpha\beta + \beta^2) - 28(\alpha+\beta). \end{aligned}$$

  It remains to prove that

$$16(\alpha^2 - \alpha\beta + \beta^2) \leq 28$$

  or

$$\alpha^2 - \alpha\beta + \beta^2 \leq 7/4.$$

  The latter is true, since

$$\sup_{\alpha,\beta} (\alpha^2 - \alpha\beta + \beta^2) = \sup_{\alpha\in(-1/2,0)} (\alpha^2 - \alpha + 1) = 7/4.$$

The expression $\mathcal{F}$ achieves its maximum value 0 for the pairs (α, β) equal to $(0, 0)$ and $(-1/2, 1)$.

The estimate (b) for the multiplicative preconditioner to B_{11} is analyzed in a similar way, using the notations introduced in the case of conforming finite elements. We again apply a local analysis and consider the generalized eigenvalue problem

$$S_{11:Q}\mathbf{v}_Q = \lambda_Q E_{11:Q}\mathbf{v}_Q. \tag{4.64}$$

In Lemma 3.11 it was shown that $\lambda_Q^{(2)} = \lambda_Q^{(3)} = \lambda_Q^{(4)} = 1$. Moreover, the following relation was also delivered there:

$$\lambda_Q^{(1)} = 1 - (\mu^{(2,3)})^2. \tag{4.65}$$

Here $\mu^{(2,3)} = 1 - \lambda^{(2,3)}$ are the eigenvalues introduced in the analysis of the additive preconditioner to B_{11}. We apply now (4.61) and get the estimate

$$\lambda_Q^{(1)} > \frac{8}{15}$$

from which (4.58) follows straightforwardly.

The structure and hence the computational complexity of the additive and multiplicative preconditioners for the matrix B_{11} is the same as of the related preconditioners of the matrix A_{11} for the conforming finite elements. Therefore, the optimality statement (c) follows directly from Theorem 3.13 and Theorem 3.14. □

The presented results are easily converted to the cases of the DA and the GDA splittings. The related pivot block to be considered has the form

$$\begin{bmatrix} \tilde{A}_{11} & \tilde{A}_{12} \\ \tilde{A}_{12}^T & \tilde{A}_{22} \end{bmatrix}.$$

Here, the (locally performed) elimination of the block $\tilde{A}_{11}$ gives rise to a Schur complement equal to

$$\tilde{A}_{22} - \tilde{A}_{12}^T \tilde{A}_{11}^{-1} \tilde{A}_{12} = B_{11}$$

the preconditioning of which was already discussed. The latter relation follows from direct computation, cf. [32, 107].

A detailed study of the effect of a minimum angle condition on the preconditioning of the pivot block arising from two-level splittings of Crouzeix–Raviart FE-spaces can be found in [107].

4.5 Numerical results

The presented numerical results (for further test problems see [77]) are selected to (partly) illustrate the analysis from the previous sections. We consider the model problem (1.1) on the unit square $\Omega = (0,1)\times(0,1)$ for the coefficient matrix

$$\mathbf{a}(\mathbf{x}) = a(e) \begin{bmatrix} \varepsilon & -\delta \\ -\delta & 1 \end{bmatrix}, \tag{4.66}$$

using Dirichlet boundary conditions. The discretization is by linear Crouzeix–Raviart finite elements on a uniform mesh with mesh size $h \in \{1/64, \ldots, 1/1024\}$ resulting in 8192, ..., 2097152 elements and 12416, ..., 3147776 nodes, respectively (the latter being the size of the matrix of the linear system to solve).

The computational domain is split into four subdomains, i.e., $\overline{\Omega} = \overline{\Omega}_1 \cup \ldots \cup \overline{\Omega}_4$ where $\overline{\Omega}_1 = [0, 1/2]^2$, $\overline{\Omega}_2 = [1/2, 1] \times [0, 1/2]$, $\overline{\Omega}_3 = [0, 1/2] \times [1/2, 1]$, $\overline{\Omega}_4 = [1/2, 1]^2$.

Example 4.22. The (mixed derivative) parameter δ is varied from 0 to $1/4$, which is combined with a jump of two orders of magnitude in the coefficient $a(e)$, i.e., $\alpha = 10^{-2}$ and a ratio $1 : 10$ of anisotropy, i.e., $\varepsilon = 10^{-1}$.

The experiments consist of solving the systems of linear algebraic equations, arising from this example, thereby varying various parameters including the mesh size h. Below we report the convergence results for the nonlinear algebraic multilevel iteration method (which has been discussed in Section 2.5), see also [18, 19, 72]. The reason for choosing the parameter-free (but nonlinear) variant of the AMLI method is that we do not want to distort the comparison of the performance of different splittings by employing (different) matrix polynomials whose coefficients (in particular) depend on the estimates of γ. As it has been observed in the context of various nonconforming finite elements, the nonlinear AMLI method, i.e., the self-adapting variable-step preconditioner, usually performs at least as good as the linear AMLI method [62, 77]. The stabilization of the convergence is achieved by using two inner generalized conjugate gradient (GCG) iterations in all cases. The outer iteration, which we initialize with a random start vector, is stopped as soon as the residual vector satisfies the criteria

$$\|\mathbf{r}_{(n_{\text{it}})}\| / \|\mathbf{r}_{(0)}\| \le 10^{-6},$$

where n_{it} denotes the number of iterations that we report in the table below. The coarsest-grid problem with mesh size $h = 1/16$ is always solved directly, i.e., for the sequences of discrete problems with decreasing mesh size a 3- to 7-level method is performed. The nonlinear AMLI method is employed in its multiplicative version, cf. [62, 72, 73, 77]. However, as we conclude from the estimates of the CBS constant that have been presented in the previous section, all three splittings considered here allow for a construction of optimal-order additive methods [6, 15], e.g., using third order stabilization polynomials.

In the following numerical experiments the nonlinear AMLI algorithm has been performed with an inexact solve of the linear system with the pivot block B_{11} (as obtained after the local elimination of the unknowns corresponding to the block $\tilde{A}_{11}$) on all levels; its approximate inverse (acting on a vector) was implemented via an incomplete factorization based on a drop tolerance tol, which was chosen of the same size as the parameter ε (of anisotropy).[3] This general approach provides an alternative to the previously considered robust (but more complicated from implementation point of view) additive and multiplicative preconditioners to the pivot block B_{11}.

In order to have a reasonable comparison of the efficiency of the DA, the GDA, and the FR splitting from a practical point of view we report the CPU time for the entire solution process, which includes the time for the construction of the preconditioner. As to the state of our implementation the latter is in the range of 20 to 30 per cent of the reported total CPU time. All experiments have been performed on a Linux PC with 2.4 GHz and 4 GB of physical memory.

The presented experiments combine a jump in the coefficient with moderate anisotropy and illustrate the effect of a mixed derivative term in the model problem. The results are shown in Table 4.1. As predicted by the theory we observe that each of the three splittings is robust with respect to jumps in the coefficients (if they do not occur in the interior of any element of the coarsest mesh partition). The reported CPU time demonstrates an (almost) optimal order of computational complexity for all three splittings (the CPU time typically increases by a factor less than five when the problem size increases by a factor four). At least with the FR splitting, which yields the fastest solver, the largest problem (with more than three million unknowns) still can be solved in approximately one minute on a standard PC. Note also that though raising the number of iterations to a higher (but constant) level, as compared to the GDA (or FR) splitting, the DA splitting outperforms GDA in terms of computing time when $\delta = 0$, which is caused by a (slightly) larger number of matrix entries in the GDA (as well as in the FR) transformation.

4.5.1 Concluding remarks

In this chapter we studied different hierarchical splittings of Crouzeix–Raviart finite element spaces based on the construction of so-called differences and aggregates. We showed that among all possible splittings of this (general) type the FR approach is the best in the sense that it results in the smallest CBS constant. The presented numerical results confirm the analysis and also favor the FR hierarchical basis from a CPU-time point of view.

[3]During the computation of the triangular incomplete factors of a matrix M the entries smaller in magnitude than the local drop tolerance (given by the product of the drop tolerance tol and the norm of the corresponding row i of M, i.e. tol $\cdot \|M_i\|$) are dropped from the appropriate factor.

Table 4.1: Nonlinear AMLI W-cycle: Number of iterations and CPU-time for Example 4.22

$1/h$	64		128		256		512		1024	
	n_{it}	sec	n_{it}	sec	n_{it}	sec	n_{it}	sec	n_{it}	sec
					$\delta = 0$					
DA	17	0.12	18	0.71	19	3.58	19	16.5	19	75.6
GDA	13	0.13	13	0.70	13	3.49	13	17.0	13	80.2
FR	10	0.10	10	0.51	10	2.57	10	12.5	10	59.0
					$\delta = 1/8$					
DA	20	0.16	21	0.94	22	4.86	21	22.5	22	106
GDA	13	0.13	13	0.74	13	3.72	13	17.9	13	82.3
FR	12	0.11	12	0.63	12	3.17	11	14.3	11	66.2
					$\delta = 1/4$					
DA	19	0.15	20	0.91	20	4.50	20	21.4	20	98.5
GDA	16	0.15	16	0.85	16	4.31	16	20.9	16	97.1
FR	14	0.12	14	0.69	14	3.47	14	17.0	14	78.7

Remark 4.23. The hierarchical error estimators introduced in Section 2.7.3 are not directly applicable to the case of Crouzeix–Raviart finite elements. Some ideas about the required additional analysis are briefly discussed in [30]. Due to the result of Theorem 4.11, the DA hierarchical splitting has some advantages in this context. Similar arguments hold true in a more general setting for the related DA splittings which will be considered in the following sections for some other nonconforming finite elements or discontinuous Galerkin discretizations.

5 Schur complement based multilevel preconditioners

5.1 Hierarchical versus standard nodal-basis

Hierarchical basis matrices, in general, are less sparse than standard nodal basis matrices, and, though they may be generated in the course of a mesh refinement procedure in a quite natural way, one is faced with additional computational costs when (re)constructing them from the nodal basis representation for the sake of an efficient solution of the linear system only. On the other hand, as we conclude from the previous sections, the theory of hierarchical basis block factorization methods, including various stabilization techniques, is fairly matured whereas alternative approaches, which avoid the hierarchical basis construction, have been far less frequently studied. Their rigorous analysis is only established up to a certain extent. For that reason, some numerical results are included in this chapter, which should demonstrate the potential of the latter approach.

As a matter of fact, avoiding the hierarchical basis representation, we have to meet certain supplementary requirements in order to develop robust two- and multilevel schemes. A general guideline is to achieve robustness, also with respect to perturbations of the M-matrix property, which is a crucial factor in this case.

In the following we will briefly discuss a technique, originally proposed in [73], which is based on a particular approximation of the Schur complement. Assuming access to the individual (fine-grid) element stiffness matrices, any element agglomeration procedure, see, e.g., [69, 79], can be used in order to assemble macroelement matrices that provide a basis for the computation of appropriate coarse-level matrices, which makes the hierarchical basis representation (2.33) unnecessary in the construction. However, building a bridge to the methods discussed in the previous chapters of this book subserves the analysis. As an important example, a (locally computable) condition number bound for the related sparse approximation of the Schur complement will be presented in Section 5.4.

5.2 A general two-level preconditioner

We start the description of the nodal-basis approach with the consideration of a general two-level preconditioner. Let the degrees of freedom (DOF) be partitioned into two groups, usually denoted as fine and coarse DOF. We replace then the exact block factorization

$$A = \begin{bmatrix} A_{11} & A_{12} \\ A_{21} & A_{22} \end{bmatrix} = \begin{bmatrix} I & \\ A_{21}A_{11}^{-1} & I \end{bmatrix} \cdot \begin{bmatrix} A_{11} & A_{12} \\ & S \end{bmatrix} \tag{5.1}$$

with an approximate factorization (analogously to the representation (2.29))

$$B = \begin{bmatrix} I & \\ A_{21}B_{11}^{-1} & I \end{bmatrix} \cdot \begin{bmatrix} B_{11} & A_{12} \\ & Q \end{bmatrix} = \begin{bmatrix} B_{11} & A_{12} \\ A_{21} & Q + A_{21}B_{11}^{-1}A_{12} \end{bmatrix}. \tag{5.2}$$

Herein, the matrix B_{11} is a preconditioner for the pivot block A_{11} and Q is an approximation of the exact Schur complement $S = A_{22} - A_{21}A_{11}^{-1}A_{12}$. Note that a change to the hierarchical (two-level) basis would only affect the off-diagonal blocks A_{12} and A_{21} in (5.1); the pivot block A_{11} as well as the Schur complement S are the same for both bases [111]. However, as already mentioned, we don't want to use this stabilizing modification of the off-diagonal blocks here.

In the following, $\check{\alpha}, \check{\beta}, \check{\gamma}, \hat{\alpha}, \hat{\beta}, \hat{\gamma}$ denote positive constants satisfying

$$\begin{aligned} 0 < \quad & \check{\alpha}, \check{\beta}, \check{\gamma} \quad \leq 1, \\ 1 \leq \quad & \hat{\alpha}, \hat{\beta}, \hat{\gamma} \quad < \infty. \end{aligned}$$

It is known that the spectral condition number

$$\kappa(B^{-1}A) = \frac{\lambda_{\max}(B^{-1}A)}{\lambda_{\min}(B^{-1}A)}, \tag{5.3}$$

measuring the quality of the two-level preconditioner B defined via (5.2), depends on the extremal eigenvalues of $B_{11}^{-1}A_{11}$ and $Q^{-1}S$, that is, it involves the bounds

$$\check{\alpha}\mathbf{v}_1^T A_{11}\mathbf{v}_1 \leq \mathbf{v}_1^T B_{11}\mathbf{v}_1 \leq \hat{\alpha}\mathbf{v}_1^T A_{11}\mathbf{v}_1 \quad \forall \mathbf{v}_1, \tag{5.4}$$

and

$$\check{\beta}\mathbf{v}_2^T S\mathbf{v}_2 \leq \mathbf{v}_2^T Q\mathbf{v}_2 \leq \hat{\beta}\mathbf{v}_2^T S\mathbf{v}_2 \quad \forall \mathbf{v}_2, \tag{5.5}$$

respectively. However, and this has been shown in [94], avoiding the hierarchical basis representation, a bound

$$\check{\gamma}\mathbf{v}^T A\mathbf{v} \leq \mathbf{v}^T B\mathbf{v} \leq \hat{\gamma}\mathbf{v}^T A\mathbf{v} \quad \forall \mathbf{v} \tag{5.6}$$

that is independent of the mesh size can only be obtained if B_{11}^{-1} acts nearly as an exact inverse on all vectors $A_{12}\mathbf{v}_2$ for which $\mathbf{v}_2$ is *smooth* on the coarse grid, i.e., a low energy mode of the Schur complement S. For instance, this requirement is met if the condition

$$\hat{\alpha}\mathbf{v}_2^T A_{21}B_{11}^{-1}A_{12}\mathbf{v}_2 \leq (1-\xi)\mathbf{v}_2^T A_{22}\mathbf{v}_2 + \xi\mathbf{v}_2^T A_{21}A_{11}^{-1}A_{12}\mathbf{v}_2 \quad \forall \mathbf{v}_2 \tag{5.7}$$

is fulfilled for some $\xi \leq 1$. Note that the assumption (5.7) is loosened if we let ξ be negative.

Now, inequalities (5.4) and (5.7) imply

$$\begin{aligned} \mathbf{v}_2^T (A_{22} - S)\mathbf{v}_2 &= \mathbf{v}_2^T A_{21} A_{11}^{-1} A_{12} \mathbf{v}_2 \\ &\leq \hat{\alpha} \mathbf{v}_2^T A_{21} B_{11}^{-1} A_{12} \mathbf{v}_2 \leq \mathbf{v}_2^T (A_{22} - \xi S)\mathbf{v}_2 \quad \forall \mathbf{v}_2 \end{aligned} \tag{5.8}$$

and (5.6) can be based on the bounds (5.4), (5.5) and (5.7). For details see [94] where the analysis is carried out for the case $\overline{\alpha} = 1$. We summarize the main result (without this restriction) in the theorem below, cf. [73].

Theorem 5.1. *Let A and B, as defined in (5.1) and (5.2), be symmetric nonnegative definite matrices such that A_{11} and B_{11} are invertible. Moreover, assume that (5.4) and (5.5) hold. If, in addition, condition (5.7) is satisfied for some $\xi \leq 1$, then the bound (5.6) holds, i.e.,*

$$\kappa(B^{-1}A) \leq \frac{\hat{\gamma}}{\check{\gamma}}, \tag{5.9}$$

where $\check{\gamma}$ is the smallest root of

$$\gamma^2 - \gamma(\check{\beta} + \hat{\alpha} - \xi(\hat{\alpha} - \check{\alpha})) + \check{\alpha}\check{\beta} \tag{5.10}$$

and $\hat{\gamma}$ is the largest root of

$$\gamma^2 - \gamma(\hat{\beta} + \hat{\alpha} - \xi(\hat{\alpha} - \check{\alpha})) + \check{\alpha}\hat{\beta}. \tag{5.11}$$

Proof. Scaling the matrix A with $\hat{\alpha}$, i.e., $\overline{A} := \hat{\alpha}\, A$, the bounds (5.4) and (5.5) read

$$\frac{\check{\alpha}}{\hat{\alpha}} \mathbf{v}_1^T \overline{A}_{11} \mathbf{v}_1 \leq \mathbf{v}_1^T B_{11} \mathbf{v}_1 \leq \mathbf{v}_1^T \overline{A}_{11} \mathbf{v}_1 \quad \forall \mathbf{v}_1 \tag{5.12}$$

$$\frac{\check{\beta}}{\hat{\alpha}} \mathbf{v}_2^T \overline{S} \mathbf{v}_2 \leq \mathbf{v}_2^T Q \mathbf{v}_2 \leq \frac{\hat{\beta}}{\hat{\alpha}} \mathbf{v}_2^T \overline{S} \mathbf{v}_2 \quad \forall \mathbf{v}_2 \tag{5.13}$$

where $\overline{S} = \hat{\alpha}\, S = \hat{\alpha}\,(A_{22} - A_{21} A_{11}^{-1} A_{12}) = \overline{A}_{22} - \overline{A}_{21} \overline{A}_{11}^{-1} \overline{A}_{12}$ is the scaled Schur complement. Moreover, (5.7) yields

$$\mathbf{v}_2^T \overline{A}_{21} B_{11}^{-1} \overline{A}_{12} \mathbf{v}_2 \leq (1 - \xi) \mathbf{v}_2^T \overline{A}_{22} \mathbf{v}_2 + \xi \mathbf{v}_2^T \overline{A}_{21} \overline{A}_{11}^{-1} \overline{A}_{12} \mathbf{v}_2 \quad \forall \mathbf{v}_2. \tag{5.14}$$

What follows is the proof of the left-hand side inequality of (5.6). In order to show that

$$\mathbf{v}^T (B - \check{\gamma}\, A)\, \mathbf{v} = \mathbf{v}^T \,(B - \frac{\check{\gamma}}{\hat{\alpha}} \overline{A})\, \mathbf{v} \geq 0 \quad \forall \mathbf{v}, \tag{5.15}$$

which is equivalent to $B - \check{\gamma} A$ being SPSD, we look at the elaborated expression for (5.15) which is given by

$$\begin{bmatrix} \mathbf{v}_1 \\ \mathbf{v}_2 \end{bmatrix}^T \begin{bmatrix} B_{11} - \frac{\check{\gamma}}{\check{\alpha}} \overline{A}_{11} & \left(1 - \frac{\check{\gamma}}{\check{\alpha}}\right) \overline{A}_{12} \\ \left(1 - \frac{\check{\gamma}}{\check{\alpha}}\right) \overline{A}_{21} & Q + \overline{A}_{21} B_{11}^{-1} \overline{A}_{12} - \frac{\check{\gamma}}{\check{\alpha}} \overline{A}_{22} \end{bmatrix} \begin{bmatrix} \mathbf{v}_1 \\ \mathbf{v}_2 \end{bmatrix} \geq 0. \quad (5.16)$$

First we observe that (5.10) has two positive roots and that the identity

$$\check{\gamma} = \check{\beta} - \frac{\check{\gamma}\,(\hat{\alpha} - \check{\alpha})\,(1 - \xi)}{\check{\alpha} - \check{\gamma}} \quad (5.17)$$

holds. In view of (5.12) we know that $B_{11} - \frac{\check{\gamma}}{\check{\alpha}} \overline{A}_{11}$ is SPD if $\check{\gamma} \leq \check{\alpha}$, the latter being true for the smallest root $\check{\gamma}$ of (5.10). Hence (5.16) holds if and only if the Schur complement

$$S_{B-\check{\gamma} A} = Q + \overline{A}_{21} B_{11}^{-1} \overline{A}_{12} - \frac{\check{\gamma}}{\hat{\alpha}} \overline{A}_{22} - \left(1 - \frac{\check{\gamma}}{\hat{\alpha}}\right)^2 \overline{A}_{21} \left(B_{11} - \frac{\check{\gamma}}{\hat{\alpha}} \overline{A}_{11}\right)^{-1} \overline{A}_{12}$$

is SPSD, i.e, if

$$\mathbf{v}_2^T \{Q - \frac{\check{\gamma}}{\hat{\alpha}} \overline{A}_{22} + \overline{A}_{21} \left[B_{11}^{-1} - \left(1 - \frac{\check{\gamma}}{\hat{\alpha}}\right)^2 \left(B_{11} - \frac{\check{\gamma}}{\hat{\alpha}} \overline{A}_{11}\right)^{-1} \right] \overline{A}_{12} \} \mathbf{v}_2 \geq 0. \quad (5.18)$$

Combining (5.17) and (5.14) we find

$$\begin{aligned} (\check{\beta} - \check{\gamma})\, \mathbf{v}_2^T \overline{A}_{22} \mathbf{v}_2 &= \frac{\check{\gamma}\,(\hat{\alpha} - \check{\alpha})\,(1 - \xi)}{\check{\alpha} - \check{\gamma}} \mathbf{v}_2^T \overline{A}_{22} \mathbf{v}_2 \\ &\geq \frac{\check{\gamma}\,(\hat{\alpha} - \check{\alpha})}{\check{\alpha} - \check{\gamma}} \mathbf{v}_2^T \left[\overline{A}_{21} \left(B_{11}^{-1} - \xi\, \overline{A}_{11}^{-1}\right) \overline{A}_{12} \right] \mathbf{v}_2. \quad (5.19) \end{aligned}$$

Now, using (5.12), (5.13) and (5.19) in (5.18) it follows that

$$\begin{aligned} \mathbf{v}_2^T S_{B-\check{\gamma} A} \mathbf{v}_2 &\geq \mathbf{v}_2^T \Big\{ \frac{\check{\beta} - \check{\gamma}}{\hat{\alpha}} \overline{A}_{22} - \frac{\check{\beta}}{\hat{\alpha}} \overline{A}_{21} \overline{A}_{11}^{-1} \overline{A}_{12} \\ &\quad + \overline{A}_{21} \left[B_{11}^{-1} - \left(1 - \frac{\check{\gamma}}{\hat{\alpha}}\right)^2 \left(B_{11} - \frac{\check{\gamma}}{\hat{\alpha}} \overline{A}_{11}\right)^{-1} \right] \overline{A}_{12} \Big\} \mathbf{v}_2 \\ &\geq \mathbf{v}_2^T \overline{A}_{21} \Big\{ \left(1 + \frac{\check{\gamma}\,(\hat{\alpha} - \check{\alpha})}{\hat{\alpha}\,(\check{\alpha} - \check{\gamma})}\right) B_{11}^{-1} - \left(\frac{\check{\beta}}{\hat{\alpha}} + \frac{\xi\, \check{\gamma}\,(\hat{\alpha} - \check{\alpha})}{\hat{\alpha}\,(\check{\alpha} - \check{\gamma})}\right) \overline{A}_{11}^{-1} \\ &\quad - \left(1 - \frac{\check{\gamma}}{\hat{\alpha}}\right)^2 \left(B_{11} - \frac{\check{\gamma}}{\hat{\alpha}} \overline{A}_{11}\right)^{-1} \Big\} \overline{A}_{12} \mathbf{v}_2. \quad (5.20) \end{aligned}$$

Next, due to the identities

$$1+\frac{\check{\gamma}\,(\hat{\alpha}-\check{\alpha})}{\hat{\alpha}\,(\check{\alpha}-\check{\gamma})}=\frac{\check{\alpha}\,(\hat{\alpha}-\check{\gamma})}{\hat{\alpha}\,(\check{\alpha}-\check{\gamma})}$$

and

$$\frac{1}{\hat{\alpha}}\left(\check{\beta}+\frac{\xi\,\check{\gamma}\,(\hat{\alpha}-\check{\alpha})}{(\check{\alpha}-\check{\gamma})}\right)=\frac{\check{\gamma}\,(\hat{\alpha}-\check{\gamma})}{\hat{\alpha}\,(\check{\alpha}-\check{\gamma})},$$

the latter of which is obtained by using (5.17), the right-hand side of (5.20) can be rewritten in the form

$$\begin{aligned}
&\mathbf{v}_2^T\overline{A}_{21}\left\{\tfrac{\check{\alpha}(\hat{\alpha}-\check{\gamma})}{\hat{\alpha}(\check{\alpha}-\check{\gamma})}B_{11}^{-1}-\tfrac{\check{\gamma}(\hat{\alpha}-\check{\gamma})}{\hat{\alpha}(\check{\alpha}-\check{\gamma})}\overline{A}_{11}^{-1}-\left(1-\tfrac{\check{\gamma}}{\hat{\alpha}}\right)^2(B_{11}-\tfrac{\check{\gamma}}{\hat{\alpha}}\overline{A}_{11})^{-1}\right\}\overline{A}_{12}\mathbf{v}_2\\
&=\tfrac{\hat{\alpha}-\check{\gamma}}{\hat{\alpha}(\check{\alpha}-\check{\gamma})}\mathbf{v}_2^T\overline{A}_{21}\left\{\check{\alpha}B_{11}^{-1}-\check{\gamma}\overline{A}_{11}^{-1}-\tfrac{(\hat{\alpha}-\check{\gamma})(\check{\alpha}-\check{\gamma})}{\hat{\alpha}}(B_{11}-\tfrac{\check{\gamma}}{\hat{\alpha}}\overline{A}_{11})^{-1}\right\}\overline{A}_{12}\mathbf{v}_2\\
&=\tfrac{\hat{\alpha}-\check{\gamma}}{\hat{\alpha}(\check{\alpha}-\check{\gamma})}\mathbf{v}_2^TY^T\left\{\check{\alpha}I-\check{\gamma}X^{-1}-\tfrac{(\hat{\alpha}-\check{\gamma})(\check{\alpha}-\check{\gamma})}{\hat{\alpha}}(I-\tfrac{\check{\gamma}}{\hat{\alpha}}X)^{-1}\right\}Y\mathbf{v}_2\quad\forall\mathbf{v}_2
\end{aligned}\tag{5.21}$$

where $X=B_{11}^{-1/2}\overline{A}_{11}B_{11}^{-1/2}$ and $Y=B_{11}^{-1/2}\overline{A}_{12}$. Because of (5.12) all eigenvalues of the matrix X are greater or equal to 1 and less or equal to $\hat{\alpha}/\check{\alpha}$, i.e.,

$$\lambda_i(X)=\lambda_i(B_{11}^{-1/2}\overline{A}_{11}B_{11}^{-1/2})\in\left[1,\frac{\hat{\alpha}}{\check{\alpha}}\right].\tag{5.22}$$

But then, since $\check{\gamma}\le\check{\alpha}$ it is easily seen that

$$\begin{aligned}
f(\lambda)&:=\check{\alpha}-\check{\gamma}\lambda^{-1}-\frac{(\hat{\alpha}-\check{\gamma})(\check{\alpha}-\check{\gamma})}{\hat{\alpha}}(1-\frac{\check{\gamma}}{\hat{\alpha}}\lambda)^{-1}\\
&=(1-\frac{\check{\gamma}}{\hat{\alpha}}\lambda)^{-1}\left[\frac{(\hat{\alpha}-\check{\gamma}\lambda)(\check{\alpha}-\check{\gamma}\lambda^{-1})-(\hat{\alpha}-\check{\gamma})(\check{\alpha}-\check{\gamma})}{\hat{\alpha}}\right]\\
&=(\hat{\alpha}-\check{\gamma}\lambda)^{-1}\check{\gamma}\,(\hat{\alpha}+\check{\alpha}-\hat{\alpha}\lambda^{-1}+\check{\alpha}\lambda)\\
&=(\hat{\alpha}-\check{\gamma}\lambda)^{-1}\check{\gamma}\,(1-\lambda^{-1})\,(\hat{\alpha}-\check{\alpha}\lambda)
\end{aligned}\tag{5.23}$$

is nonnegative for all $\lambda\in[1,\hat{\alpha}/\check{\alpha}]$ and thus (5.21) is always nonnegative, i.e., $S_{B-\check{\gamma}A}$ is indeed an SPSD matrix.

The proof of the right-hand side inequality of (5.6) is similar and is left as an exercise. □

For M-matrices condition (5.7) has been analyzed in [94]. It can be shown that if $B_{11}=B_{11,\mathrm{MILU}}$, i.e., B_{11} results from a modified incomplete factorization of A_{11} (where the modification vector is assumed to be $(1,1,\dots,1)^T$), then (5.7) is always satisfied for $\xi=0$. Moreover, if A arises from a linear finite element discretization of a two-dimensional second order elliptic PDE (on a triangular mesh with no interior angle larger than $\frac{\pi}{2}$) one can actually prove that (5.7) holds for $\xi = \frac{1}{2}$. In general, this assumption can be checked (and a value for ξ can be determined) based on the following considerations (which also apply to SPD non-M-matrices): First, we note that $(A_{22} - \xi S - \hat{\alpha} A_{21} B_{11}^{-1} A_{12})$ is the Schur complement of the matrix

$$M = \begin{bmatrix} \frac{1}{\hat{\alpha}} B_{11} & A_{12} \\ A_{21} & A_{22} - \xi S \end{bmatrix}.$$

Thus, since B_{11} is SPD, condition (5.7) holds if and only if M is symmetric positive semidefinite (SPSD). Further, if

$$\mathbf{v}_1^T B_{11} \mathbf{v}_1 \geq \mathbf{v}_1^T \tilde{B}_{11} \mathbf{v}_1 \quad \forall \mathbf{v}_1 \tag{5.24}$$

and

$$\mathbf{v}_2^T S \mathbf{v}_2 \leq \mathbf{v}_2^T \tilde{S} \mathbf{v}_2 \quad \forall \mathbf{v}_2 \tag{5.25}$$

for any SPSD matrices $\tilde{B}_{11}$ and $\tilde{S}$ it is clear that (5.7) follows if the matrix

$$\tilde{M} = \begin{bmatrix} \frac{1}{\hat{\alpha}} \tilde{B}_{11} & A_{12} \\ A_{21} & A_{22} - \xi \tilde{S} \end{bmatrix} \tag{5.26}$$

is SPSD. This motivates the study of the class of two-level preconditioners defined via (5.2) where B_{11} satisfies a row-sum criterion. However, for the Schur complement approximation Q we suggest its computation by assembling (exact) local Schur complements (cf. Section 5.4).

5.3 Incomplete factorization based on exact local factorization

As already mentioned in Section 5.2, a preconditioner B_{11} for the pivot block A_{11} in particular has to be accurate on a certain subspace (cf. (5.7) and (5.8)). We will compare the efficiency of two types of preconditioners $B_{11} = LU$ which satisfy the row-sum criterion

$$A_{11}\mathbf{e} = B_{11}\mathbf{e}, \qquad \mathbf{e} := (1, 1, \ldots, 1)^T. \tag{5.27}$$

The first preconditioner, denoted by $B_{11,\mathrm{MILU}}$, is obtained from classical MILU factorization [5, 51] of A_{11} where off-diagonal fill-in (for L and U) is only allowed in positions in which A_{11} is nonzero. It is well known that even for SPD matrices incomplete factorization methods can suffer from breakdown since they may yield zero (or negative) pivots when an exact factorization would give only positive pivots [52]. However, in the present context the partitioning of the matrix A (cf. (5.1) in Section 5.2) usually causes pivot blocks which prevent possible breakdowns.

For the purpose of comparison, we construct a second preconditioner, now based on local complete factorizations. Here we take advantage of the knowledge of the individual element matrices. A straightforward approach is to regard the small-sized macroelement matrices (resulting from assembling element matrices). For that reason we transfer the global ordering of nodes (and DOF) to a local ordering, i.e., we label the interior fine-grid nodes (DOF) first, followed by those on the boundary of the considered macroelement, and label the coarse-grid nodes (DOF) last, as exemplified in Figure 5.1.

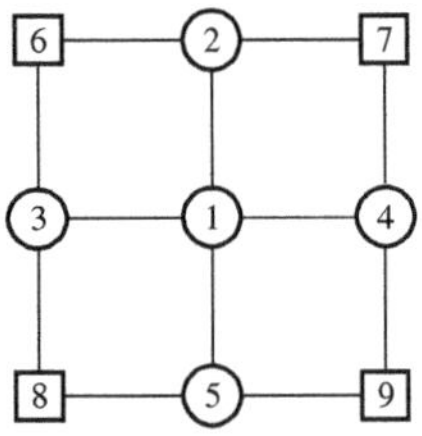

Figure 5.1: Local node numbering within macroelement (quadrilateral elements)

With respect to this local ordering the macroelement matrices take a 2×2 block form

$$A_E = \begin{bmatrix} A_{E:11} & A_{E:12} \\ A_{E:21} & A_{E:22} \end{bmatrix}, \tag{5.28}$$

where $A_{E:11}$ and $A_{E:22}$ are the blocks corresponding to fine and coarse-grid DOF, respectively.

Let us assume that $A_{E:11}$ is non-singular for all macroelements E. Then the factorization

$$A_{E:11} = L_E U_E \quad \forall E \tag{5.29}$$

is well defined where L_E and U_E can be scaled such that the diagonal of L_E is the identity, i.e., $\operatorname{diag}(L_E) = I$. Now, since

$$A_{11} = \sum_E R_E^T A_{E:11} R_E, \tag{5.30}$$

where R_E denotes a Boolean matrix representing the restriction to the macroelement degrees of freedom, the sum of upper triangular matrices U_E yields an approximate upper triangular factor of the global pivot block A_{11}, i.e.,

$$U := \sum_E R_E^T U_E R_E. \tag{5.31}$$

Consequently, a preconditioner $B_{11} = B_{11,\text{ILUE}}$ is defined in terms of

$$B_{11,\text{ILUE}} := LU, \quad U := \sum_E R_E^T U_E R_E, \quad L := U^T \operatorname{diag}(U)^{-1}. \tag{5.32}$$

The following lemma provides a Cauchy–Schwarz inequality for matrices.

Lemma 5.2. *For $i = 1, 2, \ldots, N$ let X_i be real $n \times k$ and Y_i be real $n \times m$ matrices. Then, if the $m \times m$ matrix $Z_{11} := \sum_{i=1}^{N} Y_i^T Y_i$ is invertible the following inequality holds:*

$$\sum_{i=1}^{N} X_i^T X_i - \sum_{i=1}^{N} X_i^T Y_i \left(\sum_{i=1}^{N} Y_i^T Y_i \right)^{-1} \sum_{i=1}^{N} Y_i^T X_i \geq 0. \tag{5.33}$$

Proof. We have

$$\begin{aligned} &\begin{bmatrix} \mathbf{v}_1 \\ \mathbf{v}_2 \end{bmatrix}^T \begin{bmatrix} Y_i^T Y_i & Y_i^T X_i \\ X_i^T Y_i & X_i^T X_i \end{bmatrix} \begin{bmatrix} \mathbf{v}_1 \\ \mathbf{v}_2 \end{bmatrix} \\ &= \langle (Y_i \mathbf{v}_1 + X_i \mathbf{v}_2), (Y_i \mathbf{v}_1 + X_i \mathbf{v}_2) \rangle \\ &= \| Y_i \mathbf{v}_1 + X_i \mathbf{v}_2 \|^2 \geq 0 \quad \forall \mathbf{v} = \begin{bmatrix} \mathbf{v}_1 \\ \mathbf{v}_2 \end{bmatrix} \quad \forall i. \end{aligned}$$

Hence,

$$Z := \begin{bmatrix} \sum_i Y_i^T Y_i & \sum_i Y_i^T X_i \\ \sum_i X_i^T Y_i & \sum_i X_i^T X_i \end{bmatrix} \geq 0, \tag{5.34}$$

and since $Z_{11} = \sum_{i=1}^{N} Y_i^T Y_i$ is an SPD matrix (5.34) holds if and only if the Schur complement

$$S_Z := \sum_{i=1}^{N} X_i^T X_i - \sum_{i=1}^{N} X_i^T Y_i \left(\sum_{i=1}^{N} Y_i^T Y_i \right)^{-1} \sum_{i=1}^{N} Y_i^T X_i$$

is SPSD. □

Remark 5.3. For $k = m = n = 1$ inequality (5.33) reduces to the classical discrete Cauchy–Schwarz inequality

$$\sum_{i=1}^{N} x_i^2 \sum_{i=1}^{N} y_i^2 - \left(\sum_{i=1}^{N} x_i y_i \right)^2 \geq 0.$$

Our aim is now to find constants $\hat{\alpha}$ and $\check{\alpha}$ for (5.4) that result in a proper upper bound for the relative condition number of the preconditioner B_{11} to the pivot block A_{11}, i.e., $\kappa(B_{11}^{-1} A_{11}) \leq \hat{\alpha} / \check{\alpha}$.

Note that $A_{E:11} = U_E^T \operatorname{diag}(U_E)^{-1} U_E$, and, $R_E R_E^T = I_E$. Now, by choosing $X_i \equiv X_E := R_E^T \operatorname{diag}(U_E)^{-1/2} U_E R_E$ and $Y_i \equiv Y_E := R_E^T \operatorname{diag}(U_E)^{1/2} R_E$ it follows from Lemma 5.2 that

$$\mathbf{v}_1^T B_{11} \mathbf{v}_1 \leq \mathbf{v}_1^T A_{11} \mathbf{v}_1 \quad \forall \mathbf{v}_1,$$

i.e., the right-hand side inequality in (5.4) holds with $\hat{\alpha} = 1$:

$$\begin{aligned} A_{11} - B_{11} &= \sum_E R_E^T A_{E:11} R_E \\ &\quad - \sum_E R_E^T U_E^T R_E \left(\sum_E R_E^T \operatorname{diag}(U_E) R_E \right)^{-1} \sum_E R_E^T U_E R_E \\ &= \sum_E R_E^T U_E^T \operatorname{diag}(U_E)^{-1/2} R_E R_E^T \operatorname{diag}(U_E)^{-1/2} U_E R_E \\ &\quad - \sum_E R_E^T U_E^T \operatorname{diag}(U_E)^{-1/2} R_E R_E^T \operatorname{diag}(U_E)^{1/2} R_E \\ &\quad \cdot \left(\sum_E R_E^T U_E^T \operatorname{diag}(U_E)^{1/2} R_E R_E^T \operatorname{diag}(U_E)^{1/2} R_E \right)^{-1} \\ &\quad \cdot \sum_E R_E^T U_E^T \operatorname{diag}(U_E)^{1/2} R_E R_E^T \operatorname{diag}(U_E)^{-1/2} R_E \\ &= \sum_E X_E^T X_E - \sum_E X_E^T Y_E \left(\sum_E Y_E^T Y_E \right)^{-1} \sum_E Y_E^T X_E \geq 0. \end{aligned}$$

To find a tight lower bound, i.e., a constant $\check{\alpha}$ such that the left-hand side inequality in (5.4) is as sharp as possible, is a more difficult task in general. The following constructive approach delivers insight but also a computable estimate:

Let $D_U := \operatorname{diag}(U)$ denote the diagonal matrix whose diagonal agrees with that of U, i.e., $(D_U)_{ii} = U_{ii}$, and let the (local) macroelement matrix $\tilde{B}_{E:11}$ be defined by

$$\tilde{B}_{E:11} := U_E^T R_E D_U^{-1} R_E^T U_E. \tag{5.35}$$

Moreover, let

$$\hat{\lambda} := \lambda_{\max}(\tilde{B}_{E:11}^{-1} A_{E:11}) \tag{5.36}$$

be the maximal eigenvalue of $\tilde{B}_{E:11}^{-1} A_{E:11}$. Finally, let n_E^f denote the number of faces of any macroelement E. Now, if for any two adjacent macroelements E and E', which share a face, the relation

$$\begin{aligned} &(c-1)\left(\frac{1}{n_E^f} R_E^T U_E^T R_E D_U^{-1} R_E^T U_E R_E + \frac{1}{n_{E'}^f} R_{E'}^T U_{E'}^T R_{E'} D_U^{-1} R_{E'}^T U_{E'} R_{E'} \right) \\ &+c\left(R_E^T U_E^T R_E D_U^{-1} R_{E'}^T U_{E'} R_{E'} + R_{E'}^T U_{E'}^T R_{E'} D_U^{-1} R_E^T U_E R_E \right) \geq 0 \end{aligned} \tag{5.37}$$

holds for some constant $c \geq 1$, then by choosing $\breve{\alpha} := 1/(c\,\hat{\lambda})$ we arrive at the desired estimate:

$$\begin{aligned}
\frac{1}{\breve{\alpha}}\mathbf{v}_1^T B_{11}\mathbf{v}_1 &= c\,\hat{\lambda}\,\mathbf{v}_1^T\left(\sum_E R_E^T U_E^T R_E D_U^{-1} \sum_E R_E^T U_E R_E\right)\mathbf{v}_1 \\
&\geq \hat{\lambda}\,\mathbf{v}_1^T\left(\sum_E R_E^T U_E^T R_E D_U^{-1} R_E^T U_E R_E\right)\mathbf{v}_1 \\
&= \hat{\lambda}\,\mathbf{v}_1^T\left(\sum_E R_E^T \tilde{B}_{E:11} R_E\right)\mathbf{v}_1 \\
&\geq \mathbf{v}_1^T\left(\sum_E R_E^T A_{E:11} R_E\right)\mathbf{v}_1 = \mathbf{v}_1^T A_{11}\mathbf{v}_1 \quad \forall \mathbf{v}_1.
\end{aligned}$$

Note that the constant c in the (interface) condition (5.37) can easily be determined by evaluating the corresponding (local) constant for all (different) configurations of the macroelement interfaces and then taking the maximum value (for the global c). In many cases, e.g., for the model problems considered at the end of this chapter, the condition number of $B_{11}^{-1}A_{11}$ can be estimated by

$$\kappa(B_{11}^{-1}A_{11}) \lesssim \hat{\lambda} = \lambda_{\max}(\tilde{B}_{E:11}^{-1}A_{E:11}).$$

Typically $\kappa(B_{11}^{-1}A_{11})$ also depends on certain problem parameters such as the ratio of anisotropy or the Poisson ratio (in case of linear elasticity). A feasible step towards improving the robustness of this ILU-type preconditioner based on exact local factorization is the employment of proper row-sum criteria. If we want to satisfy the row-sum criterion (5.27) the diagonal of U (as defined in (5.31)) needs a recalculation. For any two $N \times N$ matrices $C = (c_{ij})$ and $B_{11} = (b_{ij})$, the latter being a product of the form $B_{11} = U^T \operatorname{diag}(U)^{-1} U$, where $U = (u_{ij})$ is an upper triangular matrix, the criterion $C\mathbf{e} = B_{11}\mathbf{e}$ (see (5.27)), is equivalent to

$$u_{kk} = \sum_{j=1}^{N} c_{kj} - \sum_{j=k+1}^{N} u_{kj} - \sum_{i=1}^{k-1} \frac{u_{ik}}{u_{ii}} \sum_{j=i}^{N} u_{ij} \quad \forall k = 1, 2, \ldots, N. \tag{5.38}$$

This a-posteriori modification of the diagonal of U can be done efficiently using the algorithm below.

Algorithm 5.4 (A-posteriori modification of the diagonal of U).

---

for $k = 1$ **to** N

$$s_C(k) := \sum_{j=1}^{N} c_{kj}$$

$$s_U(k) := \sum_{j=k+1}^{N} u_{kj}$$

$$u_{kk} := s_C(k) - s_U(k) - \sum_{i=1}^{k-1} \frac{u_{ik} s_U(i)}{u_{ii}}$$

$$s_U(k) := s_U(k) + u_{kk}$$

end

Remark 5.5. Note that the number of operations required for the execution of Algorithm 5.4 is of order $O(n_C + n_U)$ where n_C and n_U denote the number of nonzero entries of C respectively U.

Modifying the diagonal of U according to the aforementioned algorithm yields a preconditioner, henceforth denoted by $B_{11,\mathrm{MILUE}} = LU$, where L is again defined by $L := U^T(\mathrm{diag}(U))^{-1}$.

5.4 Local Schur complements

In this section we will define a particular approximation Q to the exact Schur complement $S = A_{22} - A_{21}A_{11}^{-1}A_{12}$, and analyze its relative condition number.

The proposed approximation technique is simple: From the macroelement matrices (5.28) one computes the exact local Schur complements

$$S_E = A_{E:22} - A_{E:21}(A_{E:11})^{-1}A_{E:12} \tag{5.39}$$

for all macroelements E, which then serve as element matrices on the next coarser level. The procedure is repeated on all levels, i.e.,

$$A_e^{(k-1)} := S_E^{(k)} \quad \forall E \tag{5.40}$$

at levels $k = \ell, \ell - 1, \dots, 1$. In other words, we assemble the small-sized local Schur complement matrices to a global Schur complement approximation, i.e.,

$$Q^{(k)} := \sum_E R_E^{(k)T} S_E^{(k)} R_E^{(k)}. \tag{5.41}$$

In the following, whenever it is clear that we refer to some fixed level k, we drop the index k, writing $Q = \sum_E R_E^T S_E R_E$, for instance.

Consider now the approximation property (5.5). In the remainder of this section, we will derive a locally computable constant $\check{\beta}$, and, we will show that (5.5) holds with $\hat{\beta} = 1$. First, we recall the following energy minimization property of the Schur complement (see Lemma 2.1),

$$\begin{aligned} \mathbf{v}_2^T S \mathbf{v}_2 &= \min_{\mathbf{v}_1} \begin{bmatrix} \mathbf{v}_1 \\ \mathbf{v}_2 \end{bmatrix}^T A \begin{bmatrix} \mathbf{v}_1 \\ \mathbf{v}_2 \end{bmatrix} \\ &= \min_{\mathbf{v}_1} \begin{bmatrix} \mathbf{v}_1 \\ \mathbf{v}_2 \end{bmatrix}^T \left(\sum_E R_E^T A_E R_E \right) \begin{bmatrix} \mathbf{v}_1 \\ \mathbf{v}_2 \end{bmatrix} \\ &= \min_{\mathbf{v}_1} \sum_E \left(\begin{bmatrix} \mathbf{v}_1 \\ \mathbf{v}_2 \end{bmatrix}^T R_E^T A_E R_E \begin{bmatrix} \mathbf{v}_1 \\ \mathbf{v}_2 \end{bmatrix} \right) \quad \forall \mathbf{v}_2, \end{aligned} \tag{5.42}$$

where $\mathbf{v}_E$ denotes the restriction of an arbitrary vector $\mathbf{v}$ to any given macroelement E, i.e., $\mathbf{v}_E = R_E \mathbf{v}$. Now, using (5.41) and (5.42) it is readily seen that (5.5) holds with $\hat{\beta} = 1$:

$$\begin{aligned} \mathbf{v}_2^T Q \mathbf{v}_2 &= \mathbf{v}_2^T \left(\sum_E R_E^T S_E R_E \right) \mathbf{v}_2 \\ &= \sum_E \mathbf{v}_{E:2}^T S_E \mathbf{v}_{E:2} \\ &= \sum_E \left(\min_{\mathbf{v}_{E:1}} \begin{bmatrix} \mathbf{v}_{E:1} \\ \mathbf{v}_{E:2} \end{bmatrix}^T A_E \begin{bmatrix} \mathbf{v}_{E:1} \\ \mathbf{v}_{E:2} \end{bmatrix} \right) \\ &= \sum_E \left(\min_{\mathbf{v}_1} \begin{bmatrix} \mathbf{v}_1 \\ \mathbf{v}_2 \end{bmatrix}^T R_E^T A_E R_E \begin{bmatrix} \mathbf{v}_1 \\ \mathbf{v}_2 \end{bmatrix} \right) \\ &\le \min_{\mathbf{v}_1} \sum_E \left(\begin{bmatrix} \mathbf{v}_1 \\ \mathbf{v}_2 \end{bmatrix}^T R_E^T A_E R_E \begin{bmatrix} \mathbf{v}_1 \\ \mathbf{v}_2 \end{bmatrix} \right) = \mathbf{v}_2^T S \mathbf{v}_2 \quad \forall \mathbf{v}_2. \end{aligned} \tag{5.43}$$

Remark 5.6. Alternatively, by choosing $X_i \equiv X_E := R_E^T A_{E:11}^{-1/2} A_{E:12} R_E$ and $Y_i \equiv Y_E := R_E^T A_{E:11}^{1/2} R_E$ and using (5.33) it can also be seen that (5.5) holds with $\hat{\beta} = 1$.

Regarding the constant $\check{\beta}$ the following basic relation, cf. Lemma 2.4, is an important observation: The (hierarchical basis) coarse-level matrix A_H in (2.33) and the Schur complement approximation Q in (5.41) satisfy the inequalities

$$(1 - \gamma^2) \mathbf{v}_2^T A_H \mathbf{v}_2 \le \mathbf{v}_2^T Q \mathbf{v}_2 \le \mathbf{v}_2^T A_H \mathbf{v}_2 \quad \forall \mathbf{v}_2, \tag{5.44}$$

which result from applying Lemma 2.4 at macroelement level (for every macroelement) and using the fact that both matrices, A_H and Q, are assembled from their corresponding local counterparts. Here $\gamma := \max_E \gamma_E$ where γ_E is the local CBS constant for the splitting (2.31). Using then the right hand side inequality of Lemma 2.4 (again with $A_{22} = A_H$) we deduce from (5.44) the lower bound $\breve{\beta} = 1 - \gamma^2$. This simple derivation of a lower bound in terms of γ was presented in [9]. It shows that the approximation (5.41) is reasonable if the corresponding hierarchical basis transformation defines a proper splitting. In particular, we conclude that the condition number bound

$$\kappa(Q^{-1}S) \leq \frac{1}{1-\gamma^2}$$

holds for arbitrary coefficients in the differential operator (if they are piecewise constant, cf. Theorem 3.8), and even for degenerated elements. We summarize the facts in the following theorem.

Theorem 5.7. *Let Q be the assembly of the local (macroelement matrix) Schur complements S_E, i.e., $Q = \sum_E R_E^T S_E R_E$, and let S be the global (true) Schur complement matrix in (2.34). Then*

$$(1-\gamma^2)\mathbf{v}_2^T S \mathbf{v}_2 \leq \mathbf{v}_2^T Q \mathbf{v}_2 \leq \mathbf{v}_2^T S \mathbf{v}_2 \quad \forall \mathbf{v}_2 \tag{5.45}$$

where $\gamma := \max_E \gamma_E$.

Finally, we want to emphasize that this method can also be applied to non-selfadjoint problems. Efficient Schur complement based multilevel preconditioners for linear isostasy saddle point problems have been studied in [22].

5.5 Numerical results

In this section we will present some numerical results for the nonlinear algebraic multilevel iteration algorithm, for details see [72, 73]. This is a slightly modified version of the algorithm originally proposed in [19]. However, we will consider this method employing the components defined in Sections 5.3–5.4.

The first test problem describes anisotropic diffusion. The domain Ω is the unit square. The weak formulation is given by:

Problem 1.

$$\begin{aligned} \mathcal{A}(u,v) &:= \int_\Omega (\nabla v)^T \mathbf{a}(x) \nabla u \, dx = (f,v) \quad \forall \quad v \in H_0^1(\Omega) \\ \mathbf{a}(x) &:= \begin{bmatrix} \epsilon & 0 \\ 0 & 1 \end{bmatrix}, \quad 0 < \epsilon \leq 1. \end{aligned} \tag{5.46}$$

In the second test problem, we impose a jump on the direction of strong diffusion along the line $x_2 = 1/2$, i.e., we modify Problem 1 by changing (5.46) accordingly (cf. [15]):

Problem 2. *Replace (5.46) in Problem 1 with*

$$\mathbf{a}(x) := \begin{cases} \begin{bmatrix} \epsilon & 0 \\ 0 & 1 \end{bmatrix} & \text{if } (x_1, x_2) \in (0,1) \times (0, \tfrac{1}{2}) \\ \begin{bmatrix} 1 & 0 \\ 0 & \epsilon \end{bmatrix} & \text{if } (x_1, x_2) \in (0,1) \times (\tfrac{1}{2}, 1) . \end{cases}$$

The two-level preconditioner at level k is given by

$$B^{(k)} = \begin{bmatrix} I^{(k)} & \\ A_{21}^{(k)} (B_{11}^{(k)})^{-1} & I \end{bmatrix} \cdot \begin{bmatrix} B_{11}^{(k)} & A_{12}^{(k)} \\ & Q^{(k)} \end{bmatrix}, \tag{5.47}$$

where the Schur complement approximation from Section 5.4 defines the coarse-grid matrix, i.e.,

$$A^{(k-1)} := Q^{(k)}, \tag{5.48}$$

and the preconditioner $B_{11}^{(k)} = L^{(k)} U^{(k)}$ is obtained from (modified) incomplete factorization of $A_{11}^{(k)}$ (cf. Section 5.3).

In Tables 5.1 and 5.2 we list the number of outer generalized conjugate gradient (GCG) iterations that suffice to reduce the l^2-norm of the initial residual (corresponding to a random initial guess) by a factor 10^6.[1] We compare two iterative methods, (I) and (II), both based on nonlinear AMLI (cf. Section 2.5). In the method (I) we use a global MILU factorization for preconditioning the A_{11} block at all levels, i.e., $B_{11} = B_{11,\text{MILU}}$. In method (II) we use the preconditioner $B_{11,\text{MILUE}}$, defined via (5.32) and (5.38), instead. Starting with the first coarse level, two inner GCG iterations are (recursively) performed on every other level, which results in a cheap(er) W-cycle variant.

The numerical tests clearly indicate the potential of the considered preconditioning technique. We observe that even with a global MILU factorization (procedure (I)) of the pivot block the corresponding multilevel iteration becomes an optimal-order method if the ratio of anisotropy is not too large. However, for increasing anisotropy the convergence of method (I) deteriorates whereas method (II) yields the desired robustness. In particular, the results for Problem 2, where the parameter ϵ additionally introduces a discontinuity in the PDE coefficients (changing the direction of dominating anisotropy), demonstrate the importance of an adequate

[1] An asterisk indicates that 250 outer iterations are not sufficient to reach the convergence criterion.

Table 5.1: Number of outer GCG iterations for Problem 1

	$1/h$	8	16	32	64	128	256	512
$\epsilon = 1.0$	(I)	5	6	6	7	7	7	7
	(II)	5	6	7	7	7	7	7
$\epsilon = 0.5$	(I)	5	7	7	8	8	8	8
	(II)	5	7	8	8	8	8	8
$\epsilon = 0.25$	(I)	7	9	9	9	9	9	9
	(II)	5	7	9	9	9	9	9
$\epsilon = 0.1$	(I)	10	16	18	18	18	18	18
	(II)	5	8	10	10	11	11	11
$\epsilon = 0.01$	(I)	18	*	*	*	*	*	*
	(II)	5	7	9	10	11	11	11

Table 5.2: Number of outer GCG iterations for Problem 2

	$1/h$	8	16	32	64	128	256	512
$\epsilon = 0.5$	(I)	6	7	7	8	8	8	8
	(II)	5	7	8	8	8	8	8
$\epsilon = 0.25$	(I)	8	9	9	9	9	10	10
	(II)	5	8	9	9	10	9	10
$\epsilon = 0.1$	(I)	12	19	20	21	21	21	21
	(II)	6	8	10	11	11	11	11
$\epsilon = 0.05$	(I)	16	47	56	59	60	60	61
	(II)	6	8	10	11	11	11	11
$\epsilon = 0.01$	(I)	16	*	*	*	*	*	*
	(II)	6	8	10	11	11	11	11

preconditioner B_{11} for the pivot block. In the above examples, this requirement obviously can be met using exact local factorization, i.e., by employing $B_{11,\text{MILUE}}$.

This leads to the conclusion that, though its analysis is not complete yet, the Schur complement based algebraic multilevel preconditioning technique, presented in this section, offers an attractive alternative to the hierarchical basis approach.

6 Algebraic multigrid (AMG)

Algebraic Multigrid (AMG) was first introduced in the early 80s [36, 39, 40] and immediately attracted substantial interest [101, 102, 106]. Mainly, this is due to its robustness and applicability to various types of problems [48].

A detailed description of the classical AMG method, which was originally proposed in [39], can be found in [102]. The first AMG results for linear elasticity are discussed in [102].

In the last couple of years various new variants of AMG came up, most of which have been designed for special applications. Especially two classes of methods, namely AMG using element interpolation (AMGe) [44, 67, 69] and AMG based on smoothed aggregation [108, 109], considerably enhanced the range of applicability of classical AMG [102]. Whereas smoothed aggregation methods (applied to discretizations of elliptic problems) typically assume the knowledge of the near-nullspace that has to be preserved by the interpolation, AMGe methods capture this information implicitly by accessing the individual element stiffness matrices. Recent works on adaptive smoothed aggregation [45] and adaptive algebraic multigrid [46] try to remove the need of any assumptions on *algebraically smooth* error but, instead, use the method itself to determine near-nullspace components and adjust the coarsening process accordingly. In the following we will briefly outline the main idea of algebraic multigrid (AMG).

Let us consider the following problem: For a given $\mathbf{b} \in \mathbb{R}^n$ we seek the solution $\mathbf{x} \in \mathbb{R}^n$ to the linear system (1.47) assuming that A is sparse, and symmetric and positive definite.

Note that in general it is not necessary that A stems from discretization of a (system of) partial differential equation(s) at all. However, if A stems from finite-element discretization of a second-order elliptic operator it can be beneficial to exploit this knowledge. For instance, the individual element (stiffness) matrices can be used in order to derive a superior (element-based) interpolation.

Roughly speaking, algebraic multigrid is multigrid based on the matrix entries only, although this is not quite correct for some of the more recent approaches, as they exploit additional information, e.g., element matrices, nodal coordinates, the knowledge of (near) null-space components, etc.

The characteristic features of AMG are (a) a fixed relaxation (smoother), (b) an automatic coarse-grid selection, (c) "algebraic" grid-transfer operators, and (d) "algebraic" coarsening (typically, the coarse-grid matrices are defined via the Galerkin relation).

Some of the main advantages of AMG are that there is no rediscretization necessary, i.e., the coarsest grid of a sensible discretization can be very fine, the feasibility to handle problems without geometric background, and the robustness (grey box solver) of the method.

The essential components of an AMG method are:

- a set $\mathcal{D}$ of fine-grid degrees of freedom (DOF)
- a coarse grid, $\mathcal{D}_C$; typically a subset of $\mathcal{D}$
- a prolongation operator $P : \mathcal{D}_C \to \mathcal{D}$
- a smoother; typically Gauß–Seidel or Jacobi
- a coarse matrix given by $A_H = P^T AP$

Building a good algebraic multigrid method requires adequate (problem-adapted) coarse-grid selection, proper (kernel-preserving) interpolation and efficient smoothing that has to complement the coarse-grid correction.

In the following we will discuss the basic components of AMG in the framework of two-grid methods; in (full) algebraic multigrid this construction is used recursively. In particular, we will comment on the principle of coarse-grid correction, its interplay with relaxation, discuss element-based interpolation, prove a simple two-grid convergence result, and derive the error propagation relation based on which we will point out the similarities to AMLI methods. Finally, we give a brief overview on classical AMG and describe the basic idea of smoothed aggregation.

6.1 Two-grid and multigrid algorithms

6.1.1 Exact two-level method

Let us start the presentation with the exact solution of the 2×2 block system

$$A_{11}\mathbf{x}_1 + A_{12}\mathbf{x}_2 = \mathbf{b}_1, \tag{6.1a}$$

$$A_{21}\mathbf{x}_1 + A_{22}\mathbf{x}_2 = \mathbf{b}_2, \tag{6.1b}$$

which is given by

$$\begin{bmatrix} \mathbf{x}_1 \\ \mathbf{x}_2 \end{bmatrix} = \begin{bmatrix} A_{11}^{-1}\left(\mathbf{b}_1 - A_{12}S^{-1}(\mathbf{b}_2 - A_{21}A_{11}^{-1}\mathbf{b}_1)\right) \\ S^{-1}(\mathbf{b}_2 - A_{21}A_{11}^{-1}\mathbf{b}_1) \end{bmatrix} \tag{6.2}$$

where $S = A_{22} - A_{21}A_{11}^{-1}A_{12}$ is the Schur complement. Now let

$$M^{-1} := \begin{bmatrix} A_{11}^{-1} & 0 \\ 0 & 0 \end{bmatrix} \tag{6.3}$$

and

$$P := \begin{bmatrix} -A_{11}^{-1}A_{12} \\ I \end{bmatrix}. \tag{6.4}$$

Then the solution (6.2) of (6.1) can be obtained by executing the following algorithm: Set $\mathbf{x}_{(0)} = \mathbf{0}$ and then perform the steps (6.5a) to (6.6d):

$$\mathbf{e}_{(1)} = M^{-1}\begin{bmatrix}\mathbf{b}_1\\ \mathbf{0}\end{bmatrix} \tag{6.5a}$$

$$\mathbf{x}_{(1)} = \mathbf{x}_{(0)} + \mathbf{e}_{(1)} = \begin{bmatrix}A_{11}^{-1}\mathbf{b}_1\\ \mathbf{0}\end{bmatrix} \tag{6.5b}$$

$$\mathbf{r}_{(1)} = \mathbf{b} - A\mathbf{x}_{(1)} = \begin{bmatrix}0\\ \mathbf{b}_2 - A_{21}A_{11}^{-1}\mathbf{b}_1\end{bmatrix} \tag{6.5c}$$

$$\mathbf{r}^{\mathrm{c}}_{(1)} = P^T\mathbf{r}_{(1)} = \mathbf{b}_2 - A_{21}A_{11}^{-1}\mathbf{b}_1 \tag{6.6a}$$

$$\mathbf{e}^{\mathrm{c}}_{(2)} = S^{-1}\mathbf{r}^{\mathrm{c}}_{(1)} = S^{-1}(\mathbf{b}_2 - A_{21}A_{11}^{-1}\mathbf{b}_1) \tag{6.6b}$$

$$\mathbf{e}_{(2)} = P\mathbf{e}^{\mathrm{c}}_{(2)} = \begin{bmatrix}-A_{11}^{-1}A_{12}S^{-1}\mathbf{b}_2 + A_{11}^{-1}A_{12}S^{-1}A_{21}A_{11}^{-1}\mathbf{b}_1\\ S^{-1}\mathbf{b}_2 - S^{-1}A_{21}A_{11}^{-1}\mathbf{b}_1\end{bmatrix} \tag{6.6c}$$

$$\mathbf{x}_{(2)} = \mathbf{x}_{(1)} + \mathbf{e}_{(2)} = \begin{bmatrix}A_{11}^{-1}(\mathbf{b}_1 - A_{12}S^{-1}(\mathbf{b}_2 - A_{21}A_{11}^{-1}\mathbf{b}_1))\\ S^{-1}(\mathbf{b}_2 - A_{21}A_{11}^{-1}\mathbf{b}_1)\end{bmatrix} \tag{6.6d}$$

The above algorithm can be interpreted as a two-level method, which comprises one *smoothing step* (6.5) and one *correction step* (6.6). By choosing the smoother M^{-1} and the interpolation matrix P according to (6.3) and (6.4) we encounter an exact (direct) solution method. However, since A_{11}^{-1} is a full matrix in general – and thus far too expensive to compute – in practice one has to use sparse approximations, that is, a sparse smoother and a sparse approximation of the *harmonic extension* (the first component in (6.4)). In this way various inexact (iterative) two- and multilevel methods can be constructed. We shall now give a more general formulation of a two-grid and a multigrid algorithm and then comment in particular on the construction of the main components of these algorithms.

6.1.2 From two-grid to multigrid

An algebraic two-grid method is defined by

1. relax η_1 times on $A\mathbf{x} = \mathbf{b}$
2. correct $\mathbf{x} \leftarrow \mathbf{x} + P(P^TAP)^{-1}P^T(\mathbf{b} - A\mathbf{x})$
3. relax η_2 times on $A\mathbf{x} = \mathbf{b}$

An algebraic multigrid method recursively applies the above two-grid method to solve the linear system arising in the coarse-grid correction step (2). Diverse cycles differ in the sequence of recursion. Let $A^{(0)} := A$ be the fine-grid matrix, $\mathbf{b}^{(0)} := \mathbf{b}$ the corresponding right hand side, and let the sequence of coarse-grid operators be defined via the Galerkin relation

$$A^{(k+1)} := P^{(k)^T} A^{(k)} P^{(k)}$$

where the interpolation at level k is denoted by $P^{(k)}$. Let us further denote by $M^{(k)}$ the smoother at level k. Then the linear system (1.47) can be solved (approximately) by application of the algorithm AMG($\mathbf{x}^{(0)}, \mathbf{b}^{(0)}$) which has the following recursive structure:

Algorithm 6.1. AMG($\mathbf{x}^{(k)}, \mathbf{b}^{(k)}$)

$\mathbf{r}^{(k)} = \mathbf{b}^{(k)} - A^{(k)}\mathbf{x}^{(k)}$

apply η_1 pre-smoothing steps: $\mathbf{x}^{(k)} \leftarrow \mathbf{x}^{(k)} + M^{(k)^{-1}}\mathbf{r}^{(k)}$ (6.7)

coarse-grid correction: (6.8)

  set $\mathbf{b}^{(k+1)} = P^{(k)^T}(\mathbf{b}^{(k)} - A^{(k)}\mathbf{x}^{(k)})$

  if $k + 1 = \ell$

    solve $A^{(k+1)}\mathbf{x}^{(k+1)} = \mathbf{b}^{(k+1)}$ by a direct method

  else

    set $\mathbf{x}^{(k+1)} := \mathbf{0}$ and solve by υ applications of AMG($\mathbf{x}^{(k+1)}, \mathbf{b}^{(k+1)}$)

  end

  correct: $\mathbf{x}^{(k)} \leftarrow \mathbf{x}^{(k)} + P^{(k)}\mathbf{x}^{(k+1)}$

$\mathbf{r}^{(k)} = \mathbf{b}^{(k)} - A^{(k)}\mathbf{x}^{(k)}$

apply η_2 post-smoothing steps: $\mathbf{x}^{(k)} \leftarrow \mathbf{x}^{(k)} + M^{(k)^{-T}}\mathbf{r}^{(k)}$ (6.9)

Algorithm 6.1 performs a υ-fold W-cycle with η_1 pre- and η_2 post-smoothing steps. Note that in the formal description of AMG methods, the coarsest level is usually given the level index ℓ. This is convenient because applying the algorithm to a linear system (e.g., arising from discretization of a PDE on an unstructured mesh) one typically does not know in advance how many coarsening steps will be required to reach a coarse(st)-grid problem of acceptable dimension.

6.2 Main components of algebraic multigrid

6.2.1 Coarse-grid correction

In what follows let $A_H := P^T A P$ denote the coarse-grid matrix and $F = I - PA_H^{-1}P^T A$ the coarse-grid correction matrix. First we observe that $F^2 = F$ and that range(P) = range$(I - F)$. Moreover, since

$$\begin{aligned}
\mathbf{e}^T F^T A(I-F)\,\mathbf{v} &= \mathbf{e}^T F^T A\,P\,A_H^{-1} P^T A\,\mathbf{v} \\
&= \mathbf{e}^T [(I - A\,P\,A_H^{-1} P^T) A\,P\,A_H^{-1} P^T A]\,\mathbf{v} \\
&= \mathbf{e}^T [A\,P\,A_H^{-1} P^T A - A\,P\,A_H^{-1} P^T A\,P\,A_H^{-1} P^T A]\,\mathbf{v} \\
&= 0
\end{aligned}$$

we conclude that

$$\langle F\mathbf{e}, (I-F)\mathbf{v}\rangle_A = 0 \quad \forall \mathbf{e}\ \forall \mathbf{v}.$$

Hence

$$\begin{aligned}
\min_{\mathbf{d}\in \text{range}(P)} \|\mathbf{e} - \mathbf{d}\|_A^2 &= \min_{\mathbf{d}\in \text{range}(I-F)} \|\mathbf{e} - \mathbf{d}\|_A^2 \\
&= \min_{\mathbf{d}\in \text{range}(I-F)} \|F\mathbf{e} + (I-F)\mathbf{e} - \mathbf{d}\|_A^2 \\
&= \min_{\mathbf{d}\in \text{range}(I-F)} \|F\mathbf{e} - \mathbf{d}\|_A^2 \\
&= \min_{\mathbf{d}\in \text{range}(I-F)} [\|F\mathbf{e}\|_A^2 + \|\mathbf{d}\|_A^2] \\
&= \|F\mathbf{e}\|_A^2.
\end{aligned}$$

This means that Galerkin-based coarse-grid correction minimizes the energy norm of the error with respect to all variations in range(P).

6.2.2 Smoothing

For the class of problems considered in this book, classical stationary iterative methods, which are based on updating a current iterate at a node based on the values of the iterate at neighboring nodes, reduce the highly oscillatory error components fast, see Figure 6.1. That is why these methods are also referred to as smoothers. Then the resulting smooth error can be represented accurately using fewer degrees of freedom, i.e., on a coarse grid, see Figure 6.2.

The construction of a two- or multilevel method that takes advantage of this fact crucially depends on a proper coarse-grid (discretization) matrix.

The error propagation matrix of a two-grid method with ν_1 pre- and ν_2 post-smoothing steps is given by

$$(I - M^{-T} A)^{\nu_2} F (I - AM^{-1})^{\nu_1}.$$

Thus a two-grid method never diverges if $\|I - M^{-T}A\|_A = \|I - AM^{-1}\|_A \leq 1$.

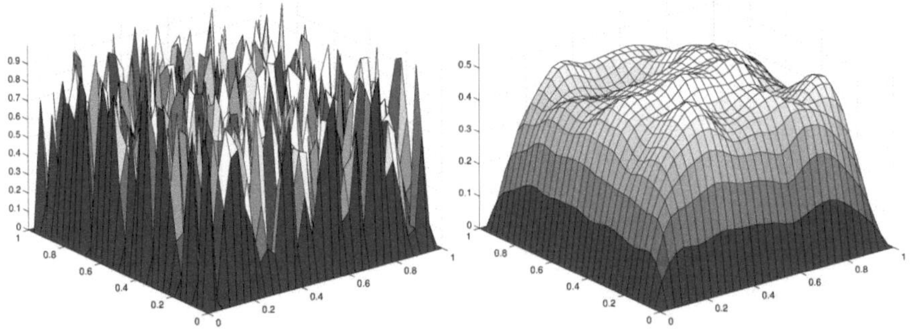

Figure 6.1: Random initial error (left) after 5 Gauß–Seidel iterations (right picture)

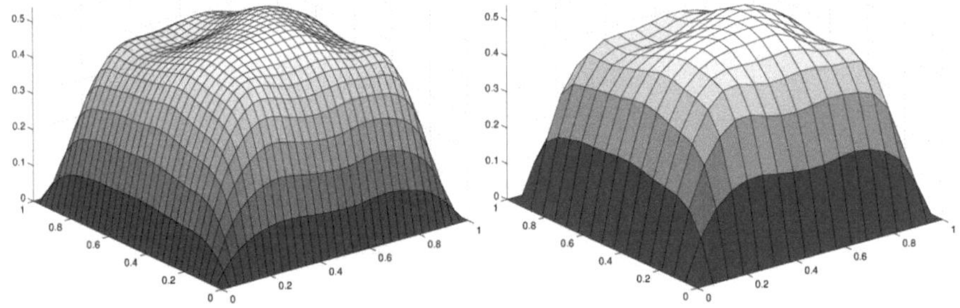

Figure 6.2: Error after 10 Gauß–Seidel iterations (left) represented on coarse mesh (right picture)

An efficient multigrid method requires relaxation and coarse-grid correction to complement each other, i.e., error not reduced by one has to be reduced by the other [44]. This necessitates that the range of interpolation should well approximate those errors not efficiently reduced by relaxation such that they can be reduced on coarser grids!

Standard relaxation schemes, like Richardson, (damped) Jacobi, or Gauß–Seidel slowly reduce low energy error, i.e., error in the direction of eigenvectors corresponding to small eigenvalues. This leads to the following heuristic:

H: *Interpolation must be able to approximate eigenvectors with error bound proportional to the size of the associated eigenvalues.*

6.2.3 Interpolation

In order to examine how well the above heuristic is satisfied for a given interpolation P let us consider a convenient linear projection Q onto range(P)

$$Q \; : \; \mathbb{R}^n \to \mathbb{R}^n$$
$$Q = PR$$

for which $RP = I$ holds. Now, if $\mathbf{e} \in$ range(P) it follows that $Q\mathbf{e} = \mathbf{e}$ and hence $(I - Q)$ can be used to measure the defect of interpolation. A simple measure for interpolation quality thus is given by

$$\mu_0(Q, \mathbf{e}) := \frac{\langle (I-Q)\mathbf{e}, (I-Q)\mathbf{e}\rangle}{\langle A\mathbf{e}, \mathbf{e}\rangle}. \tag{6.10}$$

Element- and edge-based interpolation

The key idea of AMGe is to localize this type of measure based on the knowledge of the individual element matrices A_e, i.e., one takes advantage of the fact that A is given as a sum $A = \sum_{e\in\mathcal{T}} R_e^T A_e R_e$ where $\mathcal{T}$ is a set of finite elements. Then the neighborhood Ω of a fine-grid DOF (node) i to which interpolation is desired can be defined as a (proper) set of elements (or edges) linked to this particular node; The small-sized neighborhood matrix A^Ω, which we also call an interpolation *molecule*, is assembled from the corresponding element (or edge) matrices. It admits the 2×2 block structure

$$A^\Omega(i) = A^\Omega = \begin{bmatrix} A^\Omega_{ff} & A^\Omega_{fc} \\ A^\Omega_{cf} & A^\Omega_{cc} \end{bmatrix}. \tag{6.11}$$

Consider now the small-sized (local) interpolation matrix

$$P_{A^\Omega} = P = \begin{bmatrix} P_{fc} \\ I_{cc} \end{bmatrix} \tag{6.12}$$

associated with (6.11). The $n^f_{A^\Omega} \times n^c_{A^\Omega}$ submatrix P_{fc} produces interpolation in the f-nodes; for the c-nodes P equals the identity. Under the assumption that A^Ω is SPSD the AMGe interpolation concept can be applied directly [44, 55]:

For any vector $\mathbf{e}^T = (\mathbf{e}_f^T, \mathbf{e}_c^T) \perp \ker(A^\Omega)$ we denote by

$$\mathbf{d}_f := \mathbf{e}_f - P_{fc}\mathbf{e}_c \tag{6.13}$$

the defect of (local) interpolation. With the objective of realizing the AMGe heuristic we choose P_{fc} to be the argument that minimizes

$$\max_{\mathbf{e}\perp\ker(A^\Omega)} \frac{(\mathbf{e}_f - P_{fc}\mathbf{e}_c)^T(\mathbf{e}_f - P_{fc}\mathbf{e}_c)}{\mathbf{e}^T A^\Omega \mathbf{e}}. \tag{6.14}$$

Using the substitutions (6.13) and $G := P_{fc}^T A_{ff}^\Omega P_{fc} + P_{fc}^T A_{fc}^\Omega + A_{cf}^\Omega P_{fc} + A_{cc}^\Omega$ we derive the following equivalence for (6.14):

$$
\begin{aligned}
&\max_{\mathbf{d}_f,\mathbf{e}_c} \frac{\mathbf{d}_f^T\mathbf{d}_f}{\begin{bmatrix} \mathbf{d}_f + P_{fc}\mathbf{e}_c \\ \mathbf{e}_c \end{bmatrix}^T \begin{bmatrix} A_{ff}^\Omega & A_{fc}^\Omega \\ A_{cf}^\Omega & A_{cc}^\Omega \end{bmatrix} \begin{bmatrix} \mathbf{d}_f + P_{fc}\mathbf{e}_c \\ \mathbf{e}_c \end{bmatrix}} \\
&= \max_{\mathbf{d}_f,\mathbf{e}_c} \frac{\mathbf{d}_f^T\mathbf{d}_f}{\langle A_{ff}^\Omega(\mathbf{d}_f + P_{fc}\mathbf{e}_c), \mathbf{d}_f + P_{fc}\mathbf{e}_c\rangle + 2\langle A_{fc}^\Omega \mathbf{e}_c, \mathbf{d}_f + P_{fc}\mathbf{e}_c\rangle + \langle A_{cc}^\Omega \mathbf{e}_c, \mathbf{e}_c\rangle} \\
&= \max_{\mathbf{d}_f,\mathbf{e}_c} \frac{\mathbf{d}_f^T\mathbf{d}_f}{\begin{bmatrix} \mathbf{d}_f \\ \mathbf{e}_c \end{bmatrix}^T B \begin{bmatrix} \mathbf{d}_f \\ \mathbf{e}_c \end{bmatrix}} \qquad (6.15)
\end{aligned}
$$

where

$$
B = \begin{bmatrix} A_{ff}^\Omega & A_{ff}^\Omega P_{fc} + A_{fc}^\Omega \\ P_{fc}^T A_{ff}^\Omega + A_{cf}^\Omega & G \end{bmatrix} \qquad (6.16)
$$

is SPSD. Hence,

$$
\begin{aligned}
\min_{P_{fc}} \max_{\mathbf{d}_f,\mathbf{e}_c} \frac{\mathbf{d}_f^T\mathbf{d}_f}{\begin{bmatrix} \mathbf{d}_f \\ \mathbf{e}_c \end{bmatrix}^T B \begin{bmatrix} \mathbf{d}_f \\ \mathbf{e}_c \end{bmatrix}} &= \min_{P_{fc}} \max_{\mathbf{d}_f} \frac{\mathbf{d}_f^T\mathbf{d}_f}{\min_{\mathbf{e}_c} \begin{bmatrix} \mathbf{d}_f \\ \mathbf{e}_c \end{bmatrix}^T B \begin{bmatrix} \mathbf{d}_f \\ \mathbf{e}_c \end{bmatrix}} \\
&= \min_{P_{fc}} \max_{\mathbf{d}_f} \frac{\mathbf{d}_f^T\mathbf{d}_f}{\mathbf{d}_f^T \left(A_{ff}^\Omega - (A_{ff}^\Omega P_{fc} + A_{fc}^\Omega) G^{-1} (P_{fc}^T A_{ff}^\Omega + A_{cf}^\Omega) \right) \mathbf{d}_f}. \qquad (6.17)
\end{aligned}
$$

Assuming that A_{ff}^Ω and G both are SPD the denominator of (6.17) for an arbitrary vector $\mathbf{d}_f$ is maximized and thus the minimum is attained for

$$
P_{fc} := -(A_{ff}^\Omega)^{-1} A_{fc}^\Omega, \qquad (6.18)
$$

which results in $1/(\lambda_{\min}(A_{ff}^\Omega))$. This motivates to choose the interpolation coefficients for node i to equal the i'-th row of (6.18). For a more general framework of AMG (including convergence analysis) we refer to [55].

There is also a generalization of AMGe called element-free AMGe that avoids the necessity of the individual element matrices by constructing neighborhood matrices via special extension mappings [67].

6.3 A simple convergence result

Consider a relaxation method of the form

$$\mathbf{x}^{(k+1)} = \mathbf{x}^{(k)} + M^{-1}\mathbf{r}^{(k)} \tag{6.19}$$

with error propagation

$$\mathbf{e}^{(k+1)} = (I - M^{-1}A)\mathbf{e}^{(k)}, \tag{6.20}$$

i.e., a smoother of the form $I - M^{-1}A$. Moreover, let $Q = PR$ be a projection onto range(P) for which $RP = I$ holds, e.g.,

$$Q := Q_A = P\,A_H^{-1}P^T A.$$

Then the measure

$$\mu_1(Q,\mathbf{e}) := \frac{\langle M(M+M^T-A)^{-1}M^T(I-Q)\mathbf{e}, (I-Q)\mathbf{e}\rangle}{\langle A\mathbf{e},\mathbf{e}\rangle}, \tag{6.21}$$

which involves the symmetrized smoother

$$\overline{M} := M(M+M^T-A)^{-1}M^T \tag{6.22}$$

takes into account the general smoothing process (6.19). It can be used to derive a simple convergence result for a two-grid method (using only post-smoothing) which will be given at the end of this section. First we prove the following lemma.

Lemma 6.2 (see [55]). *Let Q be any projection onto* range(P). *Assume that the following approximation property is satisfied for some constant K:*

$$\mu_1(Q,\mathbf{e}) \le K \quad \forall \mathbf{e} \in \mathbb{R}^n \setminus \{\mathbf{0}\}. \tag{6.23}$$

If $\mathbf{e} \neq \mathbf{0}$ *is A-orthogonal to* range(P), *then*

$$\|(M+M^T-A)^{1/2}M^{-1}A\mathbf{e}\|^2 \ge \frac{1}{K}\langle A\mathbf{e},\mathbf{e}\rangle. \tag{6.24}$$

Proof. Since range(Q) = range(P) and $\mathbf{e}$ is assumed to be A-orthogonal to range(P) we have

$$\langle A\mathbf{e}, Q\mathbf{v}\rangle = 0 \quad \forall \mathbf{v} \in \mathbb{R}^n. \tag{6.25}$$

Assume that (6.23) holds. From (6.25) and the Cauchy–Schwarz inequality, and using also (6.21), it follows that

$$\begin{aligned}
\langle A\mathbf{e}, \mathbf{e}\rangle &= \langle A\mathbf{e}, (I-Q)\mathbf{e}\rangle \\
&= \langle (M+M^T-A)^{1/2}M^{-1}A\mathbf{e}, (M+M^T-A)^{-1/2}M^T(I-Q)\mathbf{e}\rangle \\
&\le \|(M+M^T-A)^{1/2}M^{-1}A\mathbf{e}\|\|(M+M^T-A)^{-1/2}M^T(I-Q)\mathbf{e}\| \\
&= \|(M+M^T-A)^{1/2}M^{-1}A\mathbf{e}\|\,\mu_1(Q,\mathbf{e})^{1/2}\langle A\mathbf{e}, \mathbf{e}\rangle^{1/2} \\
&\le \|(M+M^T-A)^{1/2}M^{-1}A\mathbf{e}\|K^{1/2}\langle A\mathbf{e}, \mathbf{e}\rangle^{1/2}.
\end{aligned}$$

The result (6.24) now follows by dividing through by $\langle A\mathbf{e}, \mathbf{e}\rangle K^{1/2}$ and squaring the result. □

Theorem 6.3 (see [55]). *Assume that the approximation property (6.23) is satisfied for some constant K. Then $K \ge 1$ and*

$$\|(I-M^{-1}A)(I-Q_A)\mathbf{e}\|_A \le \left(1-\frac{1}{K}\right)^{1/2}\|\mathbf{e}\|_A. \tag{6.26}$$

Proof. We have the following identity

$$\begin{aligned}
\|(I-M^{-1}A)\mathbf{e}\|_A^2 &= \langle A\mathbf{e}, \mathbf{e}\rangle - \langle A\mathbf{e}, M^{-1}A\mathbf{e}\rangle - \langle M^{-1}A\mathbf{e}, A\mathbf{e}\rangle \\
&\quad + \langle AM^{-1}A\mathbf{e}, M^{-1}A\mathbf{e}\rangle \\
&= \langle A\mathbf{e}, \mathbf{e}\rangle - \langle (M+M^T-A)M^{-1}A\mathbf{e}, M^{-1}A\mathbf{e}\rangle.
\end{aligned}$$

Replacing $\mathbf{e}$ with $(I-Q_A)\mathbf{e}$ and applying the result of Lemma 6.2 yields

$$\begin{aligned}
\|(I-M^{-1}A)(I-Q_A)\mathbf{e}\|_A^2 &\le \left(1-\frac{1}{K}\right)\|(I-Q_A)\mathbf{e}\|_A^2 \\
&\le \left(1-\frac{1}{K}\right)\|\mathbf{e}\|_A^2.
\end{aligned}$$

To show that $K \ge 1$, note that the identity at the beginning of the proof implies (since norms are non-negative)

$$\|(M+M^T-A)^{1/2}M^{-1}A\mathbf{e}\|^2 \le \langle A\mathbf{e}, \mathbf{e}\rangle.$$

The result follows by restricting $\mathbf{e} \ne \mathbf{0}$ to be A-orthogonal to range(P) and applying Lemma 6.2. □

6.4 Error propagation of AMG and AMLI methods

Let us study the error propagation

$$\mathbf{e}_{(i+1)} = E\mathbf{e}_{(i)} \tag{6.27}$$

of a general two-level method that can be stated as

$$\mathbf{x}_{(i+1)} = \mathbf{x}_{(i)} + B^{-1}\mathbf{r}_{(i)} \tag{6.28}$$

with

$$B^{-1} := \overline{M}^{-1} + \left(I - M^{-T}A\right) P B_{H,\upsilon}^{-1} P^T \left(I - AM^{-1}\right) \tag{6.29}$$

where P denotes the prolongation operator, $\overline{M}$ is defined in (6.22), i.e.,

$$\overline{M}^{-1} = M^{-1} + M^{-T} - M^{-T}AM^{-1}, \tag{6.30}$$

and

$$B_{H,\upsilon}^{-1} := \left(I - p_\upsilon(B_H^{-1}A_H)\right) A_H^{-1} \tag{6.31}$$

is a polynomial approximation to the inverse of the Schur complement, which is based on the coarse-grid matrix

$$A_H := P^T AP$$

and the preconditioner B_H^{-1} at the coarse level. Note that the error propagation operators E and E_H corresponding to a fine- and a coarse-level update are given by

$$E = I - B^{-1}A \tag{6.32}$$

and

$$E_H = I - PB_H^{-1}P^T A, \tag{6.33}$$

respectively. The following simple identity will be useful in deriving the error propagation relation.

Lemma 6.4. *Let q_r be a polynomial of degree less or equal to r and $A_H = P^T AP$ denote the coarse-grid matrix. Then*

$$Pq_r(B_H^{-1}A_H)B_H^{-1}P^T A = q_r(PB_H^{-1}P^T A)PB_H^{-1}P^T A. \tag{6.34}$$

Proof. It suffices to prove the identity (6.34) for the case $q_r(t) = t^k$, $k \le r$. For $k = 0$ the result is trivial. For $k = 1$ we obtain

$$PB_H^{-1}A_H B_H^{-1}P^T A = PB_H^{-1}P^T APB_H^{-1}P^T A,$$

which is true since $A_H = P^T AP$. Finally, assuming that (6.34) holds true for $k = j - 1$ we find

$$\begin{aligned} P\left(B_H^{-1}A_H\right)^j B_H^{-1}P^T A &= P\left(B_H^{-1}A_H\right)^{j-1} B_H^{-1}P^T APB_H^{-1}P^T A \\ &= (PB_H^{-1}P^T A)^{j-1} PB_H^{-1}P^T APB_H^{-1}P^T A \\ &= (PB_H^{-1}P^T A)^j PB_H^{-1}P^T A. \end{aligned}$$

□

The error transfer operator for the symmetrized smoother can be written in product form, i.e.,

$$I - \overline{M}^{-1}A = \left(I - M^{-T}A\right)\left(I - M^{-1}A\right)$$

and thus, using (6.29) we find

$$\begin{aligned} I - B^{-1}A &= I - \overline{M}^{-1}A - \left(I - M^{-T}A\right) PB_{H,v}^{-1}P^T\left(I - AM^{-1}\right)A \\ &= \left(I - M^{-T}A\right)\left(I - M^{-1}A\right) \\ &- \left(I - M^{-T}A\right) PB_{H,v}^{-1}P^T A\left(I - M^{-1}A\right) \\ &= \left(I - M^{-T}A\right)\left(I - PB_{H,v}^{-1}P^T A\right)\left(I - M^{-1}A\right). \end{aligned} \quad (6.35)$$

Moreover, if $p_v(t) = 1 - q_{v-1}(t)\,t$ we can rewrite (6.31) in the form

$$\begin{aligned} B_{H,v}^{-1} &= \left(I - p_v(B_H^{-1}A_H)\right)A_H^{-1} \\ &= \left(I - \left(I - q_{v-1}(B_H^{-1}A_H)\right)B_H^{-1}A_H\right)A_H^{-1} \\ &= q_{v-1}(B_H^{-1}A_H)B_H^{-1}. \end{aligned} \quad (6.36)$$

Then by substituting (6.36) in (6.35) and using Lemma 6.4 and finally (6.33) we obtain the following representation of (6.32):

$$\begin{aligned} I - B^{-1}A &= \left(I - M^{-T}A\right)\left(I - Pq_{v-1}(B_H^{-1}A_H)B_H^{-1}P^T A\right)\left(I - M^{-1}A\right) \\ &= \left(I - M^{-T}A\right)\left(I - q_{v-1}(PB_H^{-1}P^T A)PB_H^{-1}P^T A\right)\left(I - M^{-1}A\right) \\ &= \left(I - M^{-T}A\right)\left(I - q_{v-1}(I - E_H)(I - E_H)\right)\left(I - M^{-1}A\right) \\ &= \left(I - M^{-T}A\right)\left(p_v(I - E_H)\right)\left(I - M^{-1}A\right) \\ &= \left(I - M^{-T}A\right)\tilde{p}_v(E_H)\left(I - M^{-1}A\right) \end{aligned} \quad (6.37)$$

where

$$\tilde{p}_v(t) := p_v(1 - t). \quad (6.38)$$

We shall now point out the relation between AMG and (linear) AMLI methods. Let us denote the AMLI preconditioner (at level k), as defined in (2.39), by $\tilde{B}$, i.e.,

$$\tilde{B}^{-1} = \begin{bmatrix} I & -C_{11}^{-1}A_{12} \\ 0 & I \end{bmatrix} \begin{bmatrix} C_{11}^{-1} & 0 \\ 0 & Z^{-1} \end{bmatrix} \begin{bmatrix} I & 0 \\ -A_{21}C_{11}^{-1} & I \end{bmatrix} \tag{6.39}$$

where Z^{-1} is given by (2.42) and for ease of notation we skip the level index. The off-diagonal blocks of the hierarchical two-level matrix, i.e., $\tilde{A}_{12}^{(k)}$ and $\tilde{A}_{21}^{(k)}$, are denoted by A_{12} and A_{21} accordingly. Then we can represent $\tilde{B}^{-1}$ as

$$\begin{aligned} \tilde{B}^{-1} &= \begin{bmatrix} C_{11}^{-1} & 0 \\ 0 & 0 \end{bmatrix} + \begin{bmatrix} -C_{11}^{-1}A_{12} \\ I \end{bmatrix} Z^{-1} [-A_{21}C_{11}^{-1}, I] \\ &= \begin{bmatrix} C_{11}^{-1} & 0 \\ 0 & 0 \end{bmatrix} \\ &\quad + \begin{bmatrix} I - C_{11}^{-1}A_{11} & -C_{11}^{-1}A_{12} \\ 0 & I \end{bmatrix} \begin{bmatrix} 0 \\ I \end{bmatrix} Z^{-1} [0, I] \begin{bmatrix} I - A_{11}C_{11}^{-1} & 0 \\ -A_{21}C_{11}^{-1} & I \end{bmatrix} \\ &= M^{-1} + (I - M^{-T}A)\, P Z^{-1} P^T \left(I - AM^{-1}\right) \end{aligned} \tag{6.40}$$

where

$$M^{-1} = M^{-T} = \begin{bmatrix} C_{11}^{-1} & 0 \\ 0 & 0 \end{bmatrix} \tag{6.41}$$

and

$$P = \begin{bmatrix} 0 \\ I \end{bmatrix}. \tag{6.42}$$

We note that (6.40) differs from (6.29) in the first term only. Using the symmetrized smoother $\overline{M}$ according to (6.29) we get the slightly modified AMLI preconditioner

$$B^{-1} = \begin{bmatrix} I & -C_{11}^{-1}A_{12} \\ 0 & I \end{bmatrix} \begin{bmatrix} 2C_{11}^{-1} - C_{11}^{-1}A_{11}C_{11}^{-1} & 0 \\ 0 & Z^{-1} \end{bmatrix} \begin{bmatrix} I & 0 \\ -A_{21}C_{11}^{-1} & I \end{bmatrix} \tag{6.43}$$

which satisfies the equations (6.29)–(6.30) and (2.42) where M and P are specified in (6.41) and (6.42), respectively, and the symmetrized smoother is given by

$$\overline{M}^{-1} = \begin{bmatrix} 2C_{11}^{-1} - C_{11}^{-1}A_{11}C_{11}^{-1} & 0 \\ 0 & 0 \end{bmatrix}. \tag{6.44}$$

Thus the error transfer operator of the AMLI method can be written as in (6.37), i.e.,

$$E = I - B^{-1}A = \left(I - M^{-T}A\right) \tilde{p}_\upsilon(E_H) \left(I - M^{-1}A\right). \tag{6.45}$$

We conclude that the (classical) AMLI method uses f-smoothing only. It is usually implemented based on the preconditioner (6.40), however, the variant based on (6.43), which uses the symmetrized smoother (6.44) is readily available, too. The latter exactly fits the error propagation relation of the two-level AMG method if one chooses P according to (6.42). Note that the (almost) trivial form of P in (6.40) is related to the fact that the AMLI preconditioner $\tilde{B}^{-1}$ (or B^{-1}) is defined for the hierarchical two-level matrix $\tilde{A} = J^T A J$, which already contains the coarse-level matrix as a sub-matrix in the lower right block, i.e.,

$$A_H = [0, I]\, \tilde{A} \begin{bmatrix} 0 \\ I \end{bmatrix},$$

whereas in AMG the coarse-grid matrix usually is computed from the standard (nodal basis) stiffness matrix A via the Galerkin relation

$$A_H = P^T A P,$$

using a nontrivial prolongation operator P. Thus the interpolation in AMLI is *hidden* in the basis transformation.

Regarding the polynomial $\tilde{p}_\upsilon(t)$, which determines the cycle of the multilevel method we notice that the V-cycle AMG or AMLI method corresponds to the choice

$$\tilde{p}_1(t) = p_1(1-t) := t,$$

which leads to

$$E = \left(I - M^{-T} A\right) E_H \left(I - M^{-1} A\right).$$

The standard AMG W-cycle is obtained for

$$\tilde{p}_2(t) = p_2(1-t) := t^2, \tag{6.46}$$

which leads to

$$E = \left(I - M^{-T} A\right) E_H^2 \left(I - M^{-1} A\right).$$

Note that (6.46) can also be employed in the AMLI method if a proper scaling is applied to the two-level preconditioner, i.e., if the approximation property (2.48) holds with a lower bound α, $0 < \alpha < 1$ and upper bound 1. Alternatively, if we use the polynomial (2.57), i.e.,

$$\tilde{p}_2(t) = p_2(1-t) := 1 - \left(1 + \frac{1}{\lambda}\right)(1-t) + \frac{1}{\lambda}(1-t)^2,$$

we get

$$E = \left(I - M^{-T} A\right) E_H \left(I - \frac{1}{\lambda}(I - E_H)\right) \left(I - M^{-1} A\right).$$

For a comprehensive matrix-based analysis of multilevel block factorization preconditioners we refer the reader to [114].

6.5 Classical AMG

6.5.1 Strong connections

Algebraically smooth error (with respect to M and A) satisfies

$$\|(I - M^{-1}A)\mathbf{e}\|_A \approx \|\mathbf{e}\|_A$$

and varies slowly in the direction of large (negative) connections. For (most of) the common smoothers (e.g., Gauß–Seidel) these error components can be characterized by

$$a_{ii}e_i \approx -\sum_{j \neq i} a_{ij}e_j.$$

For M-matrices this means: For each node i the error component e_i is essentially determined by those e_j for which $-a_{ij}$ is large. This motivates to introduce the following notion of strong connections:

Definition 6.5. Node i is strongly connected to node j (strongly depends on j) if

$$-a_{ij} \geq \theta \max_{k \neq i}\{-a_{ik}\}$$

for some $0 < \theta \leq 1$ (e.g., $\theta = 0.25$).

6.5.2 Coarse-grid selection

There are different reasonable ways of selecting the coarse grid nodes in AMG. Following [102] a *good* coarse grid $\mathcal{D}_c$ should satisfy two criteria:

C1: *$\mathcal{D}_c$ should be a maximum independent set, which means that no strong connections within $\mathcal{D}_c$ are allowed.*

C2: *Each node j being strongly connected to an f-node i is either contained in $\mathcal{D}_c$ or it strongly depends on at least one c-node k that itself is strongly connected to node i.*

For instance, the coarse grid can be selected in a two-stage process: First, a quick c-node choice attempts to enforce criterion (C1). Then, at a second stage, all f-nodes resulting from the first stage are tested to ensure that criterion (C2) holds, adding new c-nodes if necessary. A detailed description of the classical AMG coarse-grid selection algorithm can be found in [102]. For coarse-grid selection based on *strong edges* see also [74, 78].

6.5.3 Interpolation

The prolongation P in classical AMG can be described in a very simple way. If $\Omega(i)$ denotes the minimal neighborhood of some fine DOF (node) i, then one replaces the block A_{ff}^{Ω} in (6.11) with a modified version $\hat{A}_{ff}^{\Omega}$, which is obtained by

- adding to a_{ii} all off-diagonal entries in the i-th row that are weakly connected to i,
- second, in all rows j for DOF strongly connected to i:
  - set $a_{jj} \leftarrow \sum_{k \in \mathcal{S}_i^c} a_{jk}$,
  - set off-diagonal entries to zero.

Finally, the i-th row of the interpolation matrix P is chosen according to the i-th row of

$$-(\hat{A}_{ff}^{\Omega})^{-1} A_{fc}^{\Omega}.$$

However, for the implementation it is preferable to compute the coefficients of the interpolation matrix P in the course of the coarse-grid selection process. A detailed description of this kind of combined algorithm can be found in reference [102].

6.6 Smoothed aggregation and adaptive AMG methods

AMG based on element interpolation (AMGe) [44, 69] has been a significant progress in extending algebraic multigrid, which was originally designed having M-matrices in mind, to non-M-matrices, arising from elasticity problems, for instance. This is due to the fact that the knowledge of the element matrices carries with it implicitly the correct assignment and treatment of "strong" and "weak" connections.

Element-free AMGe [67] tries to accomplish the superior prolongation without the knowledge of the element matrices. The method uses an extension mapping to provide boundary values outside a neighborhood. In essence, this captures information that could be obtained from individual finite element stiffness matrices if they were available.

Algebraic multigrid based on smoothed aggregation [108, 109] is another approach to overcome the difficulties arising with non-M-matrices. The basic idea is to start with a simple tentative prolongator, which for instance can be constructed based on a set of node aggregates $\{\mathcal{A}_i\}$ forming a disjoint covering of the set of

DOF, i.e.,

$$\bigcup_{i=1}^{m} \mathcal{A}_i = \{1, 2, \ldots, n\}, \quad \mathcal{A}_i \cap \mathcal{A}_j = \emptyset \text{ for } i \neq j \tag{6.47}$$

where each aggregate is associated with one coarse node. Now the tentative prolongator $\hat{P} \in \mathbb{R}^{n \times n_c}$ can be defined as a discrete piecewise constant interpolation:

$$\hat{P}_{ij} = \begin{cases} 1 & \text{if } i \in \mathcal{A}_j \\ 0 & \text{otherwise.} \end{cases} \tag{6.48}$$

Note that the tentative prolongator has to be kernel preserving. This means that all zero energy modes of the principal part of the differential operator (without any boundary conditions applied) will be represented exactly by the coarse space. Let the set of such functions be denoted by $\{z_i\}_{i=1}^r$. In case of the scalar elliptic model problem (1.1) we have $r = 1$ and $z_1 = \text{const}$, i.e., $\mathbf{z}_1 = \mathbf{1}$. This implies that $\hat{P}$ has to have row sum one for all rows. In particular the interpolation weights for the piecewise constant interpolation have to be one because there is only one nonzero entry in each row of $\hat{P}$, cf. (6.47) and (6.48). For details on more general cases we refer the reader to [108]. Then the key note is to eliminate high energy components from the range of $\hat{P}$. This can be achieved by smoothing the prolongator, that is, computing the final prolongator P according to

$$P := q(D^{-1}A)\hat{P} \tag{6.49}$$

where $D \in \mathbb{R}^{n \times n}$ is a symmetric positive definite preconditioner for A, and q is a polynomial that satisfies $q(0) = 1$. The final prolongator P is used in the AMG method then in the usual fashion, i.e., each of the prolongation operators $P^{(k)}$ in Algorithm 6.1 is computed in this way. In fact, it has been shown that the use of energy minimal basis functions can even improve this approach [86, 115].

The integration of more general coarsening and smoothing processes is an important generalization of the AMG framework [55, 56, 118].

In bootstrap AMG (BAMG) methods, proposed in [38], the evolving AMG solver is used to improve its interpolation component iteratively. The selection of the coarse-grid variables can be based on a process called compatible relaxation, which has been introduced in context with highly accurate algebraic coarsening [37].

Adaptive algebraic multigrid methods, see, e.g. [46], try to remove the need of any assumptions on algebraically smooth error but, instead, use the method itself to determine near-nullspace components and to adjust the coarsening process accordingly. Adaptive smoothed aggregation [45] constructs the coarse basis functions by minimizing the sum of their energies (cf. [115]) subject to the condition that the given kernel modes are in the corresponding coarse space, and subject to restrictions on their supports.

The adaptive AMG methods break new ground in relating coarse-grid selection and interpolation directly to the relaxation process. The concept of adaptive AMG and bootstrap AMG provides additional robustness and speeds up the convergence, however, on the other hand the setup of these methods is usually more expensive.

6.7 Utilizing AMG components in AMLI

Finally we want to comment on the possibility of using the components of any AMG method in the construction of AMLI methods. Classical multigrid schemes are often optimal with a simple V-cycle [64] whereas AMLI (as described in this book) typically results in a nearly optimal order solution process in this case [112, 113, 117]. However, a smoothing procedure (similar to the ones used in classical multigrid) may compensate for this disadvantage [93]. This means in particular that the AMLI method can also be equipped with a global smoother instead of using f-smoothing only.

Moreover, and this makes the AMLI methodology even more attractive, the construction of a generalized hierarchical basis to be used for setting up an efficient AMLI algorithm can also be done completely algebraically. For instance the required two-by-two block partitioning of the matrix can be achieved by applying any independent-set ordering to the matrix graph. This yields a diagonal pivot block A_{11}, and consequently, one chooses $C_{11} = A_{11}$ in the two-level preconditioner. Then the problem reduces to find a sparse approximation to the Schur complement, which is a difficult task in general. A major drawback of this approach is that independent-set orderings in general produce a *slow coarsening* involving many approximation levels. Additionally, when computing the Schur complement approximation by neglecting in the usual Gaussian elimination process certain fill-in terms based on a numerical drop tolerance typically causes a gradual loss of sparsity for the approximate Schur complements on coarser levels, which also adds to a slow coarsening. Alternatively, one can use repeated red-black colorings or other graph-based algorithms that provide a *moderate coarsening*. In doing so, the need for a preconditioner $C_{11} \neq A_{11}$ arises. The construction of C_{11} via a (modified) incomplete LU factorization, or approximate inverse of A_{11} (satisfying certain row-sum criteria) has been discussed in a series of papers, see, e.g., [5, 72, 75, 94, 95]. More recently, it has been shown that a *fast coarsening*, as used with classical AMG, can also result in robust AMLI-type preconditioners [97].

Assume that a sequence $\{P^{(\ell-j)}\}_{j=0,1,2,\ldots,\ell-1}$ of prolongation matrices and corresponding nested (or nonnested) sequence of coarse spaces has been constructed based on AMG techniques, e.g., classical, element-based, or smoothed aggregation AMG. The coarse-space $\mathcal{V}^{(k-1)} := \{\phi_i^{(k-1)} : 1 \leq i \leq n^{(k-1)}\}$ relates to the fine-space $\mathcal{V}^{(k)} := \{\phi_i^{(k)} : 1 \leq i \leq n^{(k)}\}$ via

$$\begin{bmatrix} \phi_1^{(k-1)} \\ \phi_2^{(k-1)} \\ \vdots \\ \phi_{n^{(k-1)}}^{(k-1)} \end{bmatrix} = P^{(k)^T} \begin{bmatrix} \phi_1^{(k)} \\ \phi_2^{(k)} \\ \vdots \\ \phi_{n^{(k)}}^{(k)} \end{bmatrix}.$$

Then one can define a hierarchical (two-level) basis transformation at each level k by

$$J^{(k)} := \left[\begin{bmatrix} I \\ 0 \end{bmatrix}, P^{(k)} \right]$$

and use it in order to define the hierarchical matrices in the usual inductive way, i.e.,

$$\tilde{A}^{(k)} = \begin{bmatrix} * & * \\ * & A^{(k-1)} \end{bmatrix} := J^{(k)^T} A^{(k)} J^{(k)}$$

where

$$A^{(k-1)} = P^{(k)^T} A^{(k)} P^{(k)}$$

is the Galerkin coarse-grid matrix. In this setting practically every AMG method fits in the AMLI framework.

7 Preconditioning of Rannacher–Turek nonconforming FE systems

7.1 Rannacher–Turek nonconforming FE systems

7.1.1 The nonconforming FE problem

Let us consider the elliptic boundary value problem (1.1). As in the previous chapters, we assume that the elements of the diffusion coefficient matrix $\mathbf{a}(\boldsymbol{x})$ are piecewise smooth functions on $\overline{\Omega}$.

The weak formulation of the above problem reads as follows: given $f \in L^2(\Omega)$ find $u \in \mathcal{V} \equiv H_D^1(\Omega) = \{v \in H^1(\Omega) : v = 0 \text{ on } \Gamma_D\}$, satisfying

$$\mathcal{A}(u, v) = (f, v) \quad \forall v \in H_D^1(\Omega), \tag{7.1}$$

where

$$\mathcal{A}(u, v) = \int_\Omega \mathbf{a}(\boldsymbol{x})\nabla u(\boldsymbol{x}) \cdot \nabla v(\boldsymbol{x}) d\boldsymbol{x}.$$

We assume that the domain Ω is discretized by the partition $\mathcal{T}_h$ which is obtained by a proper refinement of a given coarser partition $\mathcal{T}_H$. For our theoretical analysis we assume also that $\mathcal{T}_H$ is aligned with the discontinuities of the elements of $\mathbf{a}(\boldsymbol{x})$ so that over each finite element $E \in \mathcal{T}_H$ the coefficients of $\mathbf{a}(\boldsymbol{x})$ are smooth functions.

Then the finite element formulation is: find $u_h \in \mathcal{V}_h$, satisfying

$$\mathcal{A}_h(u_h, v_h) = (f, v_h) \quad \forall v_h \in \mathcal{V}_h, \tag{7.2}$$

where

$$\mathcal{A}_h(u_h, v_h) = \sum_{e \in \mathcal{T}_h} \int_e \mathbf{a}(e)\nabla u_h \cdot \nabla v_h d\boldsymbol{x}.$$

Here $\mathbf{a}(e)$ is a piecewise constant symmetric positive definite matrix, defined by the integral averaged values of $\mathbf{a}(\boldsymbol{x})$ over each element from the coarser partition $\mathcal{T}_H$. We recall that in this way strong coefficient jumps across the boundaries between adjacent finite elements from $\mathcal{T}_H$ are allowed.

The resulting discrete problem to be solved is then a linear system of equations

$$A_h \mathbf{u}_h = \mathbf{f}_h, \tag{7.3}$$

with A_h and $\mathbf{f}_h$ being the corresponding global stiffness matrix and global right-hand side, and h being the discretization (meshsize) parameter of the underlying partition $\mathcal{T}_h$ of Ω.

The aim of this chapter is to present multilevel preconditioners of optimal complexity for solving the system (7.3) for the case of nonconforming Rannacher–Turek finite elements.

Nonconforming finite elements based on *rotated* multilinear shape functions were introduced by Rannacher and Turek [99] as a class of simple elements for the Stokes problem. More generally, recent activities in the development of efficient solution methods for nonconforming finite element systems are inspired by their attractive properties as a stable discretization tool for ill-conditioned problems. In this chapter we mostly pay attention to the case of robustness with respect to strong coefficient jumps. The presented AMLI methods of optimal computational complexity have their own value. However, an important additional value of the robust solvers for scalar elliptic problems is due to their applications in inner iteration procedures of composite algorithms for complicated coupled models. In the next chapter we will return to the topic of Rannacher–Turek finite elements considering some recent advances in the AMLI methods for discontinuous Galerkin FE systems.

7.1.2 Rotated bilinear elements

The unit square $[-1,1]^2$ is used as a reference element $\hat{e}$ to define the isoparametric rotated bilinear element $e \in \mathcal{T}_h$. Further, let $\Psi_e : \hat{e} \to e$ be the corresponding bilinear one-to-one transformation, and let the nodal basis functions be determined by the relations $\{\phi_i\}_{i=1}^4 = \{\hat{\phi}_i \circ \Psi_e^{-1}\}_{i=1}^4$, $\{\hat{\phi}_i\} \in \operatorname{span}\{1, x, y, x^2 - y^2\}$, where $\circ$ means the superposition of functions $\hat{\phi}_i$ and ψ_e^{-1}.

For the variant MP (mid point), $\{\hat{\phi}_i\}_{i=1}^4$ are found by the point-wise interpolation condition

$$\hat{\phi}_i(b_\Gamma^j) = \delta_{ij},$$

where b_Γ^j, $j = 1, 4$ are the midpoints of the edges of the quadrilateral $\hat{e}$. Then,

$$\hat{\phi}_1(x,y) = (1 - 2x + (x^2 - y^2))/4, \qquad \hat{\phi}_2(x,y) = (1 + 2x + (x^2 - y^2))/4,$$
$$\hat{\phi}_3(x,y) = (1 - 2y - (x^2 - y^2))/4, \qquad \hat{\phi}_4(x,y) = (1 + 2y - (x^2 - y^2))/4.$$

The variant MV (mean value) corresponds to integral mean-value interpolation condition

$$|\Gamma_{\hat{e}}^j|^{-1} \int_{\Gamma_{\hat{e}}^j} \hat{\phi}_i \, d\Gamma_{\hat{e}}^j = \delta_{ij},$$

where $\Gamma_{\hat{e}}^{j}$ are the sides of $\hat{e}$. This leads to

$$\hat{\phi}_1(x,y) = (2-4x+3(x^2-y^2))/8, \quad \hat{\phi}_2(x,y) = (2+4x+3(x^2-y^2))/8,$$
$$\hat{\phi}_3(x,y) = (2-4y-3(x^2-y^2))/8, \quad \hat{\phi}_4(x,y) = (2+4y-3(x^2-y^2))/8.$$

Let us note, that some advantages of MV variant are observed when ill-conditioned problems (including the case of strong mesh anisotropy) are discretized. As we will see later, MV is also the natural variant in the discontinuous Galerkin setting.

Consider the model anisotropic problem with diagonal coefficient matrix

$$\mathbf{a}(\boldsymbol{x}) = a(e)\begin{bmatrix} \varepsilon & 0 \\ 0 & 1 \end{bmatrix}. \tag{7.4}$$

Then, in the case of a square mesh, the element stiffness matrices, corresponding to the variants MP and MV are given by

$$A_{MP}^{(e)} = \frac{a(e)}{3}\begin{bmatrix} 1+4\varepsilon & -(2\varepsilon-1) & -(1+\varepsilon) & -(1+\varepsilon) \\ -(2\varepsilon-1) & 1+4\varepsilon & -(1+\varepsilon) & -(1+\varepsilon) \\ -(1+\varepsilon) & -(1+\varepsilon) & 4+\varepsilon & -(2-\varepsilon) \\ -(1+\varepsilon) & -(1+\varepsilon) & -(2-\varepsilon) & 4+\varepsilon \end{bmatrix},$$

and

$$A_{MV}^{(e)} = \frac{a(e)}{4}\begin{bmatrix} 3+7\varepsilon & 3-\varepsilon & -3(1+\varepsilon) & -3(1+\varepsilon) \\ 3-\varepsilon & 3+7\varepsilon & -3(1+\varepsilon) & -3(1+\varepsilon) \\ -3(1+\varepsilon) & -3(1+\varepsilon) & 7+3\varepsilon & -(1-3\varepsilon) \\ -3(1+\varepsilon) & -3(1+\varepsilon) & -(1-3\varepsilon) & 7+3\varepsilon \end{bmatrix},$$

respectively, where the node numbering is as indicated by Figure 7.3(b).

7.1.3 Rotated trilinear elements

In the 3D case, the cube $[-1,1]^3$ is used as a reference element $\hat{e}$ to define the isoparametric rotated trilinear element $e \in \mathcal{T}_h$. Now let $\psi_e : \hat{e} \to e$ be the trilinear bijective mapping between the reference element $\hat{e}$ and e. The polynomial space of shape functions $\hat{\phi}_i$ on the reference element $\hat{e}$ is defined by

$$\hat{\mathcal{P}} := \{\hat{\phi}_i : 1 \le i \le 6\} = \text{span}\{1, x, y, z, x^2-y^2, y^2-z^2\},$$

and the shape functions ϕ_i on e are computed from $\hat{\phi}_i$ via the relations

$$\{\phi_i\}_{i=1}^6 = \{\hat{\phi}_i \circ \psi_e^{-1}\}_{i=1}^6.$$

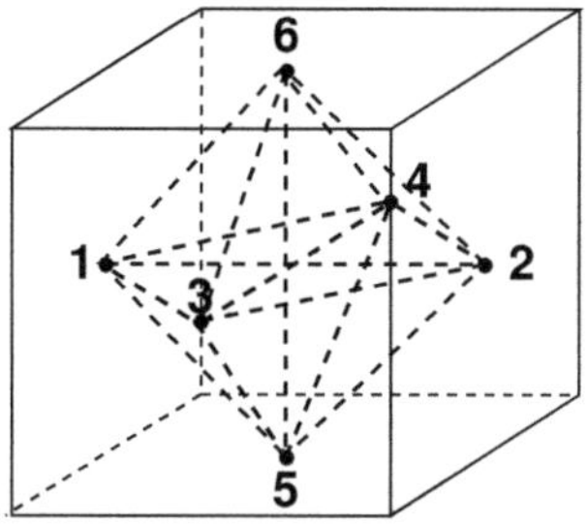

Figure 7.1: Node numbering and connectivity pattern of the reference element $\hat{e}$

For the variant MP (mid point), $\{\hat{\phi}_i\}_{i=1}^6$ are found by the interpolation condition

$$\hat{\phi}_i(b_\Gamma^j) = \delta_{ij},$$

where b_Γ^j, $j = 1, 6$ are the centers of the faces of the cube $\hat{e}$. Then,

$$\begin{aligned}
\hat{\phi}_1(x, y, z) &= (1 - 3x + 2x^2 - y^2 - z^2)/6,\\
\hat{\phi}_2(x, y, z) &= (1 + 3x + 2x^2 - y^2 - z^2)/6,\\
\hat{\phi}_3(x, y, z) &= (1 - x^2 - 3y + 2y^2 - z^2)/6,\\
\hat{\phi}_4(x, y, z) &= (1 - x^2 + 3y + 2y^2 - z^2)/6,\\
\hat{\phi}_5(x, y, z) &= (1 - x^2 - y^2 - 3z + 2z^2)/6,\\
\hat{\phi}_6(x, y, z) &= (1 - x^2 - y^2 + 3z + 2z^2)/6.
\end{aligned}$$

Alternatively, the variant MV (mean value) corresponds to the 3D integral mean-value interpolation condition

$$|\Gamma_{\hat{e}}^j|^{-1} \int_{\Gamma_{\hat{e}}^j} \hat{\phi}_i \, d\Gamma_{\hat{e}}^j = \delta_{ij},$$

where $\Gamma_{\hat{e}}^j$ are the faces of the reference element $\hat{e}$. This leads to

$$\begin{aligned}
\hat{\phi}_1(x, y, z) &= (2 - 6x + 6x^2 - 3y^2 - 3z^2)/12,\\
\hat{\phi}_2(x, y, z) &= (2 + 6x + 6x^2 - 3y^2 - 3z^2)/12,\\
\hat{\phi}_3(x, y, z) &= (2 - 3x^2 - 6y + 6y^2 - 3z^2)/12,\\
\hat{\phi}_4(x, y, z) &= (2 - 3x^2 + 6y + 6y^2 - 3z^2)/12,\\
\hat{\phi}_5(x, y, z) &= (2 - 3x^2 - 3y^2 - 6z + 6z^2)/12,\\
\hat{\phi}_6(x, y, z) &= (2 - 3x^2 - 3y^2 + 6z + 6z^2)/12.
\end{aligned}$$

Let us consider the model isotropic problem with diagonal coefficient matrix

$$\mathbf{a}(\boldsymbol{x}) = a(e)\begin{bmatrix} 1 & 0 & 0 \\ 0 & 1 & 0 \\ 0 & 0 & 1 \end{bmatrix}. \tag{7.5}$$

In what follows we will assume that all elements in the triangulation are cubes with mesh size h. Then the element stiffness matrices, corresponding to the variants MP and MV are given by

$$A_e^{MP} = a(e)\frac{2h}{9}\begin{bmatrix} 17 & -1 & -4 & -4 & -4 & -4 \\ -1 & 17 & -4 & -4 & -4 & -4 \\ -4 & -4 & 17 & -1 & -4 & -4 \\ -4 & -4 & -1 & 17 & -4 & -4 \\ -4 & -4 & -4 & -4 & 17 & -1 \\ -4 & -4 & -4 & -4 & -1 & 17 \end{bmatrix},$$

$$A_e^{MV} = a(e)2h\begin{bmatrix} 3 & 1 & -1 & -1 & -1 & -1 \\ 1 & 3 & -1 & -1 & -1 & -1 \\ -1 & -1 & 3 & 1 & -1 & -1 \\ -1 & -1 & 1 & 3 & -1 & -1 \\ -1 & -1 & -1 & -1 & 3 & 1 \\ -1 & -1 & -1 & -1 & 1 & 3 \end{bmatrix}.$$

7.2 Hierarchical two-level splittings: 2D case

In this section we provide a summary of the main results from reference [62]. Let us consider two consecutive discretizations $\mathcal{T}_H$ and $\mathcal{T}_h$. Figure 7.2 illustrates a macroelement obtained after one regular mesh-refinement step. We see that in this case the vector spaces $\mathcal{V}_H$ and $\mathcal{V}_h$ are not nested.

7.2.1 First reduce two-level splitting

To define the "first reduce" (FR) two-level splitting we apply the idea which was used in Chapter 4 in the case of Crouzeix–Raviart nonconforming elements. Following the introduced notations, let $\varphi_E = \{\phi_i(x,y)\}_{i=1}^{12}$ be the macroelement vector of the nodal basis functions and A_E be the macroelement stiffness matrix corresponding to $E \in \mathcal{T}_h$. The global stiffness matrix A_h is written as

$$A_h = \sum_{E \in \mathcal{T}_h} R_E^T A_E R_E.$$

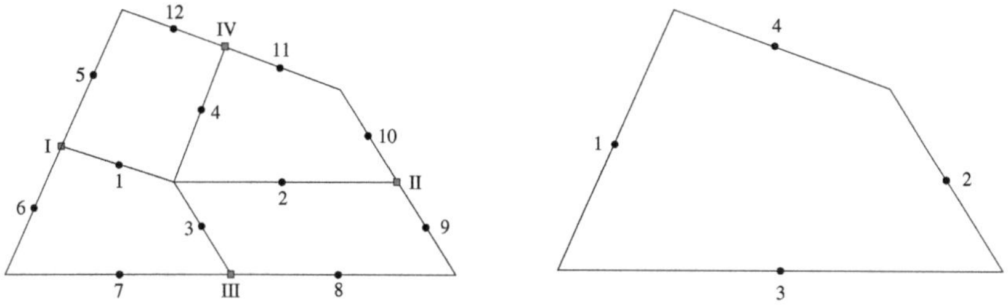

(a) One macroelement (b) One element

Figure 7.2: Uniform refinement on a general mesh

Next, we introduce the following macroelement level transformation matrix

$$J_E^T = \frac{1}{2}\begin{bmatrix} 2 &&&&&&&&&&& \\ & 2 &&&&&&&&&& \\ && 2 &&&&&&&&& \\ &&& 2 &&&&&&&& \\ &&&& 1 & -1 &&&&&& \\ &&&&&&&& 1 & -1 && \\ &&&&&& 1 & -1 &&&& \\ &&&&&&&&&& 1 & -1 \\ &&&& 1 & 1 &&&&&& \\ &&&&&&&& 1 & 1 && \\ &&&&&& 1 & 1 &&&& \\ &&&&&&&&&& 1 & 1 \end{bmatrix} \tag{7.6}$$

which defines locally a two-level hierarchical basis $\tilde{\varphi}_E$, namely, $\tilde{\varphi}_E = J_E^T \varphi_E$. The hierarchical two-level macroelement stiffness matrix is then obtained as

$$\tilde{A}_E = J_E^T A_E J_E,$$

and the related global stiffness matrix reads as

$$\tilde{A}_h = \sum_{E \in \mathcal{T}_h} \tilde{A}_E.$$

We split now the two-level stiffness matrix $\tilde{A}_h$ into 2×2 block form

$$\tilde{A}_h = \begin{bmatrix} \tilde{A}_{11} & \tilde{A}_{12} \\ \tilde{A}_{21} & \tilde{A}_{22} \end{bmatrix}, \tag{7.7}$$

where $\tilde{A}_{11}$ corresponds to the interior nodal unknowns with respect to the macro-elements $E \in \mathcal{T}_h$. As in the case of Crouzeix–Raviart finite elements, the first step of the FR algorithm is to eliminate these unknowns. For this purpose we factor $\tilde{A}_h$, i.e.,

$$\tilde{A}_h = \begin{bmatrix} \tilde{A}_{11} & 0 \\ \tilde{A}_{21} & B \end{bmatrix} \begin{bmatrix} I_1 & \tilde{A}_{11}^{-1}\tilde{A}_{12} \\ 0 & I_2 \end{bmatrix}, \tag{7.8}$$

where $B = \tilde{A}_{22} - \tilde{A}_{21}\tilde{A}_{11}^{-1}\tilde{A}_{12}$ stands for the Schur complement of this elimination step.

Next we consider a two-level splitting of the matrix B in the block form

$$B = \begin{bmatrix} B_{11} & B_{12} \\ B_{21} & B_{22} \end{bmatrix}, \tag{7.9}$$

where the first block corresponds to the half-difference basis functions. We associate the matrix B_{22} with the coarse grid. It is important to note that

$$\ker(B_{E;22}) = \ker(A_e) = \mathrm{span}\{(1,1,1,1)^T\}$$

which allows us to apply a local analysis to estimate the CBS constant γ corresponding to the splitting defined by the block partition (7.9).

For our analysis we follow the earlier established procedure, namely:

Step 1: We observe that the first left block of $\tilde{A}_h$ is a block-diagonal matrix. In this case, the diagonal entries of $\tilde{A}_{11}$ are 4×4 blocks, related to the interior points $\{1, 2, 3, 4\}$, cf. Figure 7.2, which are not connected to nodes in other macroelements. Thus, the corresponding unknowns are eliminated exactly, i.e., this is done locally. Therefore, we first compute the local Schur complements arising from static condensation of the *interior degrees of freedom* and obtain the (8×8) matrix B_E. Next we split B_E as

$$B_E = \begin{bmatrix} B_{E:11} & B_{E:12} \\ B_{E:21} & B_{E:22} \end{bmatrix} \begin{matrix} \}\ \text{two-level half-difference basis functions} \\ \}\ \text{two-level half-sum basis functions} \end{matrix}$$

written again in two-by-two block form with blocks of order (4×4).

Step 2: We are now in a position to estimate the CBS constant corresponding to the 2×2 splitting of B. As we know from the general theory, it suffices to compute the minimal eigenvalue of the generalized eigenproblem (see (2.36))

$$S_E \mathbf{v}_E = \lambda_E^{(1)} B_{E:22}\mathbf{v}_E, \quad \mathbf{v}_E \neq \mathbf{c},$$

where $S_E = B_{E:22} - B_{E:21}B_{E:11}^{-1}B_{E:12}$, and then

$$\gamma^2 \leq \max_{E\in\mathcal{T}_h} \gamma_E^2 = \max_{E\in\mathcal{T}_h} (1 - \lambda_E^{(1)}). \tag{7.10}$$

7.2.2 Two-level splitting by differences and aggregates

The "differences and aggregates" (DA) splitting is also first defined on a macroelement level. If $\phi_1, \ldots, \phi_{12}$ are the standard nodal basis functions for the macroelement, then we let

$$\begin{aligned}
\mathcal{V}(E) &= \operatorname{span}\{\phi_1, \ldots, \phi_{12}\} = \mathcal{V}_1(E) \oplus \mathcal{V}_2(E), \\
\mathcal{V}_1(E) &= \operatorname{span}\{\phi_1, \phi_2, \phi_3, \phi_4, \phi_5 - \phi_6, \phi_9 - \phi_{10}, \phi_7 - \phi_8, \phi_{11} - \phi_{12}\} \\
\mathcal{V}_2(E) &= \operatorname{span}\Big\{\phi_5 + \phi_6 + \sum_{j=1}^{4} \alpha_{1j}\phi_j, \; \phi_9 + \phi_{10} + \sum_{j=1}^{4} \alpha_{2j}\phi_j, \\
&\qquad \phi_7 + \phi_8 + \sum_{j=1}^{4} \alpha_{3j}\phi_j, \; \phi_{11} + \phi_{12} + \sum_{j=1}^{4} \alpha_{4j}\phi_j\Big\}.
\end{aligned}$$

Using the related transformation matrix J_E,

$$J_E^T = \frac{1}{2}\begin{bmatrix}
2 & & & & & & & & & & & \\
& 2 & & & & & & & & & & \\
& & 2 & & & & & & & & & \\
& & & 2 & & & & & & & & \\
& & & & 1 & -1 & & & & & & \\
& & & & & & & & 1 & -1 & & \\
& & & & & & 1 & -1 & & & & \\
& & & & & & & & & & 1 & -1 \\
\alpha_{11} & \alpha_{12} & \alpha_{13} & \alpha_{14} & 1 & 1 & & & & & & \\
\alpha_{21} & \alpha_{22} & \alpha_{23} & \alpha_{24} & & & & & 1 & 1 & & \\
\alpha_{31} & \alpha_{32} & \alpha_{33} & \alpha_{34} & & & 1 & 1 & & & & \\
\alpha_{41} & \alpha_{42} & \alpha_{43} & \alpha_{44} & & & & & & & 1 & 1
\end{bmatrix}, \tag{7.11}$$

the vector of the macroelement basis functions $\varphi_E = \{\phi_i\}_{i=1}^{12}$ is transformed to the hierarchical basis $\tilde{\varphi}_E = \{\tilde{\phi}_i\}_{i=1}^{12} = J_E^T \varphi_E$. Accordingly, the macroelement stiffness matrix is transformed into a hierarchical form

$$\tilde{A}_E = J_E^T A_E J_E = \begin{bmatrix} \tilde{A}_{E:11} & \tilde{A}_{E:12} \\ \tilde{A}_{E:21} & \tilde{A}_{E:22} \end{bmatrix} \begin{matrix} \tilde{\phi}_i \in \mathcal{V}_1(E) \\ \tilde{\phi}_i \in \mathcal{V}_2(E) \end{matrix}. \tag{7.12}$$

Following the local definitions, for the whole finite element space $\mathcal{V}_h$ with the standard nodal finite element basis $\varphi = \{\phi_h^{(i)}\}_{i=1}^{\mathcal{N}_h}$, we similarly construct the new hierarchical basis $\tilde{\varphi} = \{\tilde{\varphi}_h^{(i)}\}_{i=1}^{\mathcal{N}_h}$ and the corresponding splitting

$$\mathcal{V}_h = \mathcal{V}_1 \oplus \mathcal{V}_2. \tag{7.13}$$

The transformation J such that $\tilde{\varphi} = J^T \varphi$, can be used for transformation of the stiffness matrix A_h to hierarchical form $\tilde{A}_h = J^T A_h J$, which allows preconditioning by the two-level preconditioners based on the splitting (7.13).

Now, we are interested to analyze (i.e., to derive a uniform estimate of) the constant $\gamma = \cos(\mathcal{V}_1, \mathcal{V}_2)$ for the splitting (7.13). Again, as in the previous similar cases, we would like to perform this analysis locally, by considering the corresponding problems on macroelements. For this purpose we need to have satisfied the condition

(i) $\ker(\tilde{A}_{E:22}) = \ker(A_e)$,

which is equivalent to

$$\sum_{i=1}^{4} \alpha_{ij} = 1, \qquad \forall j \in \{1,2,3,4\}. \tag{7.14}$$

There are obviously various DA splittings satisfying the condition (i). In particular, the variant

$$[\alpha_{ij}] = \begin{bmatrix} 0 & 1 & 0 & 0 \\ 1 & 0 & 0 & 0 \\ 0 & 0 & 0 & 1 \\ 0 & 0 & 1 & 0 \end{bmatrix}$$

could be considered as a direct interpretation of the DA algorithm for Crouzeix–Raviart linear nonconforming finite elements in the present context. For further details on aggregation-based preconditioners see the review paper [29].

When the two-level algorithm is recursively generalized to the multilevel case, it is useful if

(ii) $\tilde{A}_{E:22}$ is proportional to A_e.

As was shown in Chapter 4, this property holds in a very general setting for the DA splitting of the Crouzeix–Raviart finite element space, see [31]. Unfortunately, it is rather complicated to find a parameter matrix $[\alpha_{ij}]$, which satisfies the condition (ii) in the general case of Rannacher–Turek bilinear finite elements.

7.2.3 Uniform estimates of the CBS constant for the 2D splittings

We consider here the isotropic model problem where all elements $e \in \mathcal{T}_H$ are squares and the uniform refinement is as shown in Figure 7.3. Both splitting algorithms, FR and DA, for both discretization variants, MP and MV, of rotated bilinear finite elements, are considered.

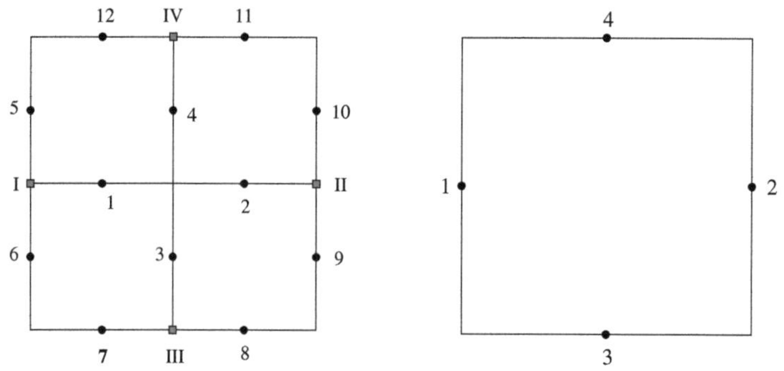

(a) One macroelement (b) One element

Figure 7.3: Uniform refinement on a square mesh

FR algorithm

Following (7.10) we compute the local CBS constants and derive the following global estimates which are uniform with respect to the size of the discrete problem and any possible jumps of the coefficients $a(e)$, $e \in \mathcal{T}_H$.

Variant MP: For the local CBS constant of the FR splitting we have

$$\lambda_E^{(1)} = \frac{5}{7}, \qquad \gamma_E^2 = 1 - \lambda_E^{(1)} = \frac{2}{7},$$

and therefore

$$\gamma_{MP}^2 \leq \frac{2}{7}. \tag{7.15}$$

Variant MV: For the FR splitting we further have

$$\lambda_E^{(1)} = \frac{5}{8}, \qquad \gamma_E^2 = 1 - \lambda_E^{(1)} = \frac{3}{8},$$

and therefore

$$\gamma_{MV}^2 \leq \frac{3}{8}. \tag{7.16}$$

Let us remind once again, that the obtained estimates hold theoretically for the two-level algorithm only. This is because the matrix B_{22} is only associated with the coarse discretization and is not proportional to the related element stiffness matrix A_e. However, as we will show in the next section, the CBS constants have a very stable behavior in the FR multilevel setting as well. The latter has been verified numerically, cf. Table 7.3 and Figure 7.4.

DA algorithm

Due to the symmetry (isotropy) of the model problem, the lower-left block of the transformation matrix J_E^T can be simplified to the form

$$\begin{bmatrix} b & c & a & a \\ c & b & a & a \\ a & a & b & c \\ a & a & c & b \end{bmatrix}. \tag{7.17}$$

The condition (i) (see (7.14)) is equivalent to

$$2a + b + c = 1.$$

Let us write the condition (ii) in the form

$$\tilde{A}_{E:22} = pA_e. \tag{7.18}$$

Then (ii) is reduced to a system of two nonlinear equations for (b, c) with a parameter p. It appears that the system for (b, c) has solutions if $p \in [p_0, \infty)$. In such a case, we can optimize the parameter p, so that the CBS constant is minimal. The obtained results are summarized below.

For the related analysis here, as well as in the 3D case (which will be presented in the next section) symbolic computations with the computer algebra program MATHEMATICA have been used [62].

<u>**Variant MP**</u>:

Lemma 7.1. *There exists a DA two-level splitting satisfying the condition (ii), if and only if,*

$$p \geq \frac{3}{7}.$$

Then, the obtained solutions for (b, c) are invariant with respect to the local CBS constant

$$\gamma_E^2 = 1 - \frac{1}{4p},$$

and for the related optimal splitting

$$\gamma_{MP}^2 \leq \frac{5}{12}. \tag{7.19}$$

Although the statements of Lemma 7.1 look very simple, the midterm derivations are rather technical, which is just illustrated by the following expressions of one of the similarly looking solutions for (b, c):

$$b = -\frac{24786 - 76160p + 2658\sqrt{\phi(p)} - 7280p\sqrt{\phi(p)} + \sqrt{\phi(p)^3}}{70(-729 + 2240p)}$$

$$c = \frac{6 - \sqrt{\phi(p)}}{70}$$

where

$$\phi(p) = -1329 + 3640p - 140\sqrt{63 - 327p + 420p^2}.$$

<u>**Variant MV**</u>: The same approach has been applied to get the estimates below.

Lemma 7.2. *There exists a DA two-level splitting satisfying the condition (ii), if and only if,*

$$p \geq \frac{2}{5}.$$

Then, the obtained solutions for (b, c) are invariant with respect to the local CBS constant

$$\gamma_E^2 = 1 - \frac{1}{4p},$$

and for the related optimal splitting

$$\gamma_{MV}^2 \leq \frac{3}{8}. \tag{7.20}$$

7.3 Hierarchical two-level splittings: 3D case

Similarly to the constructions and the analysis in the 2D case, we consider two consecutive discretizations $\mathcal{T}_H$ and $\mathcal{T}_h$. The finite element spaces $\mathcal{V}_H$ and $\mathcal{V}_h$ are not nested again, which is illustrated by Figure 7.4. Most of the results discussed in this section are originally presented in [61].

7.3.1 First reduce two-level splitting

Generalizing the 2D approach established in the previous section we denote by $\varphi_E = \{\phi_i(x, y)\}_{i=1}^{36}$ the macroelement vector of the nodal basis functions and

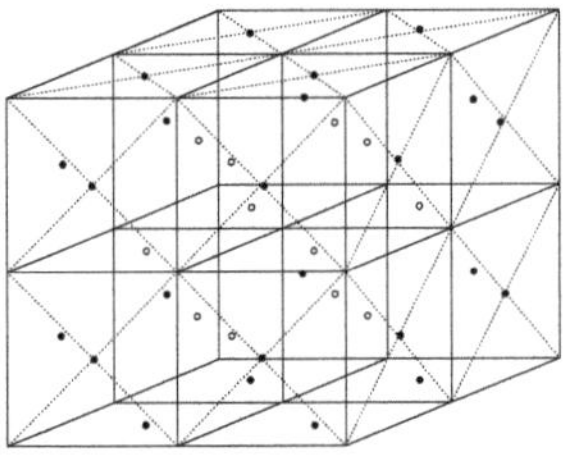

Figure 7.4: One macroelement

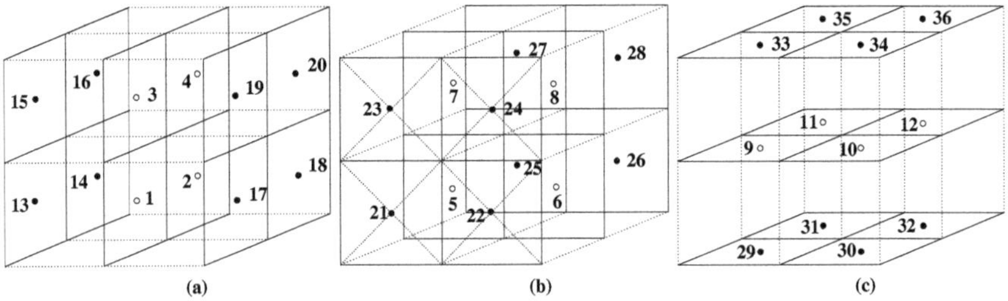

Figure 7.5: Node numbering in macroelement

by A_E the macroelement stiffness matrix corresponding to $E \in \mathcal{T}_h$. The global stiffness matrix A_h is as usual written in the form

$$A_h = \sum_{E \in \mathcal{T}_h} R_E^T A_E R_E .$$

Next, we introduce the 3D variant of (7.6). The related new macroelement level transformation matrix J_E^T is written in the form

$$J_E^T = \frac{1}{4} \begin{bmatrix} 4I & \\ & J_{E:22}^T \end{bmatrix}, \tag{7.21}$$

where I is the 12×12 identity matrix and

$$J_{E:22}^T = \begin{bmatrix} P & & & & & \\ & P & & & & \\ & & P & & & \\ & & & P & & \\ & & & & P & \\ & & & & & P \\ E_1 & E_2 & E_3 & E_4 & E_5 & E_6 \end{bmatrix}. \tag{7.22}$$

Each block E_i is a 6×4 zero matrix except for its i-th row which is composed of all ones, and

$$P = \begin{bmatrix} -1 & 1 & -1 & 1 \\ -1 & -1 & 1 & 1 \\ 1 & -1 & -1 & 1 \end{bmatrix}.$$

The matrix J_E defines locally the two-level hierarchical basis $\tilde{\varphi}_E$. Then,

$$\tilde{\varphi}_E = J_E^T \varphi_E$$

and the hierarchical two-level macroelement stiffness matrix reads as

$$\tilde{A}_E = J_E^T A_E J_E.$$

The related global two-level stiffness matrix is assembled by the macroelement once, i.e.,

$$\tilde{A}_h = \sum_{E \in \mathcal{T}_h} R_E^T \tilde{A}_E R_E.$$

We split again the two-level matrix $\tilde{A}_h$ into 2×2 block form

$$\tilde{A}_h = \begin{bmatrix} \tilde{A}_{11} & \tilde{A}_{12} \\ \tilde{A}_{21} & \tilde{A}_{22} \end{bmatrix}, \tag{7.23}$$

where $\tilde{A}_{11}$ corresponds to the interior nodal unknowns with respect to the macroelements $E \in \mathcal{T}_h$. As we know, the first step of the FR algorithm is to eliminate the first block of the unknowns. For this purpose we factor $\tilde{A}_h$, i.e.,

$$\tilde{A}_h = \begin{bmatrix} \tilde{A}_{11} & 0 \\ \tilde{A}_{21} & B \end{bmatrix} \begin{bmatrix} I_1 & \tilde{A}_{11}^{-1}\tilde{A}_{12} \\ 0 & I_2 \end{bmatrix}, \tag{7.24}$$

where $B = \tilde{A}_{22} - \tilde{A}_{21}\tilde{A}_{11}^{-1}\tilde{A}_{12}$ is the Schur complement of this elimination step.

Next we apply the two-level splitting (7.9) of the matrix B. In the 3D case this leads to the block presentation

$$B = \begin{bmatrix} B_{11} & B_{12} \\ B_{21} & B_{22} \end{bmatrix}, \tag{7.25}$$

where the first block corresponds to the differences of three different couples of basis functions from each macroelement face. The matrix B_{22} corresponds to the sum of basis functions from each macroelement face and thus is associated with the coarse grid. Let us note again that

$$\ker(B_{E;22}) = \ker(A_e) = \operatorname{span}\{(1,1,1,1,1,1)^T\}$$

which allows us to apply the standard local analysis to estimate the constant γ corresponding to the splitting defined by the block partition (7.25).

The analysis steps are completely the same as in the 2D case, namely:

Step 1: The upper-left block $\tilde{A}_{11}$ is a block-diagonal matrix. Its diagonal entries are 12×12 blocks, related to the interior points $\{1, 2, \ldots, 12\}$, cf. Figure 7.4, which are not connected to nodes in other macroelements. Thus, the corresponding unknowns are locally eliminated, and we compute the macroelement (24×24) Schur complements B_E, i.e.,

$$B_E = \begin{bmatrix} B_{E:11} & B_{E:12} \\ B_{E:21} & B_{E:22} \end{bmatrix} \begin{array}{l} \}\ \text{two-level "difference" basis functions} \\ \}\ \text{two-level "aggregated" basis functions} \end{array}$$

written again in two-by-two block form.

Step 2: Now we have to estimate the CBS constant corresponding to the 2×2 splitting of B. As we know (see, e.g., the similar 2D case), it suffices to compute the minimal eigenvalue of the 6×6 generalized eigenproblem

$$S_E \mathbf{v}_E = \lambda_E^{(1)} B_{E:22} \mathbf{v}_E, \quad \mathbf{v}_E \neq \mathbf{c} := (c, c, \ldots, c)^T, \quad c \in \mathbb{R},$$

where $S_E = B_{E:22} - B_{E:21} B_{E:11}^{-1} B_{E:12}$, and then

$$\gamma^2 \leq \max_{E \in \mathcal{T}_h} \gamma_E^2 = \max_{E \in \mathcal{T}_h} (1 - \lambda_E^{(1)}). \tag{7.26}$$

7.3.2 Two-level splitting by differences and aggregates

The splitting is first described for one macroelement, further developing the constructions from the 2D case. If $\phi_1, \ldots, \phi_{36}$ are the standard nodal basis functions for the macroelement, then we define

$$\begin{aligned}
\mathcal{V}(E) &= \operatorname{span}\{\phi_1,\dots,\phi_{36}\} = \mathcal{V}_1(E) \oplus \mathcal{V}_2(E),\\
\mathcal{V}_1(E) &= \operatorname{span}\Big\{\phi_1,\dots,\phi_{12},\\
&\qquad \phi_{14}+\phi_{16}-(\phi_{13}+\phi_{15}),\ \phi_{15}+\phi_{16}-(\phi_{13}+\phi_{14}),\\
&\qquad \phi_{13}+\phi_{16}-(\phi_{14}+\phi_{15}),\ \dots,\phi_{34}+\phi_{36}-(\phi_{33}+\phi_{35}),\\
&\qquad \phi_{35}+\phi_{36}-(\phi_{33}+\phi_{34}),\ \phi_{33}+\phi_{36}-(\phi_{34}+\phi_{35})\Big\}\\
\mathcal{V}_2(E) &= \operatorname{span}\Big\{\phi_{13}+\phi_{14}+\phi_{15}+\phi_{16}+\sum_{j=1}^{12}\beta_{1j}\phi_j,\ \dots,\\
&\qquad \phi_{33}+\phi_{34}+\phi_{35}+\phi_{36}+\sum_{j=1}^{12}\beta_{6j}\phi_j\Big\}.
\end{aligned}$$

The related transformation matrix has the form

$$J_E^T = \frac{1}{4}\begin{bmatrix} 4I & \\ J_{E:12}^T & J_{E:22}^T \end{bmatrix}, \tag{7.27}$$

where I is 12×12 identity matrix, $J_{E:22}^T$ is the same as (7.22),

$$J_{E:12}^T = \begin{bmatrix} \mathbf{0} \\ \mathcal{B} \end{bmatrix}$$

and $\mathcal{B} = (\beta_{ij})_{6\times 12}$. The vector of the macroelement basis functions $\varphi_E = \{\phi_i\}_{i=1}^{36}$ is transformed to the DA hierarchical basis

$$\tilde{\varphi}_E = \{\tilde{\phi}_i\}_{i=1}^{36} = J_E^T \varphi_E,$$

and the macroelement stiffness matrix into the hierarchical form

$$\tilde{A}_E = J_E^T A_E J_E = \begin{bmatrix} \tilde{A}_{E:11} & \tilde{A}_{E:12} \\ \tilde{A}_{E:21} & \tilde{A}_{E:22} \end{bmatrix} \begin{matrix} \tilde{\phi}_i \in \mathcal{V}_1(E) \\ \tilde{\phi}_i \in \mathcal{V}_2(E) \end{matrix}. \tag{7.28}$$

For the whole finite element space $\mathcal{V}_h$ with the standard nodal finite element basis $\varphi = \{\phi_h^{(i)}\}_{i=1}^{N_h}$ we similarly construct the new hierarchical basis (aggregating the local ones) $\tilde{\varphi} = \{\tilde{\varphi}_h^{(i)}\}_{i=1}^{N_h}$ and the corresponding splitting

$$\mathcal{V}_h = \mathcal{V}_1 \oplus \mathcal{V}_2. \tag{7.29}$$

The transformation J such that $\tilde{\varphi} = J^T\varphi$ provides also a transformation of the stiffness matrix A_h into the hierarchical form, i.e., $\tilde{A}_h = J^T A_h J$, which allows preconditioning by the two-level preconditioners based on the DA splitting (7.29).

Again, as in the previous sections, we would like to perform the analysis of the CBS constant $\gamma = \cos(\mathcal{V}_1, \mathcal{V}_2)$ locally, i.e., by considering the corresponding problems on macroelements. As we know, for this purpose we need to have satisfied the condition

(i) $\ker(\tilde{A}_{E:22}) = \ker(A_e)$,

which in this particular case is equivalent to

$$\sum_{i=1}^{6} \beta_{ij} = 1\,, \qquad \forall j \in \{1, 2, \ldots, 12\}. \tag{7.30}$$

Similarly to the situation in the 2D case, there are obviously various DA splittings satisfying the condition (i) in the 3D case.

As we know also from the previous section, it is convenient for the direct multilevel generalization of the two-level algorithm, if

(ii) $\tilde{A}_{E:22}$ is proportional to A_e.

7.3.3 Uniform estimates of the CBS constant for the 3D splittings

FR algorithm

We use (7.26) to compute the local CBS constant and thereafter to derive the following global estimates for the considered model problem. It is important to notice that the bounds are uniform with respect to the size of the discrete problem and any possible jumps of the coefficients which are aligned with the coarsest grid.

Variant MP: For the FR splitting in the 3D case we have

$$\lambda_E^{(1)} = \frac{13}{21}\,, \qquad \gamma_E^2 = 1 - \lambda_E^{(1)} = \frac{8}{21},$$

and therefore

$$\gamma_{MP}^2 \le \frac{8}{21}. \tag{7.31}$$

Variant MV: For the FR splitting in the 3D case we further have

$$\lambda_E^{(1)} = \frac{1}{2}\,, \qquad \gamma_E^2 = 1 - \lambda_E^{(1)} = \frac{1}{2},$$

and therefore

$$\gamma_{MV}^2 \le \frac{1}{2}. \tag{7.32}$$

We observe that the 3D bound (7.31) for Variant MP is slightly worse than the corresponding 2D result (7.15). For Variant MV the related FR estimates in 2D and 3D are equal.

As in the 2D case, the above FR estimates hold for the two-level algorithm only. The behavior of the CBS constants in the FR multilevel setting is studied numerically and the obtained results will be presented and discussed in the next section, cf. Table 7.4 and Figure 7.7.

DA algorithm

Now the isotropy of the model problem allows to simplify the non-zero part $\mathcal{B}$ of the lower-left block $J_{E:12}^T$ of the transformation matrix J_E^T, which in the considered 3D case has the form

$$\mathcal{B} = \begin{bmatrix} a & a & a & a & b & c & b & c & b & c & b & c \\ a & a & a & a & c & b & c & b & c & b & c & b \\ b & c & b & c & a & a & a & a & b & b & c & c \\ c & b & c & b & a & a & a & a & c & c & b & b \\ b & b & c & c & b & b & c & c & a & a & a & a \\ c & c & b & b & c & c & b & b & a & a & a & a \end{bmatrix}. \tag{7.33}$$

The condition (i) is equivalent to

$$a + b + c = 1.$$

We write again the condition (ii) in the form

$$\tilde{A}_{E:22} = pA_e, \tag{7.34}$$

and reduce (ii) to a system of three nonlinear equations for (a, b, c), with a parameter p. For the relatively less complicated 2D case, a similar approach was presented in the previous section, see for some more details in [62]. As is shown in [61], in the 3D case the system for (a, b, c) have again solutions if $p \in [p_0, \infty)$ for some $p_0 > 0$. The limit value of $p = p_0$ minimizes the CBS constant estimate. The obtained results are summarized in the next two lemmas.

<u>**Variant MP**</u>:

Lemma 7.3. *There exists a DA two-level splitting satisfying the condition (ii), if and only if,*

$$p \geq \frac{3}{14}.$$

Then, the obtained solutions for (a,b,c) *are invariant with respect to the local CBS constant*

$$\gamma_E^2 = 1 - \frac{1}{8p},$$

and for the related optimal splitting

$$\gamma_{MP}^2 \leq \frac{5}{12}. \tag{7.35}$$

The midterm derivations to get the above estimate are rather technical, which is illustrated by the presented expressions of the four different solutions for (a,b,c):

$$\left(\frac{94}{273} - \frac{2}{21}\,\xi(p) - \frac{3\,\eta(p)}{26\sqrt{2}},\ \frac{10-26\,\xi(p)}{273} + \frac{3\sqrt{2}\,\eta(p)}{52},\ \frac{5+8\,\xi(p)}{42}\right),$$

$$\left(\frac{94}{273} + \frac{2}{21}\,\xi(p) + \frac{3\,\eta(p)}{26\sqrt{2}},\ \frac{10+26\,\xi(p)}{273} - \frac{3\sqrt{2}\,\eta(p)}{52},\ \frac{5-8\,\xi(p)}{42}\right),$$

$$\left(\frac{94}{273} - \frac{2}{21}\,\xi(p) + \frac{3\,\eta(p)}{26\sqrt{2}},\ \frac{10-26\,\xi(p)}{273} - \frac{3\sqrt{2}\,\eta(p)}{52},\ \frac{5+8\,\xi(p)}{42}\right),$$

$$\left(\frac{94}{273} + \frac{2}{21}\,\xi(p) - \frac{3\,\eta(p)}{26\sqrt{2}},\ \frac{10+26\,\xi(p)}{273} + \frac{3\sqrt{2}\,\eta(p)}{52},\ \frac{5-8\,\xi(p)}{42}\right),$$

where $\xi(p) = \sqrt{-3+14\,p}$ and $\eta(p) = \sqrt{-21+104\,p}$.

Variant MV: The same approach is applied to get the estimates below.

Lemma 7.4. *There exists a DA two-level splitting satisfying the condition (ii), if and only if,*

$$p \geq \frac{1}{4}.$$

Then, the obtained solutions for (a,b,c) *are invariant with respect to the local CBS constant*

$$\gamma_E^2 = 1 - \frac{1}{8p},$$

and for the related optimal splitting

$$\gamma_{MV}^2 \leq \frac{1}{2}. \tag{7.36}$$

What we observe here is that the CBS constants in the DA approach have a reverse behavior if compared to the FR case. Here the 2D and 3D estimates for Variant MP are equal, while the 3D bound (7.36) for Variant MV is slightly larger than the bound (7.20) in 2D.

7.4 Multilevel preconditioning

Let us remind briefly some basic facts from the multilevel preconditioning theory. The AMLI methods are evolved from two-level methods. The straightforward recursive extension leads to the class of hierarchical basis (HB) methods for which the condition number grows in general exponentially with the number of levels ℓ. The AMLI preconditioners of both additive ($M_A = M_A^{(\ell)}$) or multiplicative ($M_F = M_F^{(\ell)}$) type are considered here. As we know (see Section 2.4), the purely algebraic stabilization technique is based on a properly constructed matrix polynomial P_{υ_k} of degree υ_k which is used at some (or all) of the levels $k = 1, \ldots, \ell$.

Let us denote by $C_{11}^{(k)}$ some properly scaled (see the assumption (2.75)) preconditioner for the upper left block of the hierarchical stiffness matrix

$$\tilde{A}^{(k)} = \begin{bmatrix} \tilde{A}_{11}^{(k)} & \tilde{A}_{12}^{(k)} \\ \tilde{A}_{21}^{(k)} & \tilde{A}_{22}^{(k)} \end{bmatrix}$$

at level k. Starting from the coarsest mesh (level 0) with $M_A^{(0)} = M_F^{(0)} = A^{(0)}$, the related two-level preconditioner is applied recursively at all levels $k = 1, 2, \ldots, \ell$ of mesh refinement to get the AMLI algorithms. They can be written in the following form, which in the multiplicative case is equivalent to the linear multilevel algorithm introduced in Section 2.4.

Additive AMLI:

$$M_A^{(k)} = \begin{bmatrix} C_{11}^{(k)} & 0 \\ 0 & C_{22}^{(k)} \end{bmatrix} \tag{7.37}$$

where the matrix $C_{22}^{(k)}$ is implicitly defined by the equation

$$C_{22}^{(k)^{-1}} = \left[I - P_{\upsilon_k}\left(M_A^{(k-1)^{-1}} \tilde{A}^{(k-1)} \right) \right] A^{(k-1)^{-1}}. \tag{7.38}$$

Multiplicative AMLI:

$$M_F^{(k)} = \begin{bmatrix} C_{11}^{(k)} & 0 \\ \tilde{A}_{21}^{(k)} & C_{22}^{(k)} \end{bmatrix} \begin{bmatrix} I & C_{11}^{(k)^{-1}} \tilde{A}_{12}^{(k)} \\ 0 & I \end{bmatrix} \tag{7.39}$$

where $C_{22}^{(k)}$ is again defined by (7.38).

Then, as known from the theory [16, 17], a properly scaled and shifted Chebyshev polynomial P_{υ_k} of degree υ_k, can be used in order to stabilize the condition number of the linear AMLI preconditioners. The main result from this analysis (see Section 2.6) is that the AMLI preconditioners have optimal computational complexity, if $\upsilon_k = \upsilon$,

$$\sqrt{\frac{1+\gamma}{1-\gamma}} < \upsilon < \varrho, \qquad \text{for additive AMLI,} \tag{7.40}$$

and

$$\frac{1}{\sqrt{1-\gamma^2}} < \upsilon < \varrho, \qquad \text{for multiplicative AMLI,} \tag{7.41}$$

where $\varrho \approx \frac{n_{k+1}}{n_k}$ is the reduction factor of the number of degrees of freedom.

Based on (7.40) and (7.41), the optimality conditions for the stabilization polynomial degree υ in the case of the DA splitting are summarized in the tables below, including the related CBS constant estimates.

Table 7.1: DA splitting: 2D cases

	MP	MV
γ^2	5/12	3/8
Additive AMLI	$\upsilon = 3$	$\upsilon = 3$
Multiplicative AMLI	$\upsilon \in \{2, 3\}$	$\upsilon \in \{2, 3\}$

Table 7.2: DA splitting: 3D cases

	MP	MV
γ^2	5/12	1/2
Additive AMLI	$\upsilon \in \{3, 4, \dots, 7\}$	$\upsilon \in \{3, 4, \dots, 7\}$
Multiplicative AMLI	$\upsilon \in \{2, 3, \dots, 7\}$	$\upsilon \in \{2, 3, \dots, 7\}$

Now, we turn back to the FR case. The multilevel behavior of the CBS constant is studied numerically. This means, that at each current coarsening step, the role of the element stiffness matrix is played by the related last obtained block $B_{E:22}$. The numerical results for the 2D and 3D cases are shown below in tabular and graphical form.

The computed (local) estimates for γ^2 for the FR algorithm are always smaller than the related ones for the DA algorithm. Strictly following (7.40) and (7.41), the optimality conditions for the polynomial degree υ in the case of the FR splitting are the same as for the DA splitting, see the Tables 7.1 and 7.2.

One can also observe a nice one-side convergence to the value of $\theta \approx 0.3170$ in the 2D case and $\theta \approx 0.39238$ in the 3D case for both variants, MP and MV, see Figure 7.4 and Figure 7.7. This in particular explains why in the presented numerical tests the additive AMLI in 2D is stabilized even with polynomial degree $\upsilon = 2$.

Table 7.3: Multilevel behavior of γ^2 for FR algorithm: 2D case

variant	ℓ	$\ell-1$	$\ell-2$	$\ell-3$	$\ell-4$	$\ell-5$
MP	0.2857	0.3101	0.3156	0.3167	0.3169	0.3170
MV	0.3750	0.3261	0.3187	0.3173	0.3171	0.3170

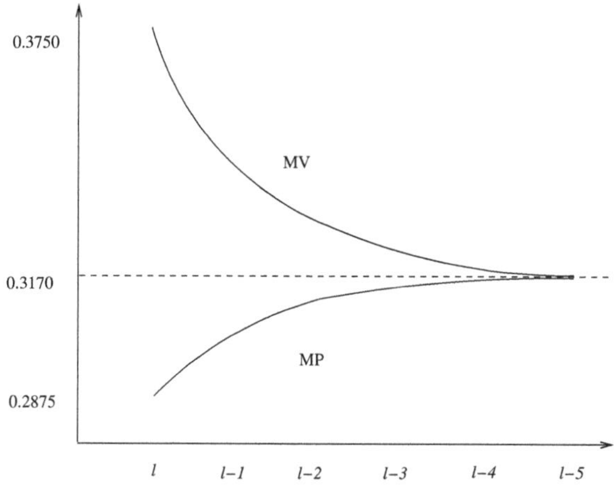

Figure 7.6: Multilevel behavior of γ^2 for FR algorithm: 2D case

Table 7.4: Multilevel behavior of γ^2 for FR algorithm: 3D case

variant	ℓ	$\ell-1$	$\ell-2$	$\ell-3$	$\ell-4$	$\ell-5$
MP	0.38095	0.39061	0.39211	0.39234	0.39237	0.39238
MV	0.5	0.4	0.39344	0.39253	0.39240	0.39238

The general conclusion of the considerations in this section is that the DA splitting provides better opportunities for a systematic theoretical analysis. However, the counterpart approach FR could have serious advantages from a practical point of view.

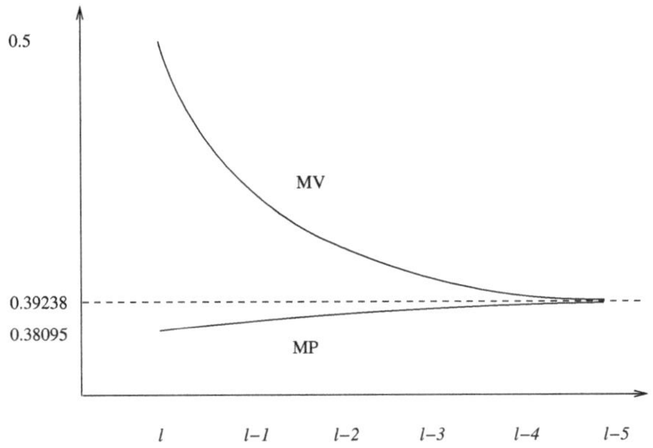

Figure 7.7: Multilevel behavior of γ^2 for FR algorithm: 3D case

7.5 Numerical tests

7.5.1 Additive and multiplicative AMLI preconditioners in 2D

Here we compare the convergence properties of the PCG method using either the additive or the multiplicative AMLI preconditioners based on either DA or FR splitting for MP and MV discretization. We solve the system of linear algebraic (finite element) equations (7.3). The square domain $\Omega = [0,1]^2$ is divided into four subdomains, namely, $\Omega = \Omega_1 \cup \ldots \cup \Omega_4$, where $\Omega_1 = [0, 1/2]^2$, $\Omega_2 = [1/2, 1] \times [0, 1/2]$, $\Omega_3 = [0, 1/2] \times [1/2, 1]$, and $\Omega_4 = [1/2, 1]^2$. The piecewise constant diffusion coefficient is given by $a(e) = 1$ on subdomains Ω_1 and Ω_4, $a(e) = \varepsilon$ on Ω_2, and $a(e) = \varepsilon^{-1}$ on Ω_3. The first Table 7.5 summarizes the number of PCG iterations that reduce the residual norm by factor of 10^6 when performing a single V-cycle AMLI. In the second Table 7.6 we list the corresponding results for the linear AMLI W-cycle, employing second-order stabilization polynomials (see also the related notice in the next Section 7.5.2). The results for the multiplicative variant are put in parentheses in each case. Though the number of iterations approximately doubles in most cases when switching from multiplicative to additive preconditioning the CPU-time (in most situations) increases by about 10 to 50 per cent. This is due to the lower computational complexity per application of the additive AMLI. This is illustrated in Figure 7.8, which depicts the logarithm ($\log_2$) of the CPU-time in milliseconds measured on a 2 GHz Linux PC for the case of the DA splitting and MV discretization. The (almost) linear profile of the diagrams gives a clear idea about the good scalability of the developed codes.

In full agreement with the theoretical analysis presented in this chapter, both preconditioners, the additive as well as the multiplicative AMLI, are perfectly robust with respect to jump discontinuities in the coefficient $a(e)$ as can be seen from the almost identical results shown in the respective rows for $\varepsilon = 1$ and $\varepsilon = 0.01$ of Tables 7.5–7.6.

Table 7.5: AMLI V-cycle: Number of PCG iterations; 2D tests

	$1/h$	32	64	128	256	512	1024
DA/MP	$\varepsilon = 1$	15 (8)	21 (10)	29 (13)	39 (16)	49 (19)	61 (22)
	$\varepsilon = 0.01$	15 (8)	22 (10)	29 (13)	39 (16)	50 (19)	63 (22)
FR/MP	$\varepsilon = 1$	11 (6)	15 (8)	19 (9)	24 (11)	28 (12)	34 (14)
	$\varepsilon = 0.01$	11 (6)	15 (8)	20 (9)	24 (11)	30 (12)	36 (14)
DA/MV	$\varepsilon = 1$	14 (8)	20 (10)	28 (13)	36 (16)	45 (18)	56 (21)
	$\varepsilon = 0.01$	14 (8)	20 (11)	28 (13)	37 (16)	47 (18)	59 (22)
FR/MV	$\varepsilon = 1$	12 (7)	16 (9)	21 (10)	26 (12)	31 (14)	37 (16)
	$\varepsilon = 0.01$	13 (7)	17 (9)	22 (10)	27 (12)	33 (14)	39 (16)

Table 7.6: Linear AMLI W-cycle: Number of PCG iterations; 2D tests

	$1/h$	32	64	128	256	512	1024
DA/MP	$\varepsilon = 1$	15 (8)	16 (8)	17 (8)	18 (8)	18 (8)	18 (8)
	$\varepsilon = 0.01$	15 (8)	17 (8)	17 (8)	18 (8)	18 (8)	19 (8)
FR/MP	$\varepsilon = 1$	11 (6)	12 (6)	12 (6)	13 (6)	13 (6)	13 (6)
	$\varepsilon = 0.01$	11 (6)	12 (6)	13 (6)	13 (6)	13 (6)	13 (6)
DA/MV	$\varepsilon = 1$	14 (8)	15 (9)	16 (9)	16 (9)	16 (9)	16 (9)
	$\varepsilon = 0.01$	14 (8)	16 (10)	16 (9)	16 (9)	17 (9)	17 (9)
FR/MV	$\varepsilon = 1$	12 (7)	14 (7)	14 (7)	14 (7)	14 (7)	14 (7)
	$\varepsilon = 0.01$	13 (7)	14 (7)	15 (7)	15 (7)	15 (7)	15 (7)

7.5.2 Problems with jumping coefficients in 3D

Two kinds of numerical tests are presented and discussed in this section (following [61]). The coefficient jumps are aligned with the coarsest mesh at the beginning – the case which was theoretically analyzed in this chapter. After that we consider the case of randomly distributed coefficients, that is, the situation in which the jumps (due to the oscillatory coefficient) can only be resolved on the finest mesh. This class of challenging problems is also referred to as *high-frequency and high-contrast problems*. Such a terminology is in particular well fit to the case of (micro) μ-FEM analysis of structures, based on voxel computer tomography images.

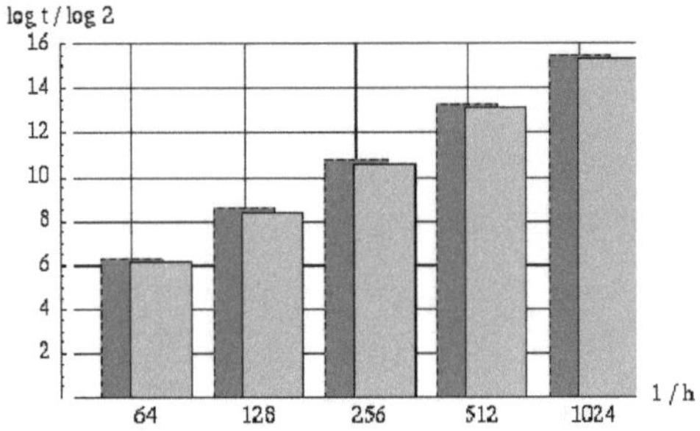

Figure 7.8: CPU-time for additive (dark) and multiplicative (light) preconditioning (logarithmic scale)

Jump in coefficients aligned with coarse mesh

The computational domain is $\Omega = (0,1)^3$. The mesh is uniform with mesh size varied in the range $h = 1/8$ to $h = 1/128$ resulting in 512 to 2 097 157 finite elements with 1 728 to 6 340 608 nodes, respectively. The matrix $\mathbf{a}(e)$ in (7.3) is defined by $\mathbf{a}(e) := \alpha(e) \cdot I$, where the following jumps in the coefficient $\alpha = \alpha(e)$ are considered:

$$\alpha(e) = \left\{ \begin{array}{ccc} 1 & \text{in} & \begin{array}{c} (I_1 \times I_1 \times I_1) \cup (I_2 \times I_2 \times I_1) \\ \cup (I_2 \times I_1 \times I_2) \cup (I_1 \times I_2 \times I_2) \end{array} \\ \varepsilon & & \text{elsewhere} \end{array} \right\},$$

where $I_1 = (0, 0.5]$ and $I_2 = (0.5, 1)$, and $\varepsilon = 10^{-3}$. The number of (outer) iterations shown in Tables 7.7–7.9 reduce the residual norm by a factor of 10^8.

Table 7.7: AMLI V-cycle: Number of PCG iterations; 3D tests

MP	h^{-1}	8	16	32	64	128
DA	$\varepsilon = 1$	9	12	16	20	24
	$\varepsilon = 10^{-3}$	9	12	16	20	25
FR	$\varepsilon = 1$	8	11	14	18	22
	$\varepsilon = 10^{-3}$	8	11	14	18	22
MV	h^{-1}	8	16	32	64	128
DA	$\varepsilon = 1$	12	17	22	29	38
	$\varepsilon = 10^{-3}$	12	17	22	30	39
FR	$\varepsilon = 1$	10	14	17	21	26
	$\varepsilon = 10^{-3}$	10	14	17	21	26

Table 7.7 presents results for the AMLI V-cycle preconditioner. Let us note, that the observed logarithmic growth of the iteration count (with respect to the problem size) is more well known for the case of 2D problems but is not covered by the theory for 3D problems. The results in Table 7.8 refer to the linear AMLI W-cycle preconditioner choosing a matrix stabilization polynomial $Q_1(t) = (1 - P_2(t))/t = q_0 + q_1 t$. As for some other numerical tests included in this book, the coefficients

$$q_0 = \frac{2}{\sqrt{1-\gamma^2}}, \quad q_1 = -\frac{1}{1-\gamma^2} \tag{7.42}$$

are used, cf. [17]. It is notable that this choice, although theoretically founded for the situation of exact inversion of the pivot block A_{11} only, still yields satisfying results in this case where an approximate inversion of A_{11}, i.e., an incomplete factorization based on a drop tolerance (ILU(tol)) is used. In the experiments presented here, the drop tolerance is set to tol $= 10^{-3}$.

In Table 7.9 we list the results for the (variable-step) nonlinear AMLI method stabilized by two inner generalized conjugate gradient iterations at every intermediate level (and using a direct solve on the coarsest mesh with mesh size $h^{-1} = 4$, as in the other tests).

Table 7.8: Linear AMLI W-cycle: Number of PCG iterations; 3D tests

MP	h^{-1}	8	16	32	64	128
DA	$\varepsilon = 1$	9	10	10	10	10
	$\varepsilon = 10^{-3}$	9	10	10	10	10
FR	$\varepsilon = 1$	8	9	9	9	9
	$\varepsilon = 10^{-3}$	8	9	9	9	9
MV	h^{-1}	8	16	32	64	128
DA	$\varepsilon = 1$	12	15	15	16	16
	$\varepsilon = 10^{-3}$	12	15	16	16	16
FR	$\varepsilon = 1$	10	12	12	12	12
	$\varepsilon = 10^{-3}$	10	12	12	12	12

As the theoretical estimates presented in this chapter predict, the AMLI preconditioners are perfectly robust with respect to jump discontinuities of the coefficients $\mathbf{a}(e)$ if they do not occur inside any element of the coarsest mesh partition. The results slightly favor the FR approach, and, they illustrate well the optimal convergence rate of the PCG solvers, when using a W-cycle AMLI as preconditioner, for all considered cases and variants.

We also observe, that the nonlinear AMLI slightly outperforms the linear one, which is better expressed in the MV case.

Table 7.9: Nonlinear AMLI W-cycle: Number of (outer) GCG iterations; 3D tests

MP	h^{-1}	8	16	32	64	128
DA	$\varepsilon = 1$	9	9	9	9	9
	$\varepsilon = 10^{-3}$	9	10	10	10	10
FR	$\varepsilon = 1$	8	9	9	9	9
	$\varepsilon = 10^{-3}$	8	9	9	9	9
MV	h^{-1}	8	16	32	64	128
DA	$\varepsilon = 1$	12	12	12	12	12
	$\varepsilon = 10^{-3}$	12	12	12	12	12
FR	$\varepsilon = 1$	10	11	11	11	11
	$\varepsilon = 10^{-3}$	10	11	11	11	11

Random distribution of jump in coefficients

The remaining experiments deal with examples where the coefficient functions are rough in the sense that their jumps are resolved on the finest mesh only. In these tests we use the FR basis transformation in combination with the nonlinear AMLI W-cycle method, i.e., two inner GCG iterations at all coarser levels. The number of outer iterations that we report reduce the residual by a factor of 10^6. The coefficient $\alpha(e)$ is constant elementwise only. For the considered case of "binary material", it is initialized randomly, taking either of the values 1 or ε, where 1 occurs with some fixed probability p. Finally, the last row of Table 7.10 shows the corresponding results for a problem where the coefficient on each element is a random number (uniformly distributed) in the interval $(0, 1)$, i.e., $\alpha(e) \in (0, 1)$.

By comparing the results shown in Tables 7.9 and 7.10, we observe that in general, this AMLI solver is not robust with respect to the considered jump discontinuities of high frequency. However, for a fixed value of ε its convergence rate is still of optimal order no matter how large the jumps are. Here, it is important to emphasize also the advantage of the nonlinear AMLI compared to polynomial stabilization of the condition number, which is more difficult to achieve for problems with coefficient jumps on the finest mesh.

The topic of robustness of the discretization and the related PCG/GCG solvers for problems with coefficient jumps of high frequency and high contrast is currently of a strongly growing interest. We will come back to such problems and to one challenging real-life application in Chapter 8 and Chapter 10.

Table 7.10: Nonlinear AMLI W-cycle: Number of (outer) GCG iterations for problem with random coefficients: 3D tests

$p = 1/2$					
FR-MV h^{-1}	8	16	32	64	128
$\varepsilon = 10^{-1}$	9	9	9	9	9
$\varepsilon = 10^{-2}$	21	22	22	21	21
$\varepsilon = 10^{-3}$	42	59	58	56	54
$p = 1/10$					
FR-MV h^{-1}	8	16	32	64	128
$\varepsilon = 10^{-1}$	9	9	9	9	9
$\varepsilon = 10^{-2}$	17	22	22	22	22
$\varepsilon = 10^{-3}$	29	60	55	50	50
Random coefficient:					
FR-MV h^{-1}	8	16	32	64	128
$\alpha(e) \in (0, 1)$	22	39	43	34	35

8 AMLI algorithms for discontinuous Galerkin FE problems

8.1 Introduction to discontinuous Galerkin FEM

Consider a second order elliptic boundary value problem on a polygonal domain $\Omega \subset \mathbb{R}^d$, $d = 2, 3$:

$$\begin{aligned} -\nabla \cdot (a(\boldsymbol{x})\nabla u) &= f(\boldsymbol{x}) \quad \text{in } \Omega \\ u(\boldsymbol{x}) &= 0 \quad \text{on } \Gamma_D \\ \partial_N u(\boldsymbol{x}) \equiv a\nabla u \cdot \boldsymbol{n} &= 0 \quad \text{on } \Gamma_N . \end{aligned} \tag{8.1}$$

For the formulation below we shall need the existence of the traces of u and $a\nabla u \cdot \boldsymbol{n}$ on certain interfaces in Ω. Thus, the solution u is assumed to have the required regularity. To simplify our exposition we assume that the set Γ_D is not empty and its $\mathbb{R}^{d-1}$-dimensional measure is nonzero.

Let $\mathcal{T}$ be a partitioning of Ω into a finite number of open subdomains (finite elements) K with boundaries ∂K. We assume that the partition is quasi uniform and regular. For each finite element we denote by h_K its size and further $h = \max_{K \in \mathcal{T}} h_K$. Let $e = \overline{K}_L \cap \overline{K}_R$ be the interface of two adjacent subdomains K_L and K_R (see Figure 8.1). The set of all such interfaces is denoted by $\mathcal{F}_0$. Note that these interfaces are inside Ω. Further, $\mathcal{F}_D$ and $\mathcal{F}_N$ will be the sets of faces/edges of finite elements on the boundary Γ_D and Γ_N, respectively. Finally, $\mathcal{F}$ will be the set of all faces/edges: $\mathcal{F} = \mathcal{F}_0 \cup \mathcal{F}_D \cup \mathcal{F}_N$. Here we allow finite elements of polygonal or polyhedral shape, with hanging nodes etc. The important assumption is that if e is a side or a face of a finite element $K \in \mathcal{T}$ then $|e| \approx h_K$ for $d = 2$ and $|e|^{\frac{1}{2}} \approx h_K$ for $d = 3$. In other words we do not allow very small edges or faces, i.e., strong mesh anisotropy is excluded.

On the partition $\mathcal{T}$ we define the finite element space

$$\mathcal{V} := \mathcal{V}(\mathcal{T}) := \{v \in L^2(\Omega) : v|_K \in P_r(K),\ K \in \mathcal{T}\},$$

where P_r is the set of polynomials of degree $r \geq 0$. For each $e = \overline{K}_L \cap \overline{K}_R \in \mathcal{F}_0$ we define the jump $[\![v]\!]$ of any function $v \in \mathcal{V}$ as the vector

$$[\![v]\!]_e := \begin{cases} v|_{K_L}\boldsymbol{n}_L + v|_{K_R}\boldsymbol{n}_R, & e = \overline{K}_L \cap \overline{K}_R, \text{ i.e., } e \in \mathcal{F}_0, \\ v|_K \boldsymbol{n}, & e = \overline{K} \cap \Gamma_D, \text{ i.e., } e \in \mathcal{F}_D. \end{cases}$$

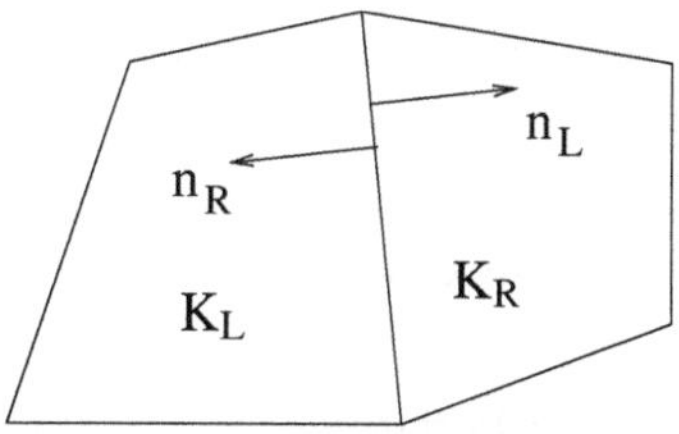

Figure 8.1: Two adjacent subdomains (elements) K_L and K_R

Here $\boldsymbol{n}_L$ and $\boldsymbol{n}_R$ are the external unit vectors to K_L and K_R, respectively.

We shall also need the following notation for the average value of a vector function $\boldsymbol{v} \in \mathcal{V}^d$ on $e = \overline{K}_L \cap \overline{K}_R$, that is,

$$\{\boldsymbol{v}\}|_e := \begin{cases} \frac{1}{2}(\boldsymbol{v}|_{K_L} + \boldsymbol{v}|_{K_R}) & \text{for } e = \overline{K}_L \cap \overline{K}_R, \text{ i.e., } e \in \mathcal{F}_0, \\ \boldsymbol{v}|_K, & \text{for } e = \overline{K} \cap \Gamma_D, \text{ i.e., } e \in \mathcal{F} \setminus \mathcal{F}_0 \end{cases}$$

and the piecewise constant function $h_{\mathcal{F}}$ defined on $\mathcal{F}$ as

$$h_{\mathcal{F}} = h_{\mathcal{F}}(\boldsymbol{x}) = \begin{cases} |e| & \text{for } \boldsymbol{x} \in e \in \mathcal{F}, d = 2, \\ |e|^{\frac{1}{2}} & \text{for } \boldsymbol{x} \in e \in \mathcal{F}, d = 3. \end{cases}$$

Further denote by

$$(a\nabla v, \nabla v)_h := \sum_{K \in \mathcal{T}} \int_K a\nabla u \, \nabla v dK,$$

$$\langle h_{\mathcal{F}}^{-1} [\![u]\!], [\![v]\!] \rangle_{\mathcal{F}_0 \cup \mathcal{F}_D} := \sum_{e \in \mathcal{F}_0 \cup \mathcal{F}_D} \int_e h_{\mathcal{F}}^{-1} [\![u]\!] \cdot [\![v]\!] ds.$$

Finally, we introduce the following norm on $\mathcal{V}$:

$$|||v|||_h^2 = (a\nabla v, \nabla v)_h + \alpha \langle h_{\mathcal{F}}^{-1} [\![v]\!], [\![v]\!] \rangle_{\mathcal{F}_0 \cup \mathcal{F}_D} . \tag{8.2}$$

Let us consider the symmetric interior penalty discontinuous Galerkin (IP-DG) finite element method (see, e.g. [3]): Find $u_h \in \mathcal{V}$ such that

$$\mathcal{A}_h(u_h, v) = (f, v), \quad \forall\, v \in \mathcal{V}, \tag{8.3}$$

where

$$\begin{aligned} \mathcal{A}_h(u_h, v) \equiv \; & (a\nabla u_h, \nabla v)_h + \alpha \langle h_{\mathcal{F}}^{-1} [\![u_h]\!], [\![v]\!] \rangle_{\mathcal{F}_0 \cup \mathcal{F}_D} \\ & - \langle \{a\nabla u_h\}, [\![v]\!] \rangle_{\mathcal{F}_0 \cup \mathcal{F}_D} - \langle [\![u_h]\!], \{a\nabla v\} \rangle_{\mathcal{F}_0 \cup \mathcal{F}_D} . \end{aligned} \tag{8.4}$$

It is well known that if α is sufficiently large then the bilinear from (8.4) is coercive and bounded on $\mathcal{V}$ equipped with the norm (8.2), (see, e.g. [3]).

Another symmetric discontinuous Galerkin scheme can be derived by using the approach developed in [54]. In this case we get a bilinear from

$$\begin{aligned} \mathcal{A}_h(u_h, v) \equiv \; & (a\nabla u_h, \nabla v)_h + \alpha \left\langle h_{\mathcal{F}}^{-1} [\![u_h]\!], [\![v]\!] \right\rangle_{\mathcal{F}_0 \cup \mathcal{F}_D} \\ & - \langle \{a\nabla u_h\}, [\![v]\!] \rangle_{\mathcal{F}_0 \cup \mathcal{F}_D} - \langle [\![u_h]\!], \{a\nabla v\} \rangle_{\mathcal{F}_0 \cup \mathcal{F}_D} \\ & - \tfrac{1}{4}\alpha^{-1} \langle h_{\mathcal{F}} [\![a\nabla u_h \cdot \boldsymbol{n}]\!], [\![a\nabla v \cdot \boldsymbol{n}]\!] \rangle_{\mathcal{F}_0}, \end{aligned} \tag{8.5}$$

which is also coercive for sufficiently large α. Note that the corresponding DG scheme is slightly different from the IP-DG method (8.3)–(8.4). We summarize the main results regarding the discontinuous Galerkin method (8.3) in the following lemma:

Lemma 8.1. *Assume that the finite element partition $\mathcal{T}$ is regular and locally quasi uniform. Then the bilinear form $\mathcal{A}_h(\cdot, \cdot)$ defined by (8.4) or (8.5) is coercive and bounded in $\mathcal{V}$ equipped with the norm (8.2) for any sufficiently large $\alpha > 0$ and the discontinuous Galerkin method (8.3) has a unique solution.*

In the next sections we present some recent results on robust AMLI preconditioners for IP-DG finite element systems. It will be shown also, that in the considered cases, the methods are stabilized for relatively small values of the parameter α.

8.2 Element-based approach: bilinear DG systems

The essential components of the AMLI algorithm for DG systems can be constructed in different ways. We start our presentation with an element-based approach, see [81, 80], which will be described in the following for bilinear DG systems. The use of trilinear shape functions (in 3D) is discussed in detail in [80] and also the extension of this approach to higher-order shape functions and/or nonconforming elements, e.g., based on rotated multilinear shape functions is possible.

Consider a general element K with all its faces internal. Let its neighboring elements, which share a face with this element, be denoted by K_1^+, K_2^+, K_3^+, and K_4^+. Here $\cdot^+$ represents the neighboring element and digits $1, \ldots, 4$ represent the face number with which the neighboring element is attached. This arrangement is depicted in Figure 8.2(a). The bilinear form for this element reads:

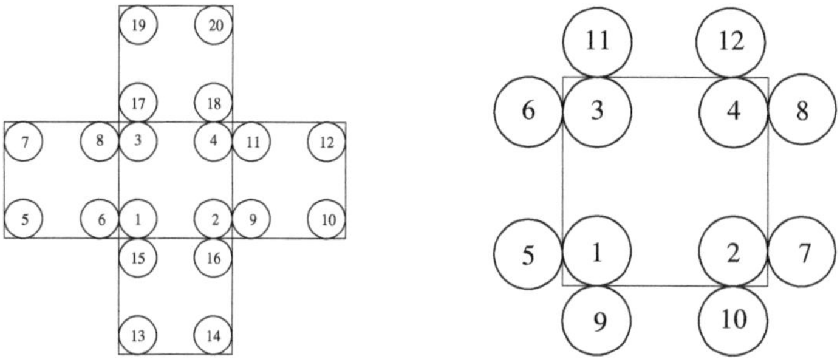

(a) Standard arrangement (b) Specific arrangement

Figure 8.2: Elemental DOF

$$
\begin{aligned}
\mathcal{A}_K(u_h, v) &= \int_K a\nabla u_h \cdot \nabla v \, dK + \alpha h_{\mathcal{F}}^{-1} \sum_{e\in\partial K} \int_e [\![u_h]\!] \cdot [\![v]\!] ds \\
&\quad - \sum_{e\in\partial K} \int_e (\{a\nabla u_h\} \cdot [\![v]\!] + [\![u_h]\!] \cdot \{a\nabla v\}) \, ds. \qquad (8.6)
\end{aligned}
$$

Using the definition of the trace operators $\{\cdot\}$ and $[\![\cdot]\!]$ and, as a standard procedure, collecting the terms with the test function for the element K we get

$$
\begin{aligned}
\mathring{\mathcal{A}}_K(u_h, v) &= \int_K a\nabla u_h \cdot \nabla v \, dK \\
&\quad + \alpha h_{\mathcal{F}}^{-1} \sum_{i=1}^{4} \int_{e_i} v_K \boldsymbol{n}_K \cdot \left(u_K \boldsymbol{n}_K + u_{K_i^+} \boldsymbol{n}_{K_i^+}\right) ds \\
&\quad - \frac{1}{2} \sum_{i=1}^{4} \int_{e_i} \Big(v_K \boldsymbol{n}_K \cdot \left(a\nabla u_K + a\nabla u_{K_i^+}\right) \\
&\quad + a\nabla v_K \cdot \left(u_K \boldsymbol{n}_K + u_{K_i^+} \boldsymbol{n}_{K_i^+}\right)\Big) \, ds. \qquad (8.7)
\end{aligned}
$$

The size of the resulting matrix is 4×20 since, as depicted in Figure 8.2(a), the DOF of the element K are connected with all the DOF of its neighboring elements K_i^+.

Since this arrangement is not well suited for setting up a local hierarchical basis transformation, we use an alternative element-based assembling procedure. Here we split the terms in the elemental bilinear form (8.6) in such a way that the ∇u_{K^+} terms are used only in the computation of the respective element K^+, and thus, only the ∇u_K terms are associated with the element K.

Moreover, to avoid the singularity of the resulting stiffness matrix arising in this arrangement, we need to consider all parts of the stabilization term in (8.6). Since this would double the contribution from the respective terms in the global stiffness matrix we take them with the weight $1/2$. The resulting elemental bilinear form is then given by

$$\begin{aligned}\mathcal{A}_K(u_h, v) &= \int_K a\nabla u_h \cdot \nabla v \, dK \\ &-\frac{1}{2}\sum_{i=1}^{4}\int_{e_i}\Big(\big(v_K \boldsymbol{n}_K + v_{K_i^+}\boldsymbol{n}_{K_i^+}\big)\cdot a\nabla u_K \\ &+a\nabla v_K \cdot \big(u_K \boldsymbol{n}_K + u_{K_i^+}\boldsymbol{n}_{K_i^+}\big)\Big)\, ds \\ &+\frac{\alpha h_{\mathcal{F}}^{-1}}{2}\sum_{i=1}^{4}\int_{e_i}\big(v_K \boldsymbol{n}_K + v_{K_i^+}\boldsymbol{n}_{K_i^+}\big)\big(u_K \boldsymbol{n}_K + u_{K_i^+}\boldsymbol{n}_{K_i^+}\big)\, ds.\end{aligned} \tag{8.8}$$

In this approach, as depicted in Figure 8.2(b), the DOF of the element K are connected with only those DOF of its neighboring elements K_i^+ which are at the common face. The resulting matrix, which is of the size 12×12, is denoted by A_K.

Remark 8.2. It is important to note that this specific splitting of terms in the bilinear form is possible only for some of the DG methods proposed in literature, e.g., the symmetric IP method [70], the method of Baumann and Oden [26], its stabilized version NIPG [100], and the method of Babuška–Zlamal [21].

Now let $N = 4N_K$ denote the total number of DOF in the system. Using the piecewise polynomial (bilinear) approximation in the weak form (8.3), with elemental bilinear form (8.8), we get the following linear system of equations

$$A\mathbf{x} = \mathbf{b}, \tag{8.9}$$

where $\mathbf{x} \in \mathbb{R}^N$, $A \in \mathbb{R}^{N\times N}$ with N_K^2 blocks of size 4×4, and $\mathbf{b} \in \mathbb{R}^N$, denote the solution vector, the global stiffness matrix, and the right-hand side data vector, respectively.

Let us next introduce the hierarchical basis transformation generating two levels of a discrete DG system; the multilevel extension is obtained by a recursive application of the two-level transformation to the coarse-level system. For that reason we assume that we have given a hierarchy of partitions $\mathcal{T}_\ell \subset \mathcal{T}_{\ell-1} \subset \ldots \subset \mathcal{T}_1 \subset \mathcal{T}_0$ of Ω, where the notation $\mathcal{T}_k = \mathcal{T}_h \subset \mathcal{T}_H = \mathcal{T}_{k-1}$ means that for any element K of the fine(r) partition $\mathcal{T}_h$ there is an element E of the coarse(r) mesh partition $\mathcal{T}_H$ such that $K \subset E$.

Consider now the linear system (8.9) resulting from the IP-DG approximation of the basic problem (8.1). The partitioning of variables (and corresponding equations) into a *fine* and a *coarse* (sub-) set, indicated by the subscripts 1 and 2, respectively, is induced by a regular mesh refinement at every level $(k-1) = 0, 1, \dots, \ell - 1$. This means that by halving the meshsize, i.e., $h = H/2$, each element is subdivided into four elements of similar shape, herewith producing the mesh at levels $k = 1, 2, \dots, \ell$. Hence, the linear system (8.9) can be represented in the 2×2 block form as

$$\begin{bmatrix} A_{11} & A_{12} \\ A_{21} & A_{22} \end{bmatrix} \begin{bmatrix} \mathbf{x}_1 \\ \mathbf{x}_2 \end{bmatrix} = \begin{bmatrix} \mathbf{b}_1 \\ \mathbf{b}_2 \end{bmatrix} \tag{8.10}$$

where $A_{21} = A_{12}^T$. Now, if we make use of the two-level transformation matrix

$$J = \begin{bmatrix} I & P_{12} \\ 0 & I \end{bmatrix}, \tag{8.11}$$

the system to be solved in the (generalized) hierarchical basis reads

$$\widehat{A}\,\widehat{\mathbf{x}} = \widehat{\mathbf{b}}. \tag{8.12}$$

The matrix $\widehat{A}$ and its submatrices $\widehat{A}_{11}$, $\widehat{A}_{12}$ $\widehat{A}_{21}$, $\widehat{A}_{22}$ are given by

$$\begin{aligned} \widehat{A} &= J^T A J = \begin{bmatrix} \widehat{A}_{11} & \widehat{A}_{12} \\ \widehat{A}_{21} & \widehat{A}_{22} \end{bmatrix}, \\ \widehat{A}_{11} &= A_{11}, \quad \widehat{A}_{12} = A_{11} P_{12} + A_{12}, \quad \widehat{A}_{21} = P_{12}^T A_{11} + A_{21}, \\ \widehat{A}_{22} &= P_{12}^T A_{11} P_{12} + A_{21} P_{12} + P_{12}^T A_{12} + A_{22}. \end{aligned}$$

The vectors $\widehat{\mathbf{x}}$ and $\widehat{\mathbf{b}}$ are transformed then from hierarchical basis to a nodal basis via $\mathbf{x} = J\,\widehat{\mathbf{x}}$ and $\mathbf{b} = \left(J^T\right)^{-1} \widehat{\mathbf{b}}$, and the following relations hold

$$\mathbf{x}_1 = \widehat{\mathbf{x}}_1 + P_{12}\widehat{\mathbf{x}}_2, \quad \mathbf{x}_2 = \widehat{\mathbf{x}}_2, \tag{8.13a}$$

$$\mathbf{b}_1 = \widehat{\mathbf{b}}_1, \quad \mathbf{b}_2 = \widehat{\mathbf{b}}_2 - P_{12}^T \mathbf{b}_1. \tag{8.13b}$$

The general macroelement we are using to define the local interpolation P_E is depicted in Figure 8.3. It is important to note that the nonzero pattern and the entries of P_E have to be defined in such a way that the local interpolation (for neighboring macroelements) is compatible, i.e., the stencil and the coefficients have to agree for fine nodes shared by adjacent macroelements in their respective local interpolation matrices P_E.

The local macroelement two-level transformation matrix J_E is given by

$$J_E = \begin{bmatrix} I & P_E \\ 0 & I \end{bmatrix}, \tag{8.14}$$

where the macroelement interpolation matrix P_E is of size 20×12. For symmetry (isotropy) reasons we consider transformations that can be characterized by the matrix P_E given below:

$$P_E = \begin{bmatrix}
p_1 & p_2 & p_2 & p_3 & 0 & 0 & 0 & 0 & 0 & 0 & 0 & 0 \\
p_2 & p_1 & p_3 & p_2 & 0 & 0 & 0 & 0 & 0 & 0 & 0 & 0 \\
p_2 & p_3 & p_1 & p_2 & 0 & 0 & 0 & 0 & 0 & 0 & 0 & 0 \\
p_3 & p_2 & p_2 & p_1 & 0 & 0 & 0 & 0 & 0 & 0 & 0 & 0 \\
q_1 & 0 & q_2 & 0 & q_3 & q_4 & 0 & 0 & 0 & 0 & 0 & 0 \\
q_2 & 0 & q_1 & 0 & q_4 & q_3 & 0 & 0 & 0 & 0 & 0 & 0 \\
0 & q_1 & 0 & q_2 & 0 & 0 & q_3 & q_4 & 0 & 0 & 0 & 0 \\
0 & q_2 & 0 & q_1 & 0 & 0 & q_4 & q_3 & 0 & 0 & 0 & 0 \\
q_1 & q_2 & 0 & 0 & 0 & 0 & 0 & 0 & q_3 & q_4 & 0 & 0 \\
q_2 & q_1 & 0 & 0 & 0 & 0 & 0 & 0 & q_4 & q_3 & 0 & 0 \\
0 & 0 & q_1 & q_2 & 0 & 0 & 0 & 0 & 0 & 0 & q_3 & q_4 \\
0 & 0 & q_2 & q_1 & 0 & 0 & 0 & 0 & 0 & 0 & q_4 & q_3 \\
q_3 & 0 & q_4 & 0 & q_1 & q_2 & 0 & 0 & 0 & 0 & 0 & 0 \\
q_4 & 0 & q_3 & 0 & q_2 & q_1 & 0 & 0 & 0 & 0 & 0 & 0 \\
0 & q_3 & 0 & q_4 & 0 & 0 & q_1 & q_2 & 0 & 0 & 0 & 0 \\
0 & q_4 & 0 & q_3 & 0 & 0 & q_2 & q_1 & 0 & 0 & 0 & 0 \\
q_3 & q_4 & 0 & 0 & 0 & 0 & 0 & 0 & q_1 & q_2 & 0 & 0 \\
q_4 & q_3 & 0 & 0 & 0 & 0 & 0 & 0 & q_2 & q_1 & 0 & 0 \\
0 & 0 & q_3 & q_4 & 0 & 0 & 0 & 0 & 0 & 0 & q_1 & q_2 \\
0 & 0 & q_4 & q_3 & 0 & 0 & 0 & 0 & 0 & 0 & q_2 & q_1
\end{bmatrix} \tag{8.15}$$

In particular, we study here two choices of P_E which are based on simple averaging:

1. $J^{(1)}$ is induced by (8.14), P_E is given by (8.15), and $p_1 = p_2 = p_3 = 1/4$, $q_1 = q_2 = 1/2$, $q_3 = q_4 = 0$.

2. $J^{(2)}$ is induced by (8.14), P_E is given by (8.15), and $p_1 = p_2 = p_3 = 1/4$, $q_1 = q_2 = q_3 = q_4 = 1/4$.

Of course, the second scheme needs a modification at the boundaries such that the sum of interpolation coefficients always equals one.[1] The connections for a few internal and face DOF are depicted in Figure 8.3, where continuous and dashed lines denote the weight $1/4$ and $1/2$, respectively. A similar treatment applies to other internal and face DOF.

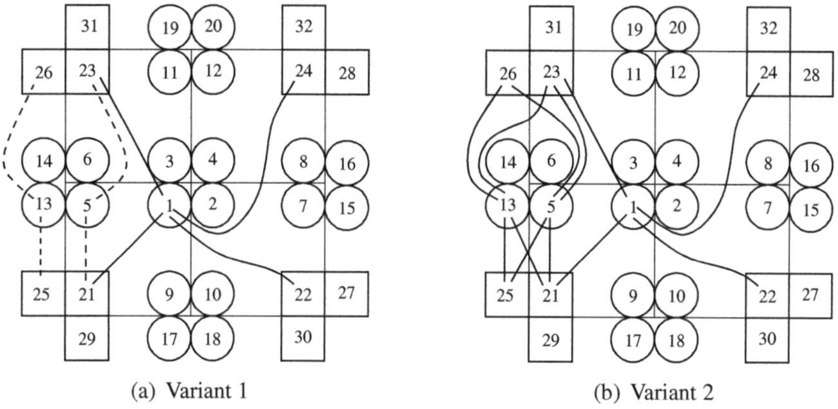

Figure 8.3: The connectivity in the local interpolation for a macroelement

Let us next consider the macroelement stiffness matrix $\widehat{A}_E$ obtained from assembling the element matrices for all elements K contained in E in the (local) hierarchical basis. Evidently, the global two-level stiffness matrix $\widehat{A}$ can be assembled from the macroelement two-level stiffness matrices, i.e.,

$$\widehat{A} = J^T A J = \sum_{E \in \mathcal{T}_H} R_E^T \widehat{A}_E R_E = \sum_{E \in \mathcal{T}_H} R_E^T J_E^T A_E J_E R_E.$$

Like the global matrix, the local matrices are also of the following 2×2 block form

$$\widehat{A}_E = \begin{bmatrix} \widehat{A}_{E:11} & \widehat{A}_{E:12} \\ \widehat{A}_{E:21} & \widehat{A}_{E:22} \end{bmatrix} = J_E^T \begin{bmatrix} A_{E:11} & A_{E:12} \\ A_{E:21} & A_{E:22} \end{bmatrix} J_E, \tag{8.16}$$

where J_E is defined by (8.14) and the local Schur complement is given by

$$S_E = \widehat{A}_{E:22} - \widehat{A}_{E:21} \widehat{A}_{E:11}^{-1} \widehat{A}_{E:12} = A_{E:22} - A_{E:21} A_{E:11}^{-1} A_{E:12}. \tag{8.17}$$

As we know from the general framework of two-level block factorization methods, it suffices to compute the minimal eigenvalue $\lambda_{E;\min}$ of the generalized eigenproblem

$$S_E \mathbf{v}_{E:2} = \lambda_E \widehat{A}_{E:22} \mathbf{v}_{E:2}, \quad \mathbf{v}_{E:2} \neq \mathbf{c} \tag{8.18}$$

[1]For instance, a macroelement touching the boundary with its left face – when DOF 13, 14, 25, 26 are missing – has the correct local interpolation coefficients $1/2$ for each of the couplings between DOF 5 and 6 and the coarse DOF 21 and 23, which is the same as for the interpolation scheme 1, cf. Figure 8.3.

in order to conclude the following upper bound for the CBS constant γ:

$$\gamma^2 \le \max_{E \in \mathcal{T}_H} \gamma_E^2 = \max_{E \in \mathcal{T}_H} (1 - \lambda_{E;\min}). \tag{8.19}$$

This relation then implies condition number estimates for the corresponding two-level preconditioner. For a proof of the following theorem see [81].

Theorem 8.3. *Consider the IP-DG formulation (8.4) of problem (8.1) based on piecewise bilinear shape functions where the diffusion coefficient $a(\boldsymbol{x}) = a$ is assumed to be piecewise constant on the coarsest mesh partition $\mathcal{T}_0$. Then for Variant 1 of the two-level basis transformation associated with the macroelement matrix which is obtained from assembling four equal element matrices $A_K(\alpha)$ the CBS constant γ has the following upper bound*

$$\gamma \le \gamma_E^{(1)} \le \sqrt{\frac{3}{4} + \frac{1}{8\alpha}} \qquad \forall \alpha \ge \frac{9}{8}. \tag{8.20}$$

In order to derive an estimate which holds true for all levels of mesh refinement and corresponding factorization steps, we need to repeat the calculation of γ_E, thereby replacing $A_K(\alpha)$ with $\widehat{A}_{E:22}$ when assembling the new macroelement matrix (at each subsequent level). In case of Variant 1 the element matrices at successive levels fulfill the relation

$$\widehat{A}_{E:22} = A_K(2\alpha). \tag{8.21}$$

In other words, the coarse-grid matrix corresponds to the same kind of IP-DG discrete problem (on a mesh with spacing $H = 2h$) but using double the value for the stabilization parameter α. Even though this results in a smaller upper bound for $\gamma_E^{(1)}$, showing that inequality (8.20) is valid for all subsequent levels if it is valid for the first one, an additional difficulty arises, which is related to the preconditioning of the A_{11}-block.

Lemma 8.4. *For our model problem with piecewise constant coefficients, i.e., for $a = a_E$, where the coefficient $a_E > 0$ can have arbitrary large jumps on the macroelement interfaces, the condition number of A_{11} is of order $\mathcal{O}(\alpha)$.*

Proof. We note that the condition number of A_{11} can be estimated locally, i.e., if

$$0 < \underline{c} \le \mathbf{v}_{E:1}^T A_{E:11} \mathbf{v}_{E:1} \le \overline{c}\alpha \quad \forall \mathbf{v}_{E:1} \forall E, \tag{8.22}$$

then the same relation holds also globally, which implies $\kappa(A_{11}) = \mathcal{O}(\alpha)$. The proof of (8.22) is as follows. Assembling any macroelement matrix $A_{E:11}$, where without loss of generality we may assume that $a_E = 1$, from the element matrix $A_K(\alpha)$, we find that $A_{E:11}$ has the representation

$$A_{E:11} = A_{E:11}(\alpha) = A_0 + A_\alpha = A_0 + \alpha A_1 \tag{8.23}$$

with constant matrices A_0 and A_1 being indefinite and SPSD, respectively. Moreover, $\overline{A} := A_0 + A_1$ is SPD. Hence, the minimal and maximal eigenvalues of the matrices $\overline{A}$ and $A_\alpha = \alpha A_1$ satisfy

$$\begin{aligned} \lambda_{\min}(\overline{A}) &= \underline{c}, \qquad & \lambda_{\max}(\overline{A}) &= c, \\ \lambda_{\min}(A_\alpha) &= 0, \qquad & \lambda_{\max}(A_\alpha) &= \mathcal{O}(\alpha), \end{aligned}$$

for some constants $\underline{c}$ and c, $0 < \underline{c} \leq c$. But then, since

$$\mathbf{v}_{E:1}^T A_{E:11}(\alpha) \mathbf{v}_{E:1} = \mathbf{v}_{E:1}^T [\overline{A} + (\alpha - 1) A_1] \mathbf{v}_{E:1} \leq \mathbf{v}_{E:1}^T [\overline{A} + A_\alpha] \mathbf{v}_{E:1} \quad \forall \mathbf{v}_{E:1} \forall E,$$

we have $\lambda_{\min}(A_{E:11}(\alpha)) \geq \underline{c}$ for all $\alpha \geq 1$ and $\lambda_{\max}(A_{E:11}(\alpha)) \leq \overline{c}\alpha$ for all $\alpha \geq 1$. In particular, by choosing $\overline{c} = \lambda_{\max}(A_1) + \lambda_{\max}(\overline{A})$ and $\underline{c} = \lambda_{\min}(\overline{A})$ the estimate (8.22) holds for all $\alpha \geq 1$. □

Thus, the block factorization in the hierarchical basis, when constructed using the transformation Variant 1, with increasing level number, in general, makes the solution of the sub-systems with A_{11} more and more difficult.

Remark 8.5. The bound (8.20) is slightly too weak in order to guarantee that the condition number of the multiplicative preconditioner (with exact inversion of the A_{11}-block) can be stabilized using Chebyshev polynomials of degree two. Combining the optimality conditions (2.78) with (8.20) we obtain $\alpha > 9/10$, which is to be satisfied for a condition number that can be uniformly bounded in the number of levels if one uses optimal third-order polynomial stabilization. Note that, however, the stability of the DG discretization scheme requires a larger α anyway, see [104].

Let us next focus on Variant 2. The following lemma (for a proof see [81]) provides some information on the element matrix $A_K^{(j)}(\alpha)$ after j regular coarsening (unrefinement) steps, starting with the element matrix

$$A_K^{(0)}(\alpha) = A_K(\alpha) \tag{8.24}$$

at the fine-grid level ℓ.

Lemma 8.6. *If we neglect boundary effects, the element matrix $A_K^{(j)}(\alpha)$ after j coarsening steps, $1 \le j < \ell$, has the representation*

$$A_K^{(j)}(\alpha) = X_K(\alpha) + jY_K \tag{8.25}$$

where Y_K is SPSD with three fourfold eigenvalues, given by $11/48$, $1/16$, and 0; and the matrix $X_K(\alpha)$ is SPSD if and only if $\alpha > 11/8$. Further, $X_K(\alpha)$ is monotonically increasing in α which means that $X_K(\alpha) - X_K(\alpha')$ is SPSD, i.e., $X_K(\alpha) - X_K(\alpha') \ge 0$ if $\alpha - \alpha' \ge 0$.

In the local analysis of the multilevel procedure we want to estimate the abstract angle between the coarse space and its complementary space in the decomposition of the space at level $(\ell - j)$, $j = 0, 1, \ldots, \ell - 1$; the corresponding element matrices are given by $A_K^{(j)}(\alpha)$, and $A_K^{(j+1)}(\alpha)$, which can be associated with the fine and coarse space at level $(\ell - j)$, respectively.

Unless otherwise specified the parameter α is always assumed to be greater than or equal to one. By construction (cf. (8.16), (8.24)–(8.25)) we have

$$\widehat{A}_{E:22}^{(j)}(\alpha) = A_K^{(j+1)}(\alpha), \tag{8.26}$$

and thus (8.17) can be written as

$$S_E^{(j)}(\alpha) = A_K^{(j+1)}(\alpha) - \widehat{A}_{E:21}^{(j)}(\alpha)(\widehat{A}_{E:11}^{(j)}(\alpha))^{-1}\widehat{A}_{E:12}^{(j)}(\alpha). \tag{8.27}$$

Furthermore, the solution of the local generalized eigenproblem (8.18) is equivalent to finding

$$\begin{aligned} \lambda_{E;\min} &:= \min_{\mathbf{v}\neq\mathbf{c}} \frac{\mathbf{v}^T S_E^{(j)}(\alpha)\,\mathbf{v}}{\mathbf{v}^T A_K^{(j+1)}(\alpha)\,\mathbf{v}} \\ &= 1 - \max_{\mathbf{v}\neq\mathbf{c}} \frac{\mathbf{v}^T \{\widehat{A}_{E:21}^{(j)}(\alpha)(\widehat{A}_{E:11}^{(j)}(\alpha))^{-1}\widehat{A}_{E:12}^{(j)}(\alpha)\}\,\mathbf{v}}{\mathbf{v}^T A_K^{(j+1)}(\alpha)\,\mathbf{v}}. \end{aligned} \tag{8.28}$$

Now, since it is sufficient to compute a lower bound λ for (8.28) we consider the inequality

$$(1-\lambda)A_K^{(j+1)}(\alpha) - R_E^{(j)}(\alpha) \ge 0 \tag{8.29}$$

where

$$R_E^{(j)}(\alpha) = \widehat{A}_{E:21}^{(j)}(\alpha)(\widehat{A}_{E:11}^{(j)}(\alpha))^{-1}\widehat{A}_{E:12}^{(j)}(\alpha). \tag{8.30}$$

Then (8.29) has to be fulfilled with a preferably large $\lambda > 0$. In this regard we first observe the monotonicity of $R_E^{(j)}(\alpha)$ and $A_K^{(j+1)}(\alpha)$.

Lemma 8.7. *For the transformation Variant 2, the matrix-valued function $R_E^{(j)}$: $\alpha \mapsto R_E^{(j)}(\alpha)$, defined by (8.30), is monotonically decreasing, which means that $R_E^{(j)}(\alpha') - R_E^{(j)}(\alpha) \geq 0$ if $1 \leq \alpha' \leq \alpha < \infty$. Furthermore, for fixed $\alpha \geq 1$ the function $R_E(\alpha) : j \mapsto R_E^{(j)}(\alpha)$ decreases in j, i.e., $R_E^{(j)}(\alpha) - R_E^{(j+1)}(\alpha) \geq 0$ if $j \in \{1, 2, \ldots\}$. By contrast, the matrix $A_K^{(j+1)}(\alpha)$ is monotonically increasing in α and j.*

Proof. From Lemma 8.6 we know that $X_K(\alpha)$ is monotonically increasing and since Y_K is SPSD (and does not depend on α) we conclude that $A_K^{(j)}(\alpha)$ is monotonically increasing in both parameters, α and j.

Then it can be easily checked that for $\alpha \geq 1$ the macroelement pivot matrix $\widehat{A}_{E:11}^{(j)}(\alpha)$ is SPD. Thus we have $(\widehat{A}_{E:11}^{(j)}(\alpha))^{-1} \geq 0$ showing that $R_E^{(j)}(\alpha) \geq 0$. Moreover, since $\widehat{A}_{E:12}^{(j)}(\alpha) = \widehat{A}_{E:12}$ and $\widehat{A}_{E:21}^{(j)}(\alpha) = \widehat{A}_{E:12}^T$ are invariant with respect to α and j, it suffices to show that $\widehat{A}_{E:11}^{(j)}(\alpha)$ is monotonically increasing in both parameters. The latter, however, follows from the fact that $\widehat{A}_{E:11}^{(j)}(\alpha)$ is a special linear combination of three symmetric matrices $A'_{E:11}$, $\overline{A}_{E:11}$ and $\overline{\overline{A}}_{E:11}$, i.e.,

$$\widehat{A}_{E:11}^{(j)}(\alpha) = A'_{E:11} + j\overline{A}_{E:11} + \alpha\overline{\overline{A}}_{E:11}$$

in which $\overline{A}_{E:11}$ and $\overline{\overline{A}}_{E:11}$ are positive semidefinite. Note that neither $\overline{A}_{E:11}$ nor $\overline{\overline{A}}_{E:11}$ depend on α or j! □

The multilevel block factorization in the hierarchical basis, (implicitly) generated by recursive application of the two-level transformation Variant 2, yields a recursive splitting of the related DG-FE spaces for which the following estimate holds.

Theorem 8.8 (see [81]). *Consider the elliptic problem (8.1) with constant coefficients discretized by the IP-DG method using bilinear elements on a uniform mesh and assume that $\alpha \geq \underline{\alpha} \geq 1$. Then the constant γ in the CBS inequality for the splitting at level $\ell - j$, $j \in \{0, 1, \ldots, \ell - 1\}$, (associated with the $(j+1)$-th coarsening step) can be estimated locally according to (8.19). In particular, if $\alpha \geq \underline{\alpha} = 4$ then the local estimate*

$$\gamma^2 \leq [\gamma_E]^2 \leq \begin{cases} 1 - \left(\frac{9}{14}\right)^2 \approx 0.586735 & \textit{for } j = 0 \\ 1 - \left(\frac{5}{7}\right)^2 \approx 0.489796 & \textit{for } j > 0 \end{cases} \tag{8.31}$$

holds.

Proof. The proof is based on the inequality (8.29). Let us first consider the case $j > 0$. We note that the maximum $\lambda > 0$ satisfying inequality (8.29) is given by (8.28). Next, using the assumption $\alpha \geq \underline{\alpha} \geq 1$ and taking into account the monotonicity properties established by Lemma 8.7 we get

$$(1-\lambda)A_K^{(j+1)}(\alpha) - R_E^{(j)}(\alpha) \geq (1-\lambda)A_K^{(2)}(\underline{\alpha}) - R_E^{(1)}(\underline{\alpha}) \quad \forall \lambda > 0,$$

and thus it follows that any $\underline{\lambda} > 0$ satisfying the inequality

$$(1-\underline{\lambda})A_K^{(2)}(\underline{\alpha}) - R_E^{(1)}(\underline{\alpha}) \geq 0$$

yields a lower bound for the desired minimum eigenvalue $\lambda_{E;\min}$, i.e.,

$$0 < \underline{\lambda} \leq \lambda_{E;\min}.$$

In particular, by choosing $\underline{\alpha} = 4$ (and $j = 1$), and solving the corresponding generalized local eigenvalue problem (8.18), this proves the estimate (8.31) for $j > 0$.

In the case $j = 0$ the bound can be verified by using similar monotonicity arguments (here only the monotonicity with respect to α is required), and finally, choosing again $\alpha = \underline{\alpha}(= 4)$. □

Remark 8.9. Comparing the bounds (8.31) and (8.20) it becomes obvious that Variant 2 of the basis transformation is preferable for stabilization of the condition number at low(er) costs. In this case both inequalities (2.78) can also be met by employing second-order (instead of third-order) Chebyshev polynomials in the (linear) AMLI cycle.

Remark 8.10. When the stabilization parameter α tends to infinity both upper bounds for γ decrease, which shows that the corresponding angles improve. While the limit for Variant 1 of the hierarchical basis is $\sqrt{3/4}$, the limiting value for Variant 2 is given by $\sqrt{3/8}$, which equals the corresponding value for the finite element spaces generated by conforming bilinear elements (cf. [98]), i.e.,

$$\lim_{\alpha \to \infty} \gamma_E(\alpha, j) = \sqrt{3/8} \quad \forall j = 0, 1, 2, \ldots .$$

Note that the same limit is obtained for any fixed α as the number of levels tends to infinity, i.e.,

$$\lim_{j \to \infty} \gamma_E(\alpha, j) = \sqrt{3/8} \quad \forall \alpha > 1,$$

which of course is not of practical relevance.

A detailed analysis of the first transformation variant when applied to three-dimensional anisotropic elliptic problems discretized by the IP-DG method (8.4) based on trilinear shape functions has been presented in reference [80]. There it has been shown that this approach yields a robust method as long as the main directions of anisotropy are aligned with the coordinate axes, which is also the case for standard Galerkin FEM based on conforming bilinear or trilinear elements.

8.3 Face-based approach: Rotated bilinear DG systems

The commonly known theory of optimal order solution methods for FEM elliptic systems is restricted to the case of coefficient jumps which are aligned with the coarse(st) mesh partitioning (triangulation). Such assumptions are usually made in case of multilevel, multigrid and domain decomposition methods. There are numerical tests confirming that the convergence of these methods deteriorates if this condition is violated. However, at the same time, there are many (multiscale and multiphysics) models for strongly heterogeneous media where the strong coefficient jumps can be resolved on the finest mesh only! The hierarchical bases proposed in this section are especially designed for problems with highly varying coefficients. The robustness issue we are investigating here concerns jump discontinuities of the PDE coefficient at arbitrary element interfaces (on the fine mesh). A hierarchical basis that provides a *robust splitting* (in this situation) yields a uniform upper bound (strictly less than one) of the CBS constant that measures the cosine of the abstract angle between the coarse space and its complementary space. Though our focus is on a particular family of rotated bilinear finite elements in two space dimensions (2D) here (see [76]) the proposed rather general approach is neither limited to this particular choice of elements nor to 2D problems.

In the setting of DG discretizations the construction of face-based transformation variants seems to be a natural choice. This originates in the observation that the global stiffness matrix related to the bilinear form (8.4) can also be assembled from small-sized local stiffness matrices associated with the individual element faces $e \in \mathcal{F}$, i.e.,

$$A = \sum_{e \in \mathcal{F}} R_e^T A_e R_e$$

where summation is understood in the sense of assembling matrices.

For an interior face $e \in \mathcal{F}_0$ the matrix A_e is then associated with the local bilinear form

$$\begin{aligned}\mathcal{A}_e(u_h, v) \equiv \; & \frac{1}{2d}\left((a\nabla u_h, \nabla v)_{K^+} + (a\nabla u_h, \nabla v)_{K^-}\right) + \alpha \left\langle h_{\mathcal{E}}{}^{-1} [\![u_h]\!], [\![v]\!]\right\rangle_e \\ & -\langle\{a\nabla u_h\}, [\![v]\!]\rangle_e - \langle[\![u_h]\!], \{a\nabla v\}\rangle_e .\end{aligned}$$

A simple computation shows that in the model case of a uniform mesh (composed of square elements) the matrices corresponding to the single contributions of horizontal and vertical (interior) faces have the representation

$$A_e^{\mathrm{h}} = A_{e,0}^{\mathrm{h}} + \alpha A_{e,1}^{\mathrm{h}}, \tag{8.32}$$

$$A_e^{\mathrm{v}} = A_{e,0}^{\mathrm{v}} + \alpha A_{e,1}^{\mathrm{v}}, \tag{8.33}$$

where the matrix terms from the right-hand side of (8.32) can be written in the form (see [76]):

$$A_{e,0}^{\mathrm{h}} = \frac{1}{8}\begin{bmatrix} 5a^+ & -3a^+ & 3a^+ & a^+ & 0 & -6a^+ & 0 & 0 \\ -3a^+ & 5a^+ & -a^+ & -3a^+ & 0 & 2a^+ & 0 & 0 \\ 3a^+ & -a^+ & -15a^+ & 3a^+ & -6a^- & 10(a^+ + a^-) & 2a^- & -6a^- \\ a^+ & -3a^+ & 3a^+ & 5a^+ & 0 & -6a^+ & 0 & 0 \\ 0 & 0 & -6a^- & 0 & 5a^- & 3a^- & -3a^- & a^- \\ -6a^+ & 2a^+ & 10(a^+ + a^-) & -6a^+ & 3a^- & -15a^- & -a^- & 3a^- \\ 0 & 0 & 2a^- & 0 & -3a^- & -a^- & 5a^- & -3a^- \\ 0 & 0 & -6a^- & 0 & a^- & 3a^- & -3a^- & 5a^- \end{bmatrix},$$

$$A_{e,1}^{\mathrm{h}} = \frac{1}{240}\begin{bmatrix} 23 & -3 & -3 & -17 & -23 & 3 & 3 & 17 \\ -3 & 3 & 3 & -3 & 3 & -3 & -3 & 3 \\ -3 & 3 & 243 & -3 & 3 & -243 & -3 & 3 \\ -17 & -3 & -3 & 23 & 17 & 3 & 3 & -23 \\ -23 & 3 & 3 & 17 & 23 & -3 & -3 & -17 \\ 3 & -3 & -243 & 3 & -3 & 243 & 3 & -3 \\ 3 & -3 & -3 & 3 & -3 & 3 & 3 & -3 \\ 17 & 3 & 3 & -23 & -17 & -3 & -3 & 23 \end{bmatrix}.$$

Here the isotropic (scalar) coefficient $a = a(K)$ is defined by $a(K^{\pm}) = a^{\pm}$. Moreover, assuming a uniform mesh the vertical face matrix A_e^{v} is obtained from A_e^{h} via the following permutation of rows and columns:

$$A_e^{\mathrm{v}} = S_{\mathrm{h,v}}^T A_e^{\mathrm{h}} S_{\mathrm{h,v}},$$

where

$$(S_{\mathrm{h,v}})_{i,j} = \begin{cases} 1 & \text{if } \ j = \mathbf{s}_i \\ 0 & \text{else,} \end{cases} \tag{8.34}$$

$$\mathbf{s} = (2,1,4,3,6,5,8,7)^T$$

and the numbering of nodes belonging to horizontal and vertical faces is as shown in Figure 8.4.

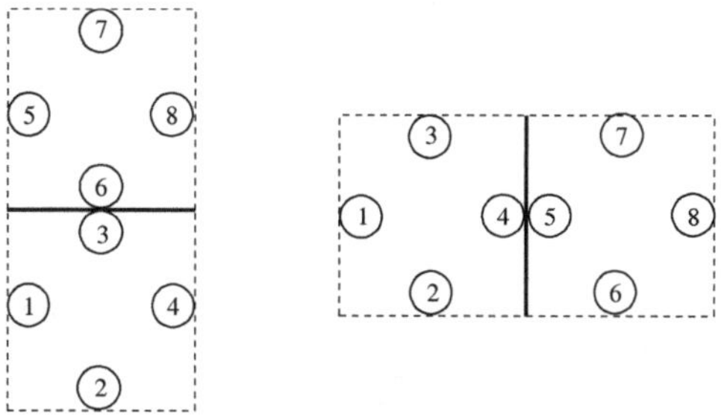

Figure 8.4: Degrees of freedom of face matrices: Horizontal face (left) and vertical face (right)

Remark 8.11. For the considered particular case of a uniform mesh of square elements the analysis of the stabilization parameter shows that when $a^+ = a^-$ the condition of Lemma 8.1 is satisfied for $\alpha > \frac{\sqrt{23329}-127}{8} \approx 3.22$.

Next we define a general so-called superelement $g \in \mathcal{G}$, which is the union of all the degrees of freedom (DOF) associated with the four faces (two horizontal and two vertical faces) that share one vertex. The characteristic macro-superelement $G \in \mathcal{M}$ is then made up of four partly overlapping superelements as shown in Figure 8.5.

Note that the construction of faces, superelements and macro-superelements is such that the global stiffness matrix can be assembled alternatively, in either way,

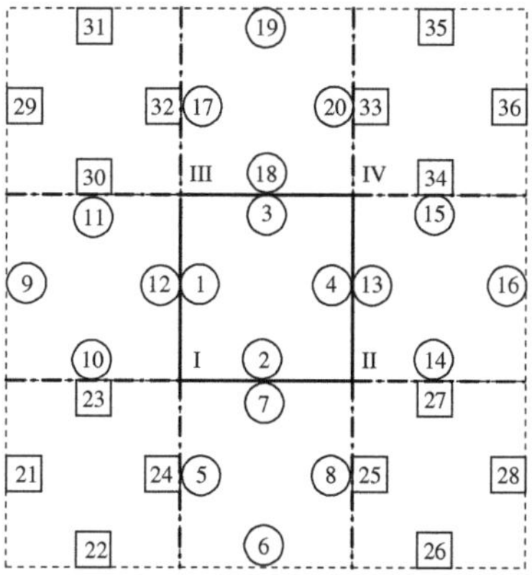

Figure 8.5: Macro-superelement G: four overlapping superelements

of the respective local matrices, i.e.,

$$A = \sum_{e \in \mathcal{F}} R_e^T A_e R_e = \sum_{g \in \mathcal{G}} R_g^T A_g R_g = \sum_{G \in \mathcal{M}} R_G^T A_G R_G.$$

Due to the overlap of superelements a proper scaling of the respective face contributions is required when assembling superelement matrices. For an interior superelement g (none of its elements touches the boundary) the correct scaling factor is $1/2$, i.e., $A_g = \sum_{e \subset g} \frac{1}{2} R_{e,g}^T A_e R_{e,g}$.

The basic construction follows now from the standard approach, as it is used for conforming bilinear elements, if one associates nodes with elements. Hence, in the present context, we consider superelements g (instead of elements) in order to compose macro-superelements G (instead of macroelements). Superelements produce overlapping (common) elements (instead of common nodes in the standard setting). Two "neighboring" macro-superelements have three elements in common, see Figure 8.6 (instead of having three common nodes in the standard situation of conforming bilinear elements). The coarse mesh hierarchy is such that the DOF associated with every other element (in x- and every other element in y-direction) belong to the coarse level. Since all DOF of a given element are either "fine" or "coarse", the number of coarse DOF is always a multiple of four (in the scalar case). The number of elements that contain the coarse DOF is approximately reduced by a factor 4 in each coarsening step (for large meshes) and thus the ratio $\varrho := N^{(k+1)}/N^{(k)} \approx 4$ where $N^{(k+1)}$ and $N^{(k)}$ denote the number of DOF at levels $k+1$ and k, respectively, see Figure 8.6.

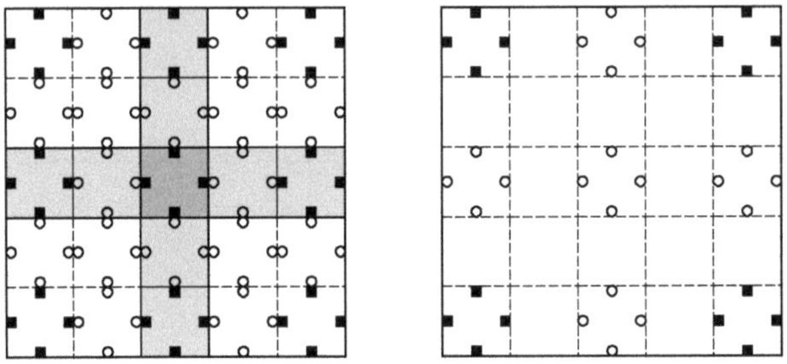

Figure 8.6: Overlap of macro-superelements and coarsening

The macro-superelement matrix associated with G is denoted by A_G. It is transformed into a hierarchical two-level basis via a local transformation

$$\hat{A}_G = J_G^T A_G J_G \tag{8.35}$$

where J_G has the form

$$J_G = \begin{bmatrix} I & P_G \\ 0 & I \end{bmatrix} \tag{8.36}$$

and P_G denotes some proper local interpolation matrix of size 20×16. Note that the local and global hierarchical bases, as presented here, do not involve so-called differences and aggregates of nodal shape functions. As opposed to the latter construction, which we studied in Chapters 4 and 7, we stay within the framework of nested meshes here.

Let us now enter into the question how to compute a local interpolation operator P_G such that the general transformation (8.36) provides a (local) minimum energy extension from the local coarse to the local fine space subject to global compatibility. We start with a so-called static condensation of the interior macro-superelement DOF with local numbers $1, 2, 3, 4$, see Figure 8.4. After this reduction step we arrive at a (in our case 32×32) local Schur complement

$$B_G = A_{G:22} - A_{G:21} A_{G:11}^{-1} A_{G:12}.$$

Here $A_{G:11}$ denotes the upper-left 4×4 submatrix corresponding to the interior DOF of G that are eliminated. Since there is no overlap of the central element (no common interior DOF) of different macro-superelements G, the exact elimination of interior unknowns in the global system can be done locally, i.e., for each macro-superelement separately. For the transformation of the local matrices B_G associated with the reduced macro-superelements, we use a local harmonic interpolation of the remaining fine DOF subject to the compatibility constraint that the fine DOF of a given element are allowed to interpolate only from the coarse DOF of its attached elements. For this purpose we assemble an auxiliary matrix C_G from those face matrices that originally produce the coupling of the remaining fine DOF (local numbers 5 to 20 in Figure 8.4) with the coarse DOF (local numbers 21 to 36). Accordingly the matrix C_G is partitioned into 2×2 blocks of size 16×16, i.e.,

$$C_G = \begin{bmatrix} C_{G:11} & C_{G:12} \\ C_{G:21} & C_{G:22} \end{bmatrix}.$$

A two-level splitting based on local energy minimization is then induced by the transformation matrix J_G^{EM} that is given as the product of two transformation steps, namely the static condensation

$$J_G^{\mathrm{SC}} = \begin{bmatrix} I & P_G^{\mathrm{SC}} & 0 \\ 0 & I & 0 \\ 0 & 0 & I \end{bmatrix} := \begin{bmatrix} I & -A_{G:11}^{-1} A_{G:12} & 0 \\ 0 & I & 0 \\ 0 & 0 & I \end{bmatrix}, \tag{8.37}$$

and the harmonic interpolation

$$J_G^{\mathrm{HI}} = \begin{bmatrix} I & 0 & 0 \\ 0 & I & P_G^{\mathrm{HI}} \\ 0 & 0 & I \end{bmatrix} := \begin{bmatrix} I & 0 & 0 \\ 0 & I & -C_{G:11}^{-1} C_{G:12} \\ 0 & 0 & I \end{bmatrix}, \tag{8.38}$$

that is,

$$J_G^{\mathrm{EM}} = J_G^{\mathrm{SC}} J_G^{\mathrm{HI}} = \begin{bmatrix} I & -A_{G:11}^{-1} A_{G:12} & A_{G:11}^{-1} A_{G:12} C_{G:11}^{-1} C_{G:12} \\ 0 & I & -C_{G:11}^{-1} C_{G:12} \\ 0 & 0 & I \end{bmatrix}. \tag{8.39}$$

Note that the block $-A_{G:11}^{-1} A_{G:12}$ in the position (1,2) of (8.39) has no effect on the angle between the coarse and its complementary space. Hence, without loss of generality, we replace this block with the matrix of all zeros. Then, the final splitting based on local energy minimization is induced by a transformation of the form (8.36), where the local interpolation operator is given by

$$P_G = P_G^{\mathrm{EM}} := \begin{bmatrix} P_G^{\mathrm{SC}} P_G^{\mathrm{HI}} \\ P_G^{\mathrm{HI}} \end{bmatrix} := \begin{bmatrix} A_{G:11}^{-1} A_{G:12} C_{G:11}^{-1} C_{G:12} \\ -C_{G:11}^{-1} C_{G:12} \end{bmatrix}. \tag{8.40}$$

In practical computations the global two-level basis transformation that is induced by the local transformation (8.36), where P_G is given by (8.40), can easily be implemented. However, the matrix (8.40) certainly depends on the size of the stabilization parameter α in (8.4) and also on the (scalar) coefficient $a = a(T)$, which is assumed to be piecewise constant with (possible) jump discontinuities at the interior element interfaces on the finest mesh, here. Let us therefore consider a general macro-superelement with piecewise constant coefficients $a_1, a_2, \ldots, a_9$, where $0 < a_i \le 1$. In the following we will define a parameter-free local two-level transformation for which it is possible to conduct a rigorous analysis that will be presented at the end of this section. However, we want to stress that this is mainly for the purpose of gaining theoretical insight.

Let us start our considerations with the building blocks of the interpolation matrix (8.40), which (for the considered model problem) have the following representation:

$$P_G^{\mathrm{SC}}(\alpha; a_1, a_2, \ldots, a_9) := -A_{G:11}^{-1} A_{G:12} = P_{G,\infty}^{\mathrm{SC}} + \frac{1}{96\alpha} P_{G,\star}^{\mathrm{SC}}(a_1, a_2, \ldots, a_9),$$

where

$$
P_{G,\infty}^{\mathrm{SC}} = \frac{1}{96}\left[\begin{array}{rrrrrrrr}
7 & -1 & 3 & -5 & 1 & -1 & -1 & 81 \\
-1 & 1 & 81 & -1 & -1 & 7 & -5 & 3 \\
-1 & 1 & 9 & -1 & -1 & -5 & 7 & 3 \\
-5 & -1 & 3 & 7 & 1 & -1 & -1 & 9
\end{array}\right.
\left.\begin{array}{rrrrrrrr}
9 & -1 & -1 & 1 & 7 & 3 & -1 & -5 \\
3 & 7 & -5 & -1 & -1 & 9 & 1 & -1 \\
3 & -5 & 7 & -1 & -1 & 81 & 1 & -1 \\
81 & -1 & -1 & 1 & -5 & 3 & -1 & 7
\end{array}\right], \tag{8.41}
$$

and the matrix $P_{G,\star}^{\mathrm{SC}}$ depends on the coefficients a_i. This implies that

$$
\lim_{\alpha\to\infty} P_G^{\mathrm{SC}}(\alpha; a_1, a_2, \ldots, a_9) = P_{G,\infty}^{\mathrm{SC}}. \tag{8.42}
$$

Hence the interpolation matrix P_G^{SC} will be close to $P_{G,\infty}^{\mathrm{SC}}$ for any (fixed) coefficient distribution if the stabilization parameter α is chosen large enough, i.e., the limit does not depend on any of the parameters a_i, $i = 1, \ldots, 9$. Dealing with the matrix P_G^{HI} the expressions for the entries are more complicated as compared to P_G^{SC}, however, again the limit for $\alpha \to \infty$ is a constant matrix, i.e.,

$$
\lim_{\alpha\to\infty} P_G^{\mathrm{HI}}(\alpha; a_1, a_2, \ldots, a_9) = P_{G,\infty}^{\mathrm{HI}} \tag{8.43}
$$

where

$$
P_{G,\infty}^{\mathrm{HI}} = \frac{1}{4}\left[\begin{array}{rrrrrrrrrrrrrrrr}
0 & 0 & 0 & 4 & 0 & 0 & 0 & 0 & 0 & 0 & 0 & 0 & 0 & 0 & 0 & 0 \\
-1 & 2 & 0 & 1 & 1 & 2 & 0 & -1 & 0 & 0 & 0 & 0 & 0 & 0 & 0 & 0 \\
-1 & 0 & 2 & 1 & 1 & 0 & 2 & -1 & 0 & 0 & 0 & 0 & 0 & 0 & 0 & 0 \\
0 & 0 & 0 & 0 & 4 & 0 & 0 & 0 & 0 & 0 & 0 & 0 & 0 & 0 & 0 & 0 \\
2 & -1 & 1 & 0 & 0 & 0 & 0 & 0 & 2 & 1 & -1 & 0 & 0 & 0 & 0 & 0 \\
0 & 0 & 4 & 0 & 0 & 0 & 0 & 0 & 0 & 0 & 0 & 0 & 0 & 0 & 0 & 0 \\
0 & 0 & 0 & 0 & 0 & 0 & 0 & 0 & 0 & 4 & 0 & 0 & 0 & 0 & 0 & 0 \\
0 & -1 & 1 & 2 & 0 & 0 & 0 & 0 & 0 & 1 & -1 & 2 & 0 & 0 & 0 & 0 \\
0 & 0 & 0 & 0 & 2 & -1 & 1 & 0 & 0 & 0 & 0 & 0 & 2 & 1 & -1 & 0 \\
0 & 0 & 0 & 0 & 0 & 0 & 4 & 0 & 0 & 0 & 0 & 0 & 0 & 0 & 0 & 0 \\
0 & 0 & 0 & 0 & 0 & 0 & 0 & 0 & 0 & 0 & 0 & 0 & 0 & 4 & 0 & 0 \\
0 & 0 & 0 & 0 & 0 & -1 & 1 & 2 & 0 & 0 & 0 & 0 & 0 & 1 & -1 & 2 \\
0 & 0 & 0 & 0 & 0 & 0 & 0 & 0 & 0 & 0 & 0 & 4 & 0 & 0 & 0 & 0 \\
0 & 0 & 0 & 0 & 0 & 0 & 0 & 0 & -1 & 2 & 0 & 1 & 1 & 2 & 0 & -1 \\
0 & 0 & 0 & 0 & 0 & 0 & 0 & 0 & -1 & 0 & 2 & 1 & 1 & 0 & 2 & -1 \\
0 & 0 & 0 & 0 & 0 & 0 & 0 & 0 & 0 & 0 & 0 & 0 & 4 & 0 & 0 & 0
\end{array}\right]. \tag{8.44}
$$

This allows us to define a two-level basis transformation of the form (8.36) via the local matrix

$$P_G = P_G^{\mathrm{LW}} = P_{G,\infty}^{\mathrm{EM}} := \begin{bmatrix} P_{G,\infty}^{\mathrm{SC}} & P_{G,\infty}^{\mathrm{HI}} \\ & P_{G,\infty}^{\mathrm{HI}} \end{bmatrix} \tag{8.45}$$

where the building blocks $P_{G,\infty}^{\mathrm{SC}}$ and $P_{G,\infty}^{\mathrm{HI}}$ are given by (8.41) and (8.44), respectively. In particular, these interpolation matrices are parameter-free and the final two-level hierarchical basis representation of the global stiffness matrix (resulting from the related transformation) is based on *limiting interpolation weights*. Since the limit for $\alpha \to \infty$ exists for all of the involved (matrix-valued) functions we have the following lemma.

Lemma 8.12. *The limit for $\alpha \to \infty$ of the transformation based on local energy minimization (see (8.40)) is given by (8.36) where P_G is chosen according to (8.45), i.e.,*

$$\lim_{\alpha\to\infty} P_G^{\mathrm{EM}}(\alpha; a_1, a_2, \ldots, a_9) = P_G^{\mathrm{LW}}. \tag{8.46}$$

The splitting obtained via the transformation (8.36) with specification (8.45) will be further studied in the remainder of this section.

Our aim is to prove for the two-level transformation with limiting interpolation weights, that the local CBS constant γ_G is uniformly bounded by some constant $c < 1$ for a general macro-superelement matrix $\hat{A}_G$ that stems from a problem with piecewise constant coefficient $a = a(T)$. Note that a on the discrete level without loss of generality can be represented by the coefficients $a_1, a_2, \ldots, a_9$, where $0 < a_i \leq 1$.

Again, the local analysis of the CBS constant follows the simple general rule to compute γ_G, see Lemma 2.4 and the relation (2.35) thereafter:

$$\gamma_G^2 = 1 - \mu_1,$$

where μ_1 is the minimal eigenvalue of the generalized eigenproblem

$$S_G \mathbf{v}_{G:2} = \mu \hat{A}_{G:22} \mathbf{v}_{G:2}, \qquad \mathbf{v}_{G:2} \neq \mathbf{c}. \tag{8.47}$$

Here $\hat{A}_{G:22}$ denotes the lower-right block of the hierarchical macro-superelement matrix (8.35) obtained via either of the local transformations J_G^{EM} or J_G^{LW}.

Let us first recall the following important property of the (local) Schur complement S_G in (8.47), which is related to energy minimizing interpolation (see Lemma (2.1)), that is, we have

$$\mathbf{v}_{G:2}^T S_G \mathbf{v}_{G:2} = \min_{\mathbf{v}_{G:1}} \begin{bmatrix} \mathbf{v}_{G:1} \\ \mathbf{v}_{G:2} \end{bmatrix}^T \begin{bmatrix} A_{G:11} & A_{G:12} \\ A_{G:21} & A_{G:22} \end{bmatrix} \begin{bmatrix} \mathbf{v}_{G:1} \\ \mathbf{v}_{G:2} \end{bmatrix} \quad \forall \mathbf{v}_{G:2} \tag{8.48}$$

where A_G represents the fine-coarse partitioned macro-superelement matrix in the standard (nodal) basis. In the following we are concerned with proving a lower bound $\underline{\mu}$ for the minimal eigenvalue μ_1 of (8.47). We therefore rewrite

$$\mathbf{v}_{G:2}^T S_G \mathbf{v}_{G:2} \geq \underline{\mu}\, \mathbf{v}_{G:2}^T \hat{A}_{G:22} \mathbf{v}_{G:2} \quad \forall \mathbf{v}_{G:2}$$

in the form

$$\min_{\mathbf{v}_{G:1}} \begin{bmatrix} \mathbf{v}_{G:1} \\ \mathbf{v}_{G:2} \end{bmatrix}^T \begin{bmatrix} A_{G:11} & A_{G:12} \\ A_{G:21} & A_{G:22} - \underline{\mu}\, \hat{A}_{G:22} \end{bmatrix} \begin{bmatrix} \mathbf{v}_{G:1} \\ \mathbf{v}_{G:2} \end{bmatrix} \geq 0 \quad \forall \mathbf{v}_{G:2}. \tag{8.49}$$

So let us denote by B the matrix in (8.49), i.e.,

$$B := \begin{bmatrix} A_{G:11} & A_{G:12} \\ A_{G:21} & A_{G:22} - \underline{\mu}\, \hat{A}_{G:22} \end{bmatrix}. \tag{8.50}$$

If we are able to show that B is symmetric positive semidefinite (SPSD) for a given constant $\underline{\mu} > 0$ then $\underline{\mu}$ provides a lower bound for μ_1. We are ready to prove the main result of this section.

Theorem 8.13. *Consider the general macro-superelement matrix $\hat{A}_G$ obtained via (8.35) where the two-level transformation J_G is based on interpolation with limiting weights, see (8.45), and the matrix A_G stems from local assembling of the face matrices (8.32) and (8.33) with piecewise constant coefficient $a(T)$ over the macro-superelement, i.e., $a_i \in (0, 1]$, $1 \leq i \leq 9$. If $\alpha \geq \alpha_0 = 25$ then*

$$\gamma_G^2 \leq \frac{3}{4}. \tag{8.51}$$

Proof. The particular construction of the interpolation (8.45) with constant interpolation weights implies that the matrix B defined in (8.50) takes the form

$$B = B(\alpha; \mu; a_1, \ldots, a_9) = \alpha\,(B_0 + \mu C_0) + \sum_{i=1}^{9} a_i (B_i + \mu C_i) \tag{8.52}$$

with constant matrices B_i and C_i for $i = 0, \ldots, 9$. The matrices B_0 and C_0 are to be found SPSD and indefinite, respectively. Moreover, a direct computation shows that $B_0 + \mu C_0$ is SPSD for any $\mu \leq \mu_0 = (821 - \sqrt{27129})/1064 \approx 0.616815$. Hence $B_0 + \mu C_0$ is SPSD for $\mu = 1/4$. If for a fixed $\mu \leq \mu_0$ and a fixed stabilization parameter $\alpha = \alpha_0$ there holds $B(\alpha_0; \mu; a_1, \ldots, a_9) \geq 0$ then clearly $B(\alpha; \mu; a_1, \ldots, a_9)$ is SPSD for all $\alpha \geq \alpha_0$ for the same value of μ. Thence it suffices to prove that $B(\alpha_0; \underline{\mu}; a_1, \ldots, a_9) \geq 0$ for a given pair $(\alpha_0, \underline{\mu})$ in order to conclude that the inequality (8.49) holds for any $\alpha \geq \alpha_0$. Now, assuming that

$\alpha = \alpha_0$ and $\mu = \underline{\mu}$ are fixed, the matrix B according to (8.52) depends linearly on the parameters $a_i \in (0, 1]$. Thus we have

$$\begin{aligned}
&\min_{\mathbf{v}_G \neq \mathbf{0}} \left(\min_{a_i \in (0,1]} \frac{\mathbf{v}_G^T B(\alpha_0; \underline{\mu}; a_1, \ldots, a_9)\mathbf{v}_G}{\mathbf{v}_G^T \mathbf{v}_G} \right) \\
&\geq \min_{\mathbf{v}_G \neq \mathbf{0}} \left(\min_{a_i \in \{0,1\}} \frac{\mathbf{v}_G^T B(\alpha_0; \underline{\mu}; a_1, \ldots, a_9)\mathbf{v}_G}{\mathbf{v}_G^T \mathbf{v}_G} \right) \\
&= \min_{a_i \in \{0,1\}} \left(\min_{\mathbf{v}_G \neq \mathbf{0}} \frac{\mathbf{v}_G^T B(\alpha_0; \underline{\mu}; a_1, \ldots, a_9)\mathbf{v}_G}{\mathbf{v}_G^T \mathbf{v}_G} \right) \\
&= \min_{a_i \in \{0,1\}} \lambda_{\min}(B(\alpha_0; \underline{\mu}; a_1, \ldots, a_9)). \qquad (8.53)
\end{aligned}$$

Finally, the minimal eigenvalue of $B(25; \frac{1}{4}; a_1, \ldots, a_9)$ equals 0 for any choice of $a_i \in \{0, 1\}$ $\forall i = 1, \ldots, 9$ which proves the bound (8.51). □

Remark 8.14. The bound (8.51) holds with strict inequality, too, because the pair $(\alpha_0, \underline{\mu}) = (25, \frac{1}{4} + \epsilon)$ can also be used in the above line of argument if $\epsilon > 0$ is sufficiently small. It should also be noted that even smaller upper bounds are obtained for γ_G when using larger values of α_0.

8.4 Two-level method and AMLI preconditioning of graph-Laplacians

Here we present a rather general approach to the solution of the IP-DG system corresponding to (8.3). The composite iterative procedure is based on a first reduction step where the two-level iteration method is applied. For DG systems it is introduced in an algebraic setting in [50] and is studied in the general algebraic framework of [56]. Together with the DG space $\mathcal{V}$ it uses an auxiliary, in general smaller, space $\mathcal{V}_o$ and proper restriction and prolongation operators. In [50] three possibilities for $\mathcal{V}_o$ are considered. Here we take one of these, namely, $\mathcal{V}_o$ is the space of piecewise constant functions over the partition $\mathcal{T}$.

The two-level method reduces the problem to a system associated with the bilinear form $\mathcal{A}_h(\cdot, \cdot)$, defined by (8.4) or (8.5), on the space $\mathcal{V}_o$. The form is simplified to the jump part only, i.e. $\alpha \left\langle h_{\mathcal{F}}^{-1} [\![u]\!], [\![v]\!] \right\rangle_{\mathcal{F}_0 \cup \mathcal{F}_D}$. Then the related matrix A_0, further called *graph-Laplacian* is defined by $\mathbf{u}^T A_0 \mathbf{v} = \left\langle h_{\mathcal{F}}^{-1} [\![u]\!], [\![v]\!] \right\rangle_{\mathcal{F}_0 \cup \mathcal{F}_D}$. Here $\mathbf{u}, \mathbf{v}$ stand for the nodal vectors corresponding to the functions $u, v \in \mathcal{V}_o$. We can associate the partition $\mathcal{T}$ with a planar graph. The finite elements are the vertices and the interfaces of the finite elements are the edges of the graph. Then taking as degrees of freedom the values of the functions in $\mathcal{V}_o$ over each finite element, we shall get a matrix that has an entry -1 at the s graph vertices connected to a chosen graph vertex and an entry s at the vertex itself. For any partitions into quadrilaterals, regardless of the shape, we get the standard 4-point stencil with $(4, -1, -1, -1, -1)$, which probably is the reason for the name *graph-Laplacian.*

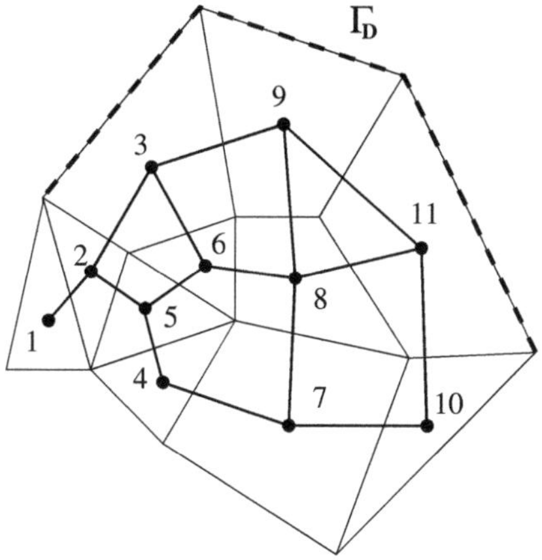

Figure 8.7: Partition $\mathcal{T}$ and related graph-Laplacian

The following graph-Laplacian A_0 corresponds to the mesh given in Figure 8.7:

$$A_0 = \begin{bmatrix} 1 & -1 & & & & & & & & & \\ -1 & 3 & -1 & & -1 & & & & & & \\ & -1 & 4 & & & -1 & & & -1 & & \\ & & & 2 & -1 & & -1 & & & & \\ & -1 & & -1 & 3 & -1 & & & & & \\ & & -1 & & -1 & 3 & & -1 & & & \\ & & & -1 & & & 3 & -1 & & -1 & \\ & & & & & -1 & -1 & 4 & -1 & & -1 \\ & & -1 & & & & & -1 & 4 & & -1 \\ & & & & & & -1 & & & 2 & -1 \\ & & & & & & & -1 & -1 & -1 & 4 \end{bmatrix}.$$

Remark 8.15. The graph-Laplacian is a symmetric M-matrix. However, this matrix does not have any approximation property on an arbitrary grid. Therefore, the multigrid theory that relies on such property cannot be used for designing robust preconditioners in this case.

We will consider now hierarchical basis splittings (HB) for the algebraic system involving the graph-Laplacian. To this end we follow the framework of HB two-level preconditioners and their multilevel (AMLI) extensions. The construction of a hierarchical decomposition for the related discontinuous Galerkin FE spaces is neither obvious nor unique. To fit the classical HB techniques, we search for a decomposition of the fine grid degrees of freedom, such that one part is associated with the degrees of freedom of the coarse grid problem. Let $A^{(\ell)} := A_0$ be the symmetric positive definite graph-Laplacian corresponding to the finest triangulation from the sequence of nested partitionings $\mathcal{T}_0 \subset \mathcal{T}_1 \subset \ldots \subset \mathcal{T}_\ell$ of the domain Ω. Further, introduce the graph-Laplacian associated with each triangulation level, i.e., $A^{(0)}, A^{(1)}, \ldots, A^{(\ell)}$.

Let us consider now two consecutive nested triangulations $\mathcal{T}_H \subset \mathcal{T}_h$ and the related graph-Laplacians A_H and A_h. We denote by χ_h the set of standard piecewise constant basis functions on the finer level and by $\tilde{\chi}_h$ the set of properly defined hierarchical basis functions. The hierarchical basis is determined by a nonsingular transformation matrix J_h^T, i.e., $\tilde{\chi}_h = J_h^T \chi_h$. As we know from the previous considerations, the hierarchical basis stiffness matrix $\tilde{A}_h$ is expressed as follows,

$$\tilde{A}_h = J_h^T A_h J_h. \tag{8.54}$$

On each finer level the matrix $\tilde{A}_h$ is partitioned into a two-by-two block form

$$\tilde{A}_h = \begin{bmatrix} \tilde{A}_{h:11} & \tilde{A}_{h:12} \\ \tilde{A}_{h:21} & \tilde{A}_{h:22} \end{bmatrix}, \tag{8.55}$$

where the lower-right diagonal block has the size of A_H. Regarding the splitting (8.55) we will suppose that it is is locally constructed so that the transformation matrix is sparse. Moreover, we will require the following relations to hold

$$\tilde{A}_{h:22} = A_H, \qquad \kappa\left(\tilde{A}_{h:11}\right) = O(1). \tag{8.56}$$

In what follows we will derive uniform estimates of the CBS constant based on the properly constructed hierarchical basis and related decomposition of the graph-Laplacian as a sum of local matrices associated with the set of edges $\mathcal{E}$ of the coarser grid $\mathcal{T}_H$. In our presentation we summarize some of the results from [83].

Mesh of triangles

Let us assume that the coarsest mesh $\mathcal{T}_0$ consists of triangles only, and each refined mesh is obtained by dividing the current triangle into four congruent triangles connecting the midpoints of its sides. Following the numbering from Figure 8.8, we

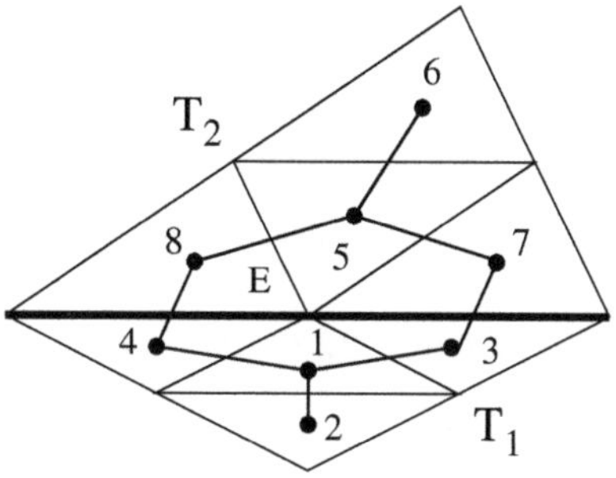

Figure 8.8: Macroelement of two adjacent triangles from $\mathcal{T}_H$

introduce the local matrix A_E in the form

$$A_E = \left[\begin{array}{cccc|cccc} 1 & -t & \tilde{t} & \tilde{t} & & & & \\ -t & t & & & & & & \\ \tilde{t} & & 1-\tilde{t} & & & -1 & & \\ \tilde{t} & & & 1-\tilde{t} & & & & -1 \\ \hline & & & & 1 & -t & \tilde{t} & \tilde{t} \\ & & & & -t & t & & \\ & & -1 & & \tilde{t} & & 1-\tilde{t} & \\ & & & -1 & \tilde{t} & & & 1-\tilde{t} \end{array}\right]. \tag{8.57}$$

This edge matrix is also associated with the macroelement $T_1 \cup T_2$ of the two adjacent triangles from $\mathcal{T}_H$ with a common side E. The role of the weight parameters $t \in (0, 1)$ and $\tilde{t} = (t-1)/2$ is to correctly distribute the contributions of the links between the interior nodes. For example, the couple (1,2) has a weight t here, but will appear also with a weight of $\tilde{t}$ in the local matrices associated with the remaining two sides of the current triangle T_1, so that the sum of weights equals one.

The hierarchical basis is introduced locally with respect to the coarser partitioning $\mathcal{T}_H$. Let us consider the triangle T_1 and the set of standard piecewise constant basis functions $\chi_{T_1} = \{\chi_{T_1:i}\}_{i=1}^4$. We introduce the related hierarchical basis

$\tilde{\chi}_{T_1} = \{\tilde{\chi}_{T_1:i}\}_{i=1}^4$ in the form

$$\begin{aligned} \tilde{\chi}_{T_1:1} &= \chi_{T_1:1} + p\chi_{T_1:2} + q\chi_{T_1:3} + q\chi_{T_1:4} \\ \tilde{\chi}_{T_1:2} &= \chi_{T_1:1} + q\chi_{T_1:2} + p\chi_{T_1:3} + q\chi_{T_1:4} \\ \tilde{\chi}_{T_1:3} &= \chi_{T_1:1} + q\chi_{T_1:2} + q\chi_{T_1:3} + p\chi_{T_1:4} \\ \tilde{\chi}_{T_1:4} &= r\,(\chi_{T_1:1} + \chi_{T_1:2} + \chi_{T_1:3} + \chi_{T_1:4}) \end{aligned} \tag{8.58}$$

where p, q are parameters to be determined later, and r is a scaling factor of the hierarchical basis function. Then the assembled transformation matrix J_E^T is as follows

$$J_E^T = \begin{bmatrix} 1 & p & q & q & & & & \\ 1 & q & p & q & & & & \\ 1 & q & q & p & & & & \\ & & & & 1 & p & q & q \\ & & & & 1 & q & p & q \\ & & & & 1 & q & q & p \\ r & r & r & r & & & & \\ & & & & r & r & r & r \end{bmatrix}, \tag{8.59}$$

and

$$\tilde{A}_E = J_E^T A_E J_E = \begin{bmatrix} \tilde{A}_{E:11} & \tilde{A}_{E:12} \\ \tilde{A}_{E:21} & \tilde{A}_{E:22} \end{bmatrix}.$$

Lemma 8.16. *Consider the hierarchical basis (8.58) for nested meshes of triangles. Then*

$$\tilde{A}_{h:22} = A_H \quad \textit{if and only if} \quad r = \frac{\sqrt{2}}{2}.$$

Proof. The definition of the last two terms in the local hierarchical basis ensures that $\tilde{A}_{E:22}$ has row-sums/column-sums equal to zero. Then, the equivalent statement

$$\tilde{A}_{E:22} = \begin{bmatrix} 1 & -1 \\ -1 & 1 \end{bmatrix}$$

simply follows from the equalities

$$\tilde{A}_{E:22}(1,1) = r^2 \sum_{i,j=1}^{4} A_E(i,j) = 2r^2\,, \quad \tilde{A}_{E:22}(2,2) = r^2 \sum_{i,j=5}^{8} A_E(i,j) = 2r^2\,.$$

□

Now, we apply Lemma 2.4 and get $\gamma_E^2 = 1 - \lambda$ where λ is the eigenvalue of the eigenproblem

$$\tilde{S}_E \mathbf{v} = \lambda \tilde{A}_{E:22} \mathbf{v}\,, \quad \mathbf{v} \neq \begin{bmatrix} 1 \\ 1 \end{bmatrix},$$

and $\tilde{S}_E^{(k)} = \tilde{A}_{E:22} - \tilde{A}_{E:21} \tilde{A}_{E:11}^{-1} \tilde{A}_{E:12}$.

Lemma 8.17. *Consider the hierarchical splitting (8.57), (8.59) with parameters $p = 1$, $q = -0.5$ and $t = 0.5$. Then the following estimate holds uniformly with respect to the refinement level k,*

$$\gamma^2 \leq \gamma_E^2 = \gamma_{TT}^2 = \frac{16}{25}\,. \tag{8.60}$$

Proof. The construction of the hierarchical basis and all related matrices are independent of the particular edge $E \in \mathcal{E}$ as well as of the current refinement level. The later holds true due to Lemma 8.16. Then, the estimate of the local CBS constant follows straightforwardly by simple computations with fixed numbers. Here, γ_{TT} indicates that the interface edge is always between two triangles. □

Mesh of quadrilaterals

We assume here, that the coarsest mesh $\mathcal{T}_0$ consists of quadrilaterals only, and each next refinement is obtained by dividing the current element in four new quadrilaterals as illustrated in Figure 8.9 (a).

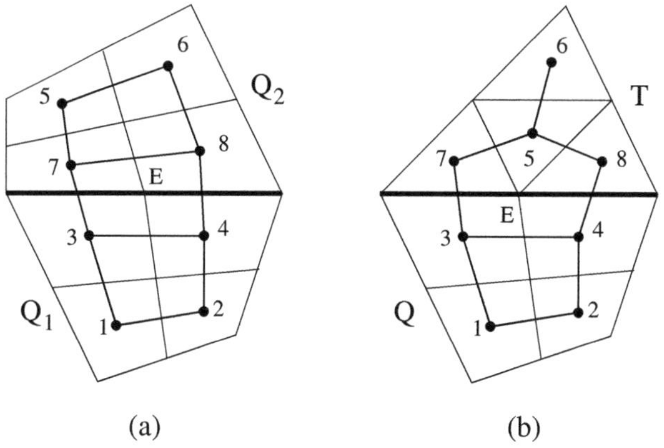

Figure 8.9: (a) Macroelement of two adjacent quadrilaterals of the mesh $\mathcal{T}_H$; (b) macroelement of adjacent triangle and quadrilateral of $\mathcal{T}_H$

Following the setting of the previous case and the node numbering from Figure 8.9, we introduce the new local matrix A_E in the form

$$A_E = \left[\begin{array}{cccc|cccc}
1/2 & -s & \tilde{s} & & & & & \\
-s & 1/2 & & \tilde{s} & & & & \\
\tilde{s} & & 3/2 & -s & & & -1 & \\
& \tilde{s} & -s & 3/2 & & & & -1 \\
\hline
& & & & 1/2 & -s & \tilde{s} & \\
& & & & -s & 1/2 & & \tilde{s} \\
& & -1 & & \tilde{s} & & 3/2 & -s \\
& & & -1 & & \tilde{s} & -s & 3/2
\end{array}\right]. \tag{8.61}$$

The weight parameters $s \in (0, 1)$ and $\tilde{s} = s - 1/2$ are again responsible for the correct distribution of the contributions of the links between the interior nodes of each quadrilateral macroelements Q_i, $i = 1, 2$, see Figure 8.9 (a).

The hierarchical basis is now introduced locally with respect to the quadrilaterals from $\mathcal{T}_H$. If we consider the macroelement Q_1, then the set of standard piecewise constant basis functions is $\boldsymbol{\chi}_{Q_1} = \{\chi_{Q_1:i}\}_{i=1}^4$, and the related hierarchical basis $\tilde{\boldsymbol{\chi}}_{Q_1} = \{\widehat{\chi}^{(k)}_{Q_1:i}\}_{i=1}^4$ is introduced in the form

$$\begin{aligned}
\tilde{\chi}_{Q_1:1} &= (\chi_{Q_1:1} + \chi_{Q_1:2}) - (\chi_{Q_1:3} + \chi_{Q_1:4} \\
\tilde{\chi}_{Q_1:2} &= (\chi_{Q_1:1} + \chi_{Q_1:3}) - (\chi_{Q_1:2} + \chi_{Q_1:4}) \\
\tilde{\chi}_{Q_1:3} &= (\chi_{Q_1:1} + \chi_{Q_1:4}) - (\chi_{Q_1:2} + \chi_{Q_1:3}) \\
\tilde{\chi}_{Q_1:4} &= r\,(\chi_{Q_1:1} + \chi_{Q_1:2} + \chi_{Q_1:3} + \chi_{Q_1:4})
\end{aligned} \tag{8.62}$$

where r is again the corresponding scaling factor. Then the assembled transformation matrix J_E^T reads as

$$J_E^T = \begin{bmatrix} 1 & 1 & -1 & -1 & & & & \\ 1 & -1 & 1 & -1 & & & & \\ 1 & -1 & -1 & 1 & & & & \\ & & & & 1 & 1 & -1 & -1 \\ & & & & 1 & -1 & 1 & -1 \\ & & & & 1 & -1 & -1 & 1 \\ r & r & r & r & & & & \\ & & & & r & r & r & r \end{bmatrix}. \quad (8.63)$$

We follow the local analysis scheme from the previous case (of a mesh of triangles) and get the next two lemmas.

Lemma 8.18. *Consider the hierarchical basis (8.62) for nested meshes of quadrilaterals. Then* $\tilde{A}_{h:22} = A_H$ *if and only if* $r = \sqrt{2}/2$.

Lemma 8.19. *The estimate*

$$\gamma^2 \leq \gamma_E^2 = \gamma_{QQ}^2 \rightarrow \frac{1}{2} \quad (8.64)$$

holds uniformly with respect to the refinement level k for the hierarchical splitting (8.62) with positive weight parameter $s \rightarrow 0^+$.

Proof. The straightforward computations lead to the following expression for the Schur complement

$$S_E = \frac{1-2s}{2(1-s)} \begin{bmatrix} 1 & -1 \\ -1 & 1 \end{bmatrix}$$

and therefore

$$\gamma_{QQ}^2 = 1 - \lambda = 1 - \frac{1-2s}{2(1-s)} = \frac{1}{2(1-s)}$$

which completes the proof. □

Here γ_{QQ} indicates that the interface edge is always between two quadrilaterals.

Mesh of quadrilaterals and triangles

The general case of a coarsest mesh $\mathcal{T}_0$ consisting of quadrilaterals and triangles is considered now. The refinement procedure is regular, and for the particular cases, it is the same as was considered in the previous two cases. What remains to be analyzed is the situation, where macroelements of different kind are adjacent as shown in Figure 8.9 (b). Combining the constructions from the previous two cases and following the node numbering from Figure 8.9 (b), we get the local matrix A_E in the form

$$A_E = \left[\begin{array}{cccc|cccc} 1/2 & -s & \tilde{s} & & & & & \\ -s & 1/2 & & \tilde{s} & & & & \\ \tilde{s} & & 3/2 & -s & & & -1 & \\ & \tilde{s} & -s & 3/2 & & & & -1 \\ \hline & & & & 1 & -t & \tilde{t} & \tilde{t} \\ & & & & -t & t & & \\ & & -1 & & \tilde{t} & & 1-\tilde{t} & \\ & & & -1 & \tilde{t} & & & 1-\tilde{t} \end{array}\right], \tag{8.65}$$

with weight parameters $s, t \in (0, 1)$, $\tilde{s} = s - 1/2$ and $\tilde{t} = (t - 1)/2$. Keeping the already introduced local definitions of hierarchical bases, we write the combined transformation matrix in the form

$$J_E^T = \begin{bmatrix} 1 & 1 & -1 & -1 & & & & \\ 1 & -1 & 1 & -1 & & & & \\ 1 & -1 & -1 & 1 & & & & \\ & & & & 1 & p & q & q \\ & & & & 1 & q & p & q \\ & & & & 1 & q & q & p \\ r & r & r & r & & & & \\ & & & & r & r & r & r \end{bmatrix}. \tag{8.66}$$

Let us stress the attention on the fact, that all locally introduced parameters are fixed for each particular triangle/quadrilateral macroelement from $\mathcal{T}_h$, independently of what kind of neighbors it has. In this respect, it is important, that $r = \sqrt{2}/2$ in both cases of triangles and quadrilaterals, see Lemma 8.16 and Lemma 8.18.

Lemma 8.20. *Consider the local matrix, corresponding to the case of an edge between a quadrilateral and a triangle, indicated below by "QT", and let* $r = \sqrt{2}/2$, $p = 1$, $q = -0.5$, $t = 0.5$, *and* $s \to 0^+$. *Then*

$$\tilde{A}_{E:22} = \begin{bmatrix} 1 & -1 \\ -1 & 1 \end{bmatrix},$$

and the relation

$$\gamma_E^2 = \gamma_{QT}^2 \to \frac{25}{43}, \tag{8.67}$$

holds uniformly with respect to the refinement level k.

Proof. Following the scheme from Lemma 8.19 we get

$$S_E = \frac{18 - 36s}{43 - 68s} \begin{bmatrix} 1 & -1 \\ -1 & 1 \end{bmatrix}$$

and therefore

$$\gamma_{QT}^2 = 1 - \lambda = 1 - \frac{18 - 36s}{43 - 68s} = \frac{25 - 32s}{43 - 68s},$$

which completes the proof. □

The next two theorems summarize the results of Lemmas 8.17, 8.19, and 8.20.

Theorem 8.21. *Consider the hierarchical splitting of the graph-Laplacian, corresponding to the general case of nested meshes, where the coarsest one $\mathcal{T}_0$ consists of quadrilaterals and triangles.*

(a) *Then $\tilde{A}_{22}^{(k+1)} = A^{(k)}$ if and only if $r = \sqrt{2}/2$;*

(b) *If $p = 1$, $q = -0.5$, $t = 0.5$, and $0 < s \leq \frac{35}{16} \approx 0.219$, then*

$$\gamma^2 \leq \max\{\gamma_{TT}^2, \gamma_{QQ}^2, \gamma_{QT}^2\} = \frac{16}{25}, \quad \textit{for all} \quad k, \quad 1 \leq k \leq \ell. \tag{8.68}$$

Theorem 8.22. *Let the parameters of the hierarchical splitting of the graph-Laplacian satisfy conditions (a)–(b) of Theorem 8.21. Then the related AMLI algorithm with acceleration polynomial of degree $\upsilon \in \{2, 3\}$ has optimal condition number and optimal total computational complexity.*

Proof. The statement follows directly from Theorem 8.21, taking into account that $\varrho = 4$ and

$$\gamma^2 \leq \frac{16}{25} < \frac{3}{4},$$

satisfying the optimality condition (2.78). □

Remark 8.23. We have assumed here, that the discontinuous Galerkin partitioning is obtained by a regular refinement of a given initial mesh, consisting of both triangles and quadrilaterals. However, the introduced approach is more generally applicable to partitionings, including pentagons, etc. The scheme is further suitable for constructing and analyzing AMLI preconditioners for discontinuous Galerkin systems in 3D.

9 AMLI methods for coupled problems

This chapter is devoted to the solution of systems of PDEs. As a general scheme, composite block iterative methods are used for the related coupled FE systems. Such methods are usually based on efficient solvers for the decoupled scalar FE elliptic problems. One such example is the Lamé system of linear elasticity where displacement decomposition methods are successfully applied if the Poisson ratio is far enough from the incompressibility limit. This means that the problem is relatively weakly coupled. When the system of PDEs is strongly coupled, specialized robust preconditioners are required. Two well-known such problems are considered in this chapter: a) The Lamé system of elasticity in the case of *almost incompressible* materials; and b) The initial-boundary value problem for the Navier–Stokes equations including the case of large Reynolds numbers. The presented AMLI methods are mostly based on materials from [33, 71, 88] for the elasticity problem, and from [27, 34] for the Navier–Stokes equations.

9.1 AMLI preconditioning of linear elasticity problems

9.1.1 Lamé system of elasticity

The target problem in this section is the following system of linear elasticity:

$$\begin{aligned} \sum_{j=1}^{2} \frac{\partial \sigma_{ij}}{\partial x_j} + f_i &= 0, \qquad \boldsymbol{x} \in \Omega, \qquad i = 1,2 \\ \boldsymbol{u} &= \boldsymbol{0}, \qquad \boldsymbol{x} \in \Gamma_D \\ \sum_{j=1}^{2} \frac{\partial \sigma_{ij}}{\partial x_j} n_j &= 0, \qquad \boldsymbol{x} \in \Gamma_N, \quad i = 1,2 \end{aligned}$$

where Ω is a polygonal domain in $\mathbb{R}^2$ and $\partial\Omega = \Gamma_D \cup \Gamma_N$ is the boundary of Ω. The stresses σ_{ij} and the strains ε_{ij} are defined by the classical Hooke's law, i.e.

$$\sigma_{ij}(\boldsymbol{u}) = \lambda \left(\sum_{k=1}^{2} \varepsilon_{kk}(\boldsymbol{u}) \right) \delta_{ij} + 2\mu\varepsilon_{ij}(\boldsymbol{u}), \qquad \varepsilon_{ij}(\boldsymbol{u}) = \frac{1}{2}\left(\frac{\partial u_i}{\partial x_j} + \frac{\partial u_j}{\partial x_i} \right).$$

We assume that the Lamé coefficients are piecewise constant in Ω. The unknowns of the problem are the displacements $\boldsymbol{u}^T = (u_1, u_2)$. A generalization to nonhomogeneous boundary condition is straightforward.

The Lamé coefficients are given by

$$\lambda = \frac{\nu E}{(1+\nu)(1-2\nu)}, \qquad \mu = \frac{E}{2(1+\nu)},$$

where E stands for the elasticity modulus, and $\nu \in [0, \frac{1}{2})$ is the Poisson ratio. We use the notion *almost incompressible* for the case $\nu = \frac{1}{2} - \delta$ ($\delta > 0$ is a small parameter). Note that the boundary value problem becomes ill-posed when $\nu = \frac{1}{2}$ (the material is *incompressible*).

For $\boldsymbol{f} = (f_1, f_2)^T \in (L_2(\Omega))^2$, the weak formulation of the boundary value problem reads:

Find $\boldsymbol{u} \in (H_0^1(\Omega))^2 = \{\boldsymbol{v} \in (H^1(\Omega))^2, \boldsymbol{v}\,|_{\partial\Omega} = 0\}$ such that

$$\mathcal{A}(\boldsymbol{u}, \boldsymbol{v}) = \int_\Omega \boldsymbol{f}^T \boldsymbol{v} dx \qquad \forall \boldsymbol{v} \in (H_0^1(\Omega))^2. \tag{9.1}$$

The bilinear form $\mathcal{A}(\boldsymbol{u}, \boldsymbol{v})$ is of the form

$$\begin{aligned} \mathcal{A}(\boldsymbol{u}, \boldsymbol{v}) &= \int_\Omega \lambda \operatorname{div}(\boldsymbol{u}) \operatorname{div}(\boldsymbol{v}) + 2\mu \sum_{i,j=1}^{2} \varepsilon_{ij}(\boldsymbol{u}) \varepsilon_{ij}(\boldsymbol{v}) dx \\ &= \int_\Omega \langle C \boldsymbol{d}(\boldsymbol{u}), \boldsymbol{d}(\boldsymbol{v}) \rangle dx, \end{aligned} \tag{9.2}$$

where

$$C = \begin{bmatrix} \lambda + 2\mu & 0 & 0 & \lambda \\ 0 & \mu & \mu & 0 \\ 0 & \mu & \mu & 0 \\ \lambda & 0 & 0 & \lambda + 2\mu \end{bmatrix}, \tag{9.3}$$

and

$$\boldsymbol{d}(\boldsymbol{u}) = \left[\frac{\partial u_1}{\partial x_1}, \frac{\partial u_1}{\partial x_2}, \frac{\partial u_2}{\partial x_1}, \frac{\partial u_2}{\partial x_2} \right]^T. \tag{9.4}$$

In case of the pure displacement problem, that is $\partial\Omega = \Gamma_D$, the following modification of the bilinear form holds true

$$\mathcal{A}(\boldsymbol{u}, \boldsymbol{v}) = \int_\Omega \langle C^s \boldsymbol{d}(\boldsymbol{u}), \boldsymbol{d}(\boldsymbol{u}) \rangle dx = \mathcal{A}^s(\boldsymbol{u}, \boldsymbol{v}) \tag{9.5}$$

where

$$C^s = \begin{bmatrix} \lambda + 2\mu & 0 & 0 & \lambda + \mu \\ 0 & \mu & 0 & 0 \\ 0 & 0 & \mu & 0 \\ \lambda + \mu & 0 & 0 & \lambda + 2\mu \end{bmatrix}. \tag{9.6}$$

Note that (9.5) holds due to homogeneous pure displacement boundary conditions as for $\boldsymbol{u}, \boldsymbol{v} \in (H_0^1(\Omega))^2$ we have

$$\int_\Omega \frac{\partial u_i}{\partial x_j}\frac{\partial v_j}{\partial x_i} d\boldsymbol{x} = \int_\Omega \frac{\partial u_i}{\partial x_i}\frac{\partial v_j}{\partial x_j} d\boldsymbol{x}.$$

The matrix C^s is positive definite. More details about the formulation of the elasticity problem and some important properties can be found e.g. in [8, 42, 43].

As in the previous chapters, we assume that the domain Ω is discretized using finite elements. The partitioning is denoted by $\mathcal{T}_\ell$ and is supposed to be obtained by ℓ regular refinement steps of a given coarse triangulation $\mathcal{T}_0$. This section is mostly focused to the robustness of the AMLI methods with respect to the Poisson ratio in the case of *almost incompressible* materials. Let us note that the condition number of the stiffness matrix deteriorates when ν approaches the incompressibility limit. More precisely, for any fixed mesh, the condition number can be estimated by $\kappa(A) = \mathcal{O}\left((1-2\nu)^{-1}\right)$.

9.1.2 On the robustness of AMLI for conforming FE elasticity systems

The first known robustness results concerning AMLI preconditioning of *almost incompressible* elasticity problems are for the case of linear triangular elements. The hierarchical splitting from Chapter 3 is assumed. Similarly to the scalar case, the CBS constant is bounded [2, 8, 87] by

$$\gamma^2 < \frac{3}{4}.$$

This estimate is uniform with respect to the Poisson ratio $\nu \in [0, 1/2)$ as well as to the mesh anisotropy and coefficient jumps which are aligned with the interfaces of the initial mesh $\mathcal{T}_0$. Then, if $\upsilon \in \{2, 3\}$, the AMLI method has optimal convergence rate. However, this is not enough to get a robust AMLI algorithm of optimal complexity. The problem is that for any fixed mesh, the condition number of the first pivot block also deteriorates with ν, that is, $\kappa(A_{11}^{(k+1)}) = \mathcal{O}\left((1-2\nu)^{-1}\right)$. Up to now, we do not know how to efficiently precondition the related blocks $A_{11}^{(k+1)}$. Unfortunately, the additive and multiplicative preconditioners developed for the scalar elliptic problems are not applicable here. The pivot block-matrices become strongly anisotropic when $\nu \to 1/2$.

What makes the problem additionally more complicated is that the directions of dominating anisotropy are different with respect to the different nodal displacements in any given mesh point.

Here we will demonstrate an alternative semi-coarsening approach [88]. Let us assume that Ω is a rectangular polygon which is discretized by some initial splitting of rectangular elements $\mathcal{T}_0$. Bilinear conforming finite elements are used to approximate the elasticity problem associated with the bilinear form (9.2) written in the form

$$\mathcal{A}(\boldsymbol{u}, \boldsymbol{v}) = \lambda \mathcal{A}_1(\boldsymbol{u}, \boldsymbol{v}) + 2\mu \mathcal{A}_2(\boldsymbol{u}, \boldsymbol{v}). \tag{9.7}$$

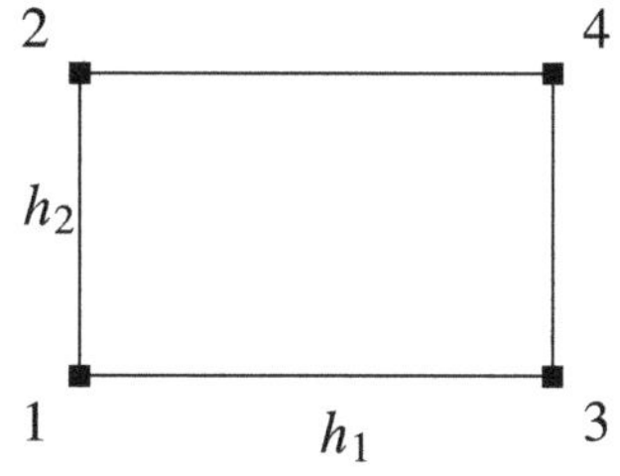

Figure 9.1: Rectangle bilinear element

Now, to get a sufficient accuracy of the numerical solution, a balanced semi-coarsening procedure is used (see Figure 9.2) to construct the nested meshes $\mathcal{T}_0 \subset \mathcal{T}_1 \subset \ldots \subset \mathcal{T}_\ell$. Associated with $\{\mathcal{T}_k\}$ are the finite element stiffness matrices $A^{(0)}, A^{(1)}, \ldots, A^{(\ell)}$ which are computed using the standard nodal basis test functions. The efficient solution of the system of equations corresponding to the finest mesh $\mathcal{T}_\ell$, namely $A\mathbf{u} = \mathbf{b}$, $A = A^{(\ell)}$, $\mathbf{b} = \mathbf{b}^{(\ell)}$ and $\mathbf{u} = \mathbf{u}^{(\ell)}$ is our goal. At the end of this introductory part we present the element stiffness matrix A_e corresponding to the studied problem. The following formulas are used for a local analysis of the constant γ in the strengthened CBS inequality. Following (9.7) we get

$$A_e = \lambda A_e^{(1)} + 2\mu A_e^{(2)} \tag{9.8}$$

where

$$A_e^{(1)} = \frac{1}{12} \begin{bmatrix} 4\epsilon & 2\epsilon & -4\epsilon & -2\epsilon & 3 & -3 & 3 & -3 \\ 2\epsilon & 4\epsilon & -2\epsilon & -4\epsilon & 3 & -3 & 3 & -3 \\ -4\epsilon & -2\epsilon & 4\epsilon & 2\epsilon & -3 & 3 & -3 & 3 \\ -2\epsilon & -4\epsilon & 2\epsilon & 4\epsilon & -3 & 3 & -3 & 3 \\ 3 & 3 & -3 & -3 & 4/\epsilon & -4/\epsilon & 2/\epsilon & -2/\epsilon \\ -3 & -3 & 3 & 3 & -4/\epsilon & 4/\epsilon & -2/\epsilon & 2/\epsilon \\ 3 & 3 & -3 & -3 & 2/\epsilon & -2/\epsilon & 4/\epsilon & -4/\epsilon \\ -3 & -3 & 3 & 3 & -2/\epsilon & 2/\epsilon & -4/\epsilon & 4/\epsilon \end{bmatrix}$$

and

$$A_e^{(2)} = \frac{1}{12}\left[\begin{array}{cccccccc}
4\epsilon+2/\epsilon & 2\epsilon-2/\epsilon & -4\epsilon+1/\epsilon & -2\epsilon-1/\epsilon & 3/2 & 3/2 & -3/2 & -3/2 \\
2\epsilon-2/\epsilon & 4\epsilon+2/\epsilon & -2\epsilon-1/\epsilon & -4\epsilon+1/\epsilon & -3/2 & -3/2 & 3/2 & 3/2 \\
-4\epsilon+1/\epsilon & -2\epsilon-1/\epsilon & 4\epsilon+2/\epsilon & 2\epsilon-2/\epsilon & 3/2 & 3/2 & -3/2 & -3/2 \\
-2\epsilon-1/\epsilon & -4\epsilon+1/\epsilon & 2\epsilon-2/\epsilon & 4\epsilon+2/\epsilon & -3/2 & -3/2 & 3/2 & 3/2 \\
3/2 & -3/2 & 3/2 & -3/2 & 2\epsilon+4/\epsilon & \epsilon-4/\epsilon & -2\epsilon+2/\epsilon & -\epsilon-2/\epsilon \\
3/2 & -3/2 & 3/2 & -3/2 & \epsilon-4/\epsilon & 2\epsilon+4/\epsilon & -\epsilon-2/\epsilon & -2\epsilon+2/\epsilon \\
-3/2 & 3/2 & -3/2 & 3/2 & -2\epsilon+2/\epsilon & -\epsilon-2/\epsilon & 2\epsilon+4/\epsilon & \epsilon-4/\epsilon \\
-3/2 & 3/2 & -3/2 & 3/2 & -\epsilon-2/\epsilon & -2\epsilon+2/\epsilon & \epsilon-4/\epsilon & 2\epsilon+4/\epsilon
\end{array}\right].$$

In the above matrices ϵ stands for the ratio of mesh anisotropy, i.e., $\epsilon = \frac{h_2}{h_1}$, the ordering of the entries corresponds to the separate displacements vector of the nodal parameters $\{u_1^1, u_1^2, u_1^3, u_1^4, u_2^1, u_2^2, u_2^3, u_2^4\}$, and to the nodes of the reference element shown in Figure 9.1. We observe the following useful relations: $A_{e:11}^{(2)} = A_{e:11}^{(1)} + \frac{1}{2}A_{e:22}^{(1)}$, $A_{e:22}^{(2)} = A_{e:22}^{(1)} + \frac{1}{2}A_{e:11}^{(1)}$, $A_{e:12}^{(2)} = A_{e:12}^{(1)} + \frac{1}{2}A_{e:21}^{(1)}$, where the 2×2 block representation corresponds to the components $\{u_1^1, u_1^2, u_1^3, u_1^4\}$ and $\{u_2^1, u_2^2, u_2^3, u_2^4\}$.

We will apply here the so called *hybrid V-cycle AMLI* where the degree υ_k of the accelerating matrix polynomial is cyclicly varied. The related AMLI optimality conditions are formulated in the next theorem.

Theorem 9.1 ([112]). *The PCG iteration method defined by the AMLI preconditioner is of optimal order if the following conditions hold:*

(i) *The properly scaled approximation $C_{11}^{(k+1)}$ is chosen such that*

$$\kappa(C_{11}^{(k+1)^{-1}} A_{11}^{(k+1)}) = \mathcal{O}(1),$$

and the action of $C_{11}^{(k+1)^{-1}}$ on a given vector $\mathbf{v}$, requires $\mathcal{O}(N_{k+1} - N_k)$ arithmetic operations.

(ii) *The degree υ_k of the accelerating polynomial p_{υ_k} satisfies the conditions:*

$$\upsilon_k = 1 \qquad\qquad \textit{if } (k \bmod k_0) \neq 0,$$

$$\frac{1}{\sqrt{1-\gamma^{(k_0)^2}}} < \upsilon_k < \varrho_{k_0} \quad \textit{if } (k \bmod k_0) = 0.$$

Here N_k is the number of the nodal unknowns, corresponding to $\mathcal{T}_k$; $\varrho_{k_0} = \frac{N_{(j+1)k_0}}{N_{jk_0}}$ stands for the mesh refinement ratio of k_0 consecutive mesh refinement steps; and $\gamma^{(k_0)}$ is the constant in the strengthened CBS inequality, corresponding to the nested FE spaces $\mathcal{V}_{(j+1)k_0}$ and $\mathcal{V}_{jk_0}$.

The proof of the theorem can be found in [112], where the *hybrid V-cycle AMLI* is introduced ($k_0 \neq 1$). The standard AMLI algorithm corresponds to the case $k_0 = 1$. In the general case, the hybrid V-cycle AMLI is a special case W-cycle algorithm with linear parts of graph length k_0.

For the local analysis of the CBS constant $\gamma^{(k_0)}$ we consider the macro-element stiffness matrix $A_E^{((j+1)k_0))}$ of a given current element $E \in \mathcal{T}_{jk_0}$, related to $\mathcal{T}_{(j+1)k_0}$, i.e., to the discretization after k_0 consecutive mesh refinement steps:

$$A_E^{((j+1)k_0)} = \begin{bmatrix} A_{E:11}^{((j+1)k_0)} & A_{E:12}^{((j+1)k_0)} \\ A_{E:21}^{((j+1)k_0)} & A_{E:22}^{((j+1)k_0)} \end{bmatrix}, \tag{9.9}$$

where $A_{E:11}^{((j+1)k_0)}$ is the block, corresponding to the nodes from $\mathcal{T}_{(j+1)k_0} \setminus \mathcal{T}_{jk_0}$. Following the well established general procedure, the CBS constant is estimated by

$$\gamma^{(k_0)} \leq \max_{E\in\mathcal{T}_{(j+1)k_0}} \{\max_j \{\gamma_{E:j}^{(k_0)}\}\}. \tag{9.10}$$

The local constant $\gamma_{E:j}^{(k_0)}$ can be computed as $\gamma_{E:j}^{(k_0)} = \sqrt{\lambda_1}$, where λ_1 is the largest eigenvalue in the generalized eigenvalue problem

$$S_E^{((j+1)k_0)}\mathbf{v} = (1-\lambda)A_e^{(jk_0)}\mathbf{v}, \qquad \mathbf{v} \in \mathbb{R}^8 \setminus \ker(A_e^{(jk_0)}), \tag{9.11}$$

$A_e^{(jk_0)}$ is the standard nodal basis element stiffness matrix, and

$$S_E^{((j+1)k_0)} = A_{E:22}^{((j+1)k_0)} - A_{E:21}^{((j+1)k_0)} A_{E:11}^{((j+1)k_0)^{-1}} A_{E:12}^{((j+1)k_0)}$$

is the Schur complement at level $((j+1)k_0)$.

There are various problems (see, e.g., in the previous chapters), where the local eigenvalue problem (9.11) can be solved (or analyzed) directly. As in the current case the local problem is relatively larger, we will use in addition the following lemmas.

Lemma 9.2. *Consider the matrix*

$$B_E^{((j+1)k_0)} = \begin{bmatrix} A_{E:11}^{((j+1)k_0)} & A_{E:12}^{((j+1)k_0)} \\ A_{E:21}^{((j+1)k_0)} & A_{E:22}^{((j+1)k_0)} - \Theta A_e^{(jk_0)} \end{bmatrix}. \tag{9.12}$$

The CBS constant $\gamma_{E:j}^{(k_0)}$ corresponding to (9.11) satisfies the estimate

$$\gamma_{E:j}^{(k_0)} \leq \sqrt{1-\Theta},$$

if and only if, $B_E^{((j+1)k_0)}$ is positive semidefinite.

Lemma 9.3. *Applying (9.8), we write the matrix B_E from Lemma 9.2 in the form*

$$B_E = \lambda B_E^{(1)} + 2\mu B_E^{(2)},$$

where the superscripts related to the refinement level are omitted. Then the matrix B_E is positive semidefinite for each $\nu \in [0, \frac{1}{2})$, if and only if, $B_E^{(1)}$ and $B_E^{(2)}$ are positive semidefinite.

The proof of the lemmas follows from the definitions of the Schur complement (for more details see [87]), and from the relations (9.2) and (9.8). These two lemmas allow us to estimate $\gamma_{E:j}^{(k_0)}$ directly, verifying that the corresponding matrices $B_E^{(1)}$ and $B_E^{(2)}$ are positive semidefinite, instead of solving the more complicated eigenvalue problem (9.11). Let us remind that such a technique was already used in the proof of Theorem 8.13.

We consider now a variant of the hybrid V-cycle AMLI algorithm with $k_0 = 2$, $C_{11}^{(k+1)} = A_{11}^{(k+1)}$ and semi-coarsening mesh refinement as shown (for $\varrho = 2$) in Figure 9.2. The mesh is refined at the even steps ($k = 2, 4, 6, \ldots$) along "x_1" direction, and respectively, at the odd steps ($k = 3, 5, 7, \ldots$) along "x_2" direction. One important property of this construction is that the blocks $A_{11}^{(k+1)}$ are band matrices with a bandwidth $2(2\varrho + 1)$. We call this algorithm *balanced semi-coarsening AMLI*, and will study the constant $\gamma^{(k_0)}$ (corresponding to two coupled refinement steps) trying to find υ_k satisfying the condition (ii) from Theorem 9.1.

The next theorem gives a solution of the problem in the case of an isotropic mesh.

Theorem 9.4. *Consider the balanced semi-coarsening AMLI with $\varrho = 2$, and square initial mesh $\mathcal{T}_0$. Then the constant in the strengthened CBS inequality is bounded uniformly with respect to the Poisson ratio, and*

$$\left(\gamma^{(k_0)}\right)^2 \leq \frac{6}{7} \tag{9.13}$$

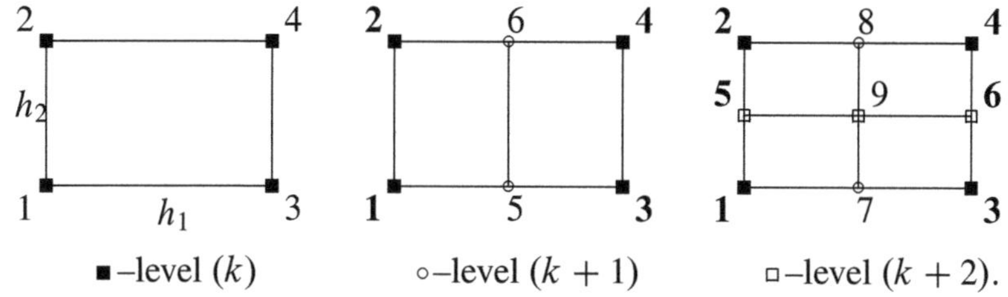

Figure 9.2: Balanced semi-coarsening

Proof. Following Lemma 9.2 and Lemma 9.3, we prove that the matrices $B_E^{(1)}$ and $B_E^{(2)}$ are positive semidefinite. Here B_E corresponds to (9.12) with $\Theta = \frac{1}{7}$. Note, that the initial mesh $\mathcal{T}_0$ is square, so that the local constant in the strengthened CBS inequality is independent of the current element and of the refinement level, that is, we will obtain the global estimate (9.13), by a local analysis with $\epsilon = 1$. We get $B_E^{(1)}$ and $B_E^{(2)}$ explicitly, applying the FEM assembling procedure to the element stiffness matrices $A_e^{(1)}$ and $A_e^{(2)}$ given by (9.8). Now we use a standard LDL^T factorization to prove the positive semi-definiteness of the above matrices. In such a way we obtain by direct (parameter-free) computations the presentations

$$B_E^{(1)} = L_E^{(1)} D_E^{(1)} {L_E^{(1)}}^T, \tag{9.14}$$

where $L_E^{(1)}$ is a lower triangular matrix and $D_E^{(1)}$ is the diagonal matrix

$$D_E^{(1)} = \operatorname{diag}\left(\frac{4}{3}, \frac{4}{3}, \frac{1}{3}, \frac{7}{12}, 0, \frac{4}{7}, \frac{19}{48}, \frac{7}{48}, \frac{7}{19}, 0, \frac{1}{112}, \frac{1}{112}, 0, 0, 0, 0, 0, 0\right),$$

and

$$B_E^{(2)} = L_E^{(2)} D_E^{(2)} {L_E^{(2)}}^T, \tag{9.15}$$

with

$$D_E^{(2)} = \operatorname{diag}\Big(14, 14, \frac{49}{8}, 7, 6, 7, \frac{539}{192}, \frac{161}{32}, \frac{60}{11},$$

$$\frac{105}{23}, \frac{719}{480}, \frac{28303}{23008}, \frac{1704}{1705}, \frac{18530}{17679}, \frac{32}{5}, 0, 0, 0\Big).$$

Since the entries of $D_E^{(1)}$ and $D_E^{(2)}$ are non negative, the matrices $B_E^{(1)}$ and $B_E^{(2)}$ are positive semi-definite, which completes the proof of the theorem. □

Combining the estimate (9.13) from Theorem 9.4 with the basic Theorem 9.1 we get the following result.

Theorem 9.5. *The hybrid V-cycle balanced semi-coarsening AMLI algorithm, defined on a square initial mesh $\mathcal{T}_0$, and with parameters $\varrho = 2$, $k_0 = 2$ and $\upsilon_{2j} = 3$, is of optimal order, uniformly with respect to the Poisson ratio $\nu \in [0, 1/2)$, and to coefficient jumps which are aligned with the initial mesh $\mathcal{T}_0$.*

Proof. Let us recall that the blocks $\tilde{A}_{11}^{(k+1)} = A_{11}^{(k+1)}$ are $2(2\varrho + 1)$-diagonal matrices, so that the condition (i) from Theorem 9.1 is satisfied by the construction of the semi-coarsening AMLI algorithm. We substitute now the current AMLI parameters in the second optimality condition (ii) and get

$$\sqrt{7} < 3 < 4,$$

which completes the proof. □

The result from Theorem 9.5 is an optimal algorithm for PCG iterative solution of FE elasticity systems. The total computational cost is proportional to the size of the discrete problem, with a proportionality constant independent of the Poisson ratio.

We present now a numerically performed analysis of the behavior of $\gamma^{(k_0)}$, varying the modified Poisson ratio $\tilde{\nu} = \nu/(1-\nu) \in [0, 1)$ and the aspect ratio ϵ, for the same balanced semi-coarsening AMLI setting, i.e. with $\varrho = 2$.

Table 9.1: Numerically computed $\kappa = (1 - (\gamma_e^{(k_0)})^2)^{-1}$: balanced semi-coarsening AMLI, $\varrho = 2$, $\frac{N_{2(j+1)}}{N_{2j}} = 4$

$\tilde{\nu}$	$\epsilon = 0.01$	$\epsilon = 0.1$	$\epsilon = 1$	$\epsilon = 10$	$\epsilon = 100$
0.7	3.99882	3.88755	3.04878	3.88755	3.99883
0.8	3.99913	3.90420	3.82352	3.90420	3.99898
0.9	3.99970	3.97029	5.00000	3.97020	3.99964
0.99	4.40685	4.89500	6.73973	4.88123	4.01299
0.999	4.14115	6.47205	6.97309	6.48445	4.22446

The presented data show first that the estimate from Theorem 5.1 is asymptotically exact. We see also that:

- The statements of Theorem 9.4 and Theorem 9.5 remain valid in the case of a rectangular initial mesh $\mathcal{T}_0$ with strongly varying mesh aspect ratio $\epsilon = h_2/h_1$.
- The hybrid V-cycle balanced semi-coarsening AMLI algorithm, defined on a rectangular initial mesh $\mathcal{T}_0$, and with parameters $\varrho = 2$, $k_0 = 2$ and $\upsilon_{2j} = 2$, is of optimal order for moderate values of the modified Poisson ratio $\tilde{\nu} \in [0, 0.8)$.

The last modification of Theorem 9.5 (with $\upsilon_{2j} = 2$) is important for the efficient application of the considered balanced semi-coarsening algorithm, as the condition $\tilde{\nu} \in [0, 0.8)$ is satisfied for many elastic materials of practical importance.

At the end of this part, let us note that the presented results concern only the efficient solution of elasticity FE systems as they arise applying standard conforming finite elements. The complete solution of the so-called *locking* phenomenon, which normally occurs in the limit case of *almost incompressible* media, requires the development of robust solution methods for *locking-free* discretization methods. One such AMLI method is presented at the end of this section.

9.1.3 Locking-free AMLI methods for Crouzeix–Raviart FE discretization of the pure displacement elasticity problem

Let us consider the pure displacement problem ($\partial\Omega = \Gamma_D$), and let us assume that Crouzeix–Raviart nonconforming finite elements are used for discretization of the variational problem (9.1). The modified bilinear form $\mathcal{A}^s$ defined by (9.5) is applied. Let us denote by $\boldsymbol{u}_h \in \mathcal{V}_h$ the approximate solution corresponding to the triangulation $\mathcal{T}_h$ where $\mathcal{V}_h$ is the related Crouzeix–Raviart FE space. Then the following error estimate holds.

Theorem 9.6 ([43]). *There exists a constant $C_{\Omega,\theta}$ (independent of $h, \lambda, \mu; \theta$ is the smallest angle in the mesh), such that*

$$\|\boldsymbol{u} - \boldsymbol{u}_h\|_h \le C_{\Omega,\theta}\, h\, \|\boldsymbol{f}\|_{[L_2(\Omega)]^2},$$

where $\|\cdot\|_h := \sqrt{\mathcal{A}_h^s(\cdot,\cdot)}$.

This theorem means that the considered nonconforming linear FE approximation is *locking-free*, with the same rate of accuracy as for the linear conforming finite elements. Let us note once again, that in the case of lower order conforming finite elements, the *locking* phenomenon appears, that is, $\lim_{\nu \to 1/2} C_{\Omega,\theta} = \infty$ for any fixed mesh (for details see, e.g., [42, 43, 57]).

Our presentation here will use the structure (to some extend), the notations, and some of the figures introduced in Section 4. Let us consider two consecutive nested meshes $\mathcal{T}_H$ and $\mathcal{T}_h$. As we already know, for Crouzeix–Raviart nonconforming linear elements, the FE spaces associated with two consecutive mesh refinements are not nested. To enable the use of the general multilevel scheme, we will apply the differences and aggregates (DA) approach to construct a hierarchical two-level decomposition of the Crouzeix–Raviart pure displacement elasticity systems. The algorithm is described on macroelement level, see Figure 4.1. Let

$\phi_1, \dots, \phi_9$ be the standard nodal nonconforming linear finite element basis functions on the macroelement E. Then for the 2D elasticity problem we use the basis functions $\phi_i^{(1)} = (\phi_i, 0)^T$ and $\phi_i^{(2)} = (0, \phi_i)^T$, $i = 1, \dots, 9$. The vector of the macroelement basis functions

$$\varphi_E = \{\Phi_i\}_{i=1}^{18} = \left(\phi_1^{(1)}, \phi_1^{(2)}, \phi_2^{(1)}, \phi_2^{(2)}, \dots, \phi_9^{(1)}, \phi_9^{(2)}\right)^T$$

is transformed into a vector of new hierarchical basis functions $\tilde{\varphi}_E = \{\tilde{\Phi}_i\}_{i=1}^{18}$. Similarly to the scalar case, the transformation matrix $J_E^T = (J_{ij})^T$ is determined by the splitting $\mathcal{V}(E) = \{\Phi_i\}_{i=1}^{18} = \tilde{\mathcal{V}}_1(E) \oplus \tilde{\mathcal{V}}_2(E)$,

$$\begin{aligned} \tilde{\mathcal{V}}_1(E) &= \operatorname{span}\{\tilde{\Phi}_i\}_{i=1}^{12} \\ &= \operatorname{span}\{\phi_1^{(k)}, \phi_2^{(k)}, \phi_3^{(k)}, \phi_4^{(k)} - \phi_5^{(k)}, \\ &\qquad \phi_6^{(k)} - \phi_7^{(k)}, \phi_8^{(k)} - \phi_9^{(k)}\}_{k=1}^{2}, \end{aligned} \tag{9.16}$$

$$\begin{aligned} \tilde{\mathcal{V}}_2(E) &= \operatorname{span}\{\tilde{\Phi}_i\}_{i=13}^{18} \\ &= \operatorname{span}\{\phi_1^{(k)} + \phi_4^{(k)} + \phi_5^{(k)}, \phi_2^{(k)} + \phi_6^{(k)} + \\ &\qquad \phi_7^{(k)}, \phi_3^{(k)} + \phi_8^{(k)} + \phi_9^{(k)}\}_{k=1}^{2}. \end{aligned} \tag{9.17}$$

Accordingly, the macroelement stiffness matrix A_E^s is transformed into a hierarchical form $\tilde{A}_E^s = J_E^T A_E^s J_E$,

$$\tilde{A}_E^s = \begin{bmatrix} \tilde{A}_{E:11}^s & \tilde{A}_{E:12}^s \\ \tilde{A}_{E:21}^s & \tilde{A}_{E:22}^s \end{bmatrix} \begin{matrix} \tilde{\phi}_i \in \tilde{\mathcal{V}}_1(E) \\ \tilde{\phi}_i \in \tilde{\mathcal{V}}_2(E) \end{matrix} .$$

The corresponding global stiffness matrix

$$\tilde{A}_h^s = \sum_{E \in \mathcal{T}_H} R_E^T \tilde{A}_E^s R_E$$

is again decomposed into 2×2 blocks

$$\tilde{A}_h^s = \begin{bmatrix} \tilde{A}_{h:11}^s & \tilde{A}_{h:12}^s \\ \tilde{A}_{h:21}^s & \tilde{A}_{h:22}^s \end{bmatrix}, \tag{9.18}$$

which are induced by the decomposition on macroelement level. The block $\tilde{A}_{h:11}^s$ corresponds to the interior nodal unknowns with respect to the macro-elements $E \in \mathcal{T}_H$ plus the differences of the nodal unknowns along the sides of E. The block $\tilde{A}_{h:22}^s$ corresponds to certain aggregates of nodal unknowns. The introduced decomposition is used to construct recursively the multilevel (AMLI) preconditioners, and as we know, their condition number can be estimated based on the corresponding CBS constant γ.

Theorem 9.7 ([33]). *Consider the pure elasticity problem where the Lamé coefficients are constant on the coarse triangles $E \in \mathcal{T}_H$. Discretization by the Crouzeix–Raviart FE and DA decomposition of the stiffness matrix are applied. Then for any element size and shape, and for any Poisson ratio $\nu \in [0, \frac{1}{2})$, there holds*

$$\gamma^2 \leq \frac{3}{4}.$$

Proof. As we know, the global constant γ can be estimated by the maximum of the local ones over the macroelements. As in the proof of Theorem 4.9, the local bilinear forms are directly analyzed, instead of solving the related local eigenvalue problems for λ_1 and then computing $\gamma_E^2 = 1 - \lambda_1$. Let us first consider the case of a right angled reference macroelement, see Figure 4.2.

Let $\tilde{\mathcal{V}}_1(\widehat{E})$, $\tilde{\mathcal{V}}_2(\widehat{E})$, be the two-level splitting for the reference macroelement $\widehat{E}$. For $\boldsymbol{u} \in \tilde{\mathcal{V}}_1(\widehat{E})$ and $\mathbf{v}v \in \tilde{\mathcal{V}}_2(\widehat{E})$ we denote by $\boldsymbol{d}^{(r)} := \boldsymbol{d}(\boldsymbol{u})\,|_{T_r}$ and $\boldsymbol{\delta}^{(r)} := \boldsymbol{d}(\boldsymbol{v})\,|_{T_r}$, $r = 1, \ldots, 4$, according to the definition (9.4). Then it is easy to show (cf. [31]) that

$$\boldsymbol{d}^{(1)} + \boldsymbol{d}^{(2)} + \boldsymbol{d}^{(3)} + \boldsymbol{d}^{(4)} \;=\; \mathbf{0}, \tag{9.19}$$

$$\boldsymbol{\delta}^{(1)} = \boldsymbol{\delta}^{(2)} = \boldsymbol{\delta}^{(3)} = -\boldsymbol{\delta}^{(4)} \;=\; \boldsymbol{\delta}. \tag{9.20}$$

Hence,

$$\begin{aligned}
\mathcal{A}^s_{\widehat{E}}(\boldsymbol{u}, \boldsymbol{v}) &= \sum_{r=1}^{4} \int_{T_r} \langle C^s \boldsymbol{d}(\boldsymbol{u}), \boldsymbol{d}(\boldsymbol{v})\rangle \, dx = \sum_{r=1}^{4} \Delta \langle C^s \boldsymbol{d}^{(r)}, \boldsymbol{\delta}^{(r)}\rangle \\
&= \Delta \langle C^s \boldsymbol{\delta}, \boldsymbol{d}^{(1)} + \boldsymbol{d}^{(2)} + \boldsymbol{d}^{(3)} - \boldsymbol{d}^{(4)}\rangle \qquad (9.21)\\
&= -2\Delta \langle C^s \boldsymbol{\delta}, \boldsymbol{d}^{(4)}\rangle \leq 2\Delta \|\boldsymbol{\delta}\|_{C^s} \|\boldsymbol{d}^{(4)}\|_{C^s}
\end{aligned}$$

where $\Delta := \text{area}(T_k)$. Further,

$$\|\boldsymbol{d}^{(4)}\|^2_{C^s} = \|\boldsymbol{d}^{(1)} + \boldsymbol{d}^{(2)} + \boldsymbol{d}^{(3)}\|^2_{C^s} \leq 3 \sum_{k=1}^{3} \|\boldsymbol{d}^{(k)}\|^2_{C^s}$$

leads to

$$\mathcal{A}^s_{\widehat{E}}(\boldsymbol{u}, \boldsymbol{u}) = \sum_{k=1}^{4} \|\boldsymbol{d}^{(k)}\|^2_{C^s} \Delta \geq \left(1 + \frac{1}{3}\right) \Delta \|\boldsymbol{d}^{(4)}\|^2_{C^s} \tag{9.22}$$

and

$$\mathcal{A}^s_{\widehat{E}}(\boldsymbol{v}, \boldsymbol{v}) = 4\Delta \|\boldsymbol{\delta}\|^2_{C^s}\,. \tag{9.23}$$

Thus,

$$\mathcal{A}^s_{\widehat{E}}(\boldsymbol{u},\boldsymbol{v}) \;\le\; \sqrt{\frac{3}{4}}\,\sqrt{\mathcal{A}^s_{\widehat{E}}(\boldsymbol{u},\boldsymbol{u})}\,\sqrt{\mathcal{A}^s_{\widehat{E}}(\boldsymbol{v},\boldsymbol{v})}\,. \tag{9.24}$$

In the case of an arbitrary shaped macroelement E we can use the affine mapping $F : \widehat{E} \to E$ for transformation of the problem to the reference macroelement, for further details see, e.g., [8]. This transformation changes the coefficient matrix C^s, but the estimate $\gamma_E \le \sqrt{\frac{3}{4}}$ still holds since the result (9.24) for the reference macroelement does not depend on the coefficient matrix C^s. □

Remark 9.8. The presented uniform estimate of the CBS constant γ is a generalization of the earlier estimate from [71], namely

$$\gamma \le \frac{\sqrt{8+\sqrt{8}}}{4} \approx 0.822,$$

which is derived in the case of a regular triangulation $\mathcal{T}_H$ obtained by a diagonal subdivision of a square mesh.

Similarly to the scalar elliptic case, the next theorem holds true.

Theorem 9.9 ([33]). *Consider again the pure displacement elasticity problem with constant Lamé coefficients on the triangles $E \in \mathcal{T}_H$, discretization by the Crouzeix–Raviart FE and the DA decomposition of the stiffness matrix. Let $\tilde{A}_{h:22}$ be the aggregated coarse grid stiffness matrix corresponding to the space $\tilde{\mathcal{V}}_{h:2}$ from the DA splitting, and let A^s_H be the stiffness matrix, corresponding to the coarser triangulation $\mathcal{T}_H$, equipped with the standard nodal finite element basis. Then*

$$\tilde{A}^s_{h:22} = 4\,A^s_H\,. \tag{9.25}$$

The proof follows almost directly from the definitions of the hierarchical basis functions $\tilde{\Phi}_i$ with value equal to one in two nodes of one of the macroelement sides and one opposite inner node and the corresponding coarse grid basis function with value equal to one in one node on the same side. This result enables the direct recursive multilevel extension of the two-level multiplicative preconditioner preserving the same estimate of the CBS constant. In particular, the general scheme of the multiplicative AMLI algorithm is straightforwardly applicable.

We consider now the construction of preconditioners $\tilde{C}^s_{h:11}$ for the coarse grid complement blocks $\tilde{A}^s_{h:11}$, see decomposition (9.18). We search for optimal preconditioners in the sense that they are spectrally equivalent to the upper-left matrix block independently on the mesh size, element shape and Poisson ratio. Moreover the cost of applying the preconditioner is aimed to be proportional to the number of degrees of freedom. Similarly to Chapter 4, we apply constructions on macroelement level and assemble the local contributions to obtain $\tilde{C}^s_{h:11}$.

Let us consider the frequently used approach where we first impose a displacement decomposition ordering, then use a block-diagonal approximation of $\tilde{A}^s_{h:11}$, and then precondition the diagonal blocks which are elliptic. Let us assume that the multiplicative preconditioner from Chapter 4 is applied to the diagonal blocks of $\tilde{A}^s_{h:11}$. Then, for homogeneous isotropic materials, the following simplified estimate holds

$$\kappa\left(\left(\tilde{C}^s_{h:11}\right)^{-1}\tilde{A}^s_{h:11}\right) \le \frac{1-\nu}{1-2\nu}\frac{15}{8}.$$

This construction is optimal with respect to mesh size and mesh anisotropy but is applicable for moderate values of $\nu \in [0, \frac{1}{2})$ only. When the material is *almost incompressible*, it is better to apply a macroelement-level static condensation of $\tilde{A}^s_{h:11}$ first, which is equivalent to the elimination of all unknowns corresponding to the interior nodes of the macroelements, see Figure 4.1.

Let us assume now that the triangulations $\mathcal{T}_H$ is obtained by diagonal subdivision of a square mesh, and let the corresponding Schur complement be approximated by its diagonal. Then the resulting preconditioner satisfies the following estimate

$$\kappa\left(\left(\tilde{C}^s_{h:11}\right)^{-1}\tilde{A}^s_{h:11}\right) \le \frac{a+\cos(\alpha)}{a-\cos(\beta)} = 8.301\ldots,$$

where

$$a = \frac{4685}{2\sqrt{5391385}}, \quad \alpha = \frac{1}{3}\arccos\left(\frac{1162562569}{1078277\sqrt{5391385}}\right), \quad \beta = \frac{\pi}{3} - \alpha.$$

It is a really good finding that the above estimate is uniform with respect to the Poisson ratio ν (for details see [71]).

The next numerical tests illustrate the robustness of the latter approach including the behavior of the FEM error as well as the convergence rate of the AMLI algorithm when the size of the discrete problem is varied and $\nu \in [0, 1/2)$ tends to the *incompressible* limit. We consider a test problem in the unit square $\Omega = (0,1)^2$ with elasticity modulus $E = 1$. The right hand side corresponds to the exact solution $\boldsymbol{u}(x,y) = (\sin(\pi x)\sin(\pi y), y(y-1)x(x-1))^T$. The relative stopping criterion for the PCG iterations with preconditioner $C = C_{\text{AMLI}}$ is

$$\frac{\langle C^{-1}\mathbf{r}_{(n_{it})}, \mathbf{r}_{(n_{it})}\rangle}{\langle C^{-1}\mathbf{r}_{(0)}, \mathbf{r}_{(0)}\rangle} < \varepsilon^2,$$

where as usual $\mathbf{r}_{(i)}$ stands for the residual at the i-th iteration step.

The relative FEM errors, given in Table 9.2, well illustrate the *locking-free* approximation. Here the number of refinement steps is $\ell = 4$, $N = 1472$, and $\varepsilon = 10^{-9}$.

Table 9.2: Relative error stability for $\nu \to 1/2$

ν	$\|\boldsymbol{u} - \boldsymbol{u}_h\|_{[L_2]^2} / \|\boldsymbol{f}\|_{[L_2]^2}$	ν	$\|\boldsymbol{u} - \boldsymbol{u}_h\|_{[L_2]^2} / \|\boldsymbol{f}\|_{[L_2]^2}$
0.4	0.3108249106503572	0.4999	0.3771889077038727
0.49	0.3695943747405575	0.49999	0.3772591195613628
0.499	0.3764879643773666	0.499999	0.3772661419401481

In Table 9.3, the number of iterations are presented as a measure of the robustness of the multilevel preconditioner. The optimal order *locking-free* convergence rate of the AMLI algorithm is well expressed. Here the degree of the stabilization polynomial is set to $\upsilon = 2$ corresponding to the derived uniform estimate of the CBS constant, and providing the total computational cost optimality of the related PCG algorithm.

Table 9.3: Number of PCG iterations with AMLI preconditioning: $\varepsilon = 10^{-3}$, $\upsilon = 2$

ℓ	N	$\nu = 0.3$	$\nu = 0.4$	$\nu = 0.49$	$\nu = 0.4999$	$\nu = 0.499999$
4	1472	13	13	12	13	13
5	6016	12	12	12	13	13
6	24320	12	12	12	13	13
7	97792	11	11	11	13	13
8	196096	11	11	11	12	13

The success of the Crouzeix–Raviart and other nonconforming finite elements can be explained in general by the fact that they produce algebraic systems that are equivalent to the Schur complement system for the Lagrange multipliers arising from the mixed finite element method for Raviart–Thomas elements.

One alternative approach for *locking-free* numerical solution of the elasticity problem with general boundary conditions is based on discontinuous Galerkin approximations using Crouzeix–Raviart or Rannacher–Turek nonconforming finite elements. The corresponding *locking-free* error analysis can be found in [65, 66]. The development of robust AMLI preconditioning algorithms for the related DG-FE systems is still an open problem.

9.2 Optimal order AMLI preconditioning of the Navier–Stokes problem

9.2.1 Crouzeix–Raviart FE discretization of the velocity field

Let us consider the Dirichlet initial-boundary value problem for the Navier–Stokes equations

$$\begin{aligned}
\frac{\partial \boldsymbol{u}}{\partial t} + (\boldsymbol{u}\cdot\nabla)\boldsymbol{u} &= -\nabla p + \frac{1}{Re}\nabla^2 \boldsymbol{u} + \boldsymbol{f} && (\boldsymbol{x},t) \in \Omega \times (0,T)\\
\nabla\cdot \boldsymbol{u} &= 0 && (\boldsymbol{x},t) \in \Omega \times (0,T)\\
\boldsymbol{u} &= \boldsymbol{0} && (\boldsymbol{x},t) \in \Gamma \times (0,T)\\
\boldsymbol{u} &= \boldsymbol{0} && (\boldsymbol{x},t) \in \Omega \times \{0\}
\end{aligned} \tag{9.26}$$

where Ω is a bounded and connected domain in $\mathbb{R}^2$, and $\Gamma = \partial\Omega$. We assume also that Ω is such that the H^2 -regularity property holds for the steady Stokes problem. As was several times noticed, a generalization to nonhomogeneous boundary condition is straightforward.

The numerical solution of the incompressible Navier–Stokes equations has been the focus of the computational fluid dynamics community for over five decades. However, the question how to construct optimal schemes, in terms of computational cost and accuracy, is still not over. What is quite clear now is that the solution via Uzawa iterations of the coupled velocity-pressure discrete systems that result from the space and time discretization of the equations is quite expensive. The most popular alternative way to build efficient approximations is known as projection approach. It is based on a stable splitting of each time step.

Let us assume that Ω is a polygonal domain, and $\mathcal{T}_h$ is a triangulation of Ω. In this section we will use also the following notations: $\mathcal{V}^c$ and $\mathcal{V}^{nc}$ are respectively the linear conforming (Courant) and linear nonconforming (Crouzeix–Raviart) FE spaces satisfying homogeneous boundary conditions; $\mathcal{Q}$ is the space of piece-wise constant pressures, i.e., $\mathcal{Q} = \{q \in L_2(\Omega) : q|_e \in P_0, \forall e \in \mathcal{T}_h, \int_\Omega q = 0\}$; $(\cdot,\cdot)$ and $(\cdot,\cdot)_e$ stand for the dot products in $L_2(\Omega)$ and $L_2(T)$; and $(\cdot,\cdot)_h = \sum_{e\in\mathcal{T}_h}(\cdot,\cdot)_e$. Uniform discretization in time is used with a time step Δt. The superscript in the presented projection schemes indicates the number of the time discretization level. For example, $(\boldsymbol{u}_h^n, p_h^n)$ are the numerically computed velocities and pressure for $t = n\Delta t$. A model 2D problem in a polygonal domain Ω covered by a uniform mesh of isosceles rectangle triangles is analyzed.

Now we will consider two projection schemes (Variant A and Variant B) which are based on Crouzeix–Raviart FE approximation of the velocities and pice-wise constant approximation of the pressure. The most significant advantage of these approximations is that the divergence of the velocity field is zero inside each element, i.e. the approximation is locally conservative. Note that this is not the case, e.g., for the alternatively applied, inf-sup stable approximation for the velocities, using rotated bilinear (Rannacher–Turek) elements.

Variant A:
This scheme is based on a complete nonconforming discretization of the velocities, and is inf-sup stable and locally conservative [49].

- *Prediction step:*
  Find $\tilde{\boldsymbol{u}}_h^{n+1} \in (\mathcal{V}^{nc})^2$ such that for all $\boldsymbol{v}_h \in (\mathcal{V}^{nc})^2$

$$\begin{aligned}\left(\frac{\tilde{\boldsymbol{u}}_h^{n+1} - \boldsymbol{u}_h^n}{\Delta t}, \boldsymbol{v}_h\right) &+ ((\tilde{\boldsymbol{u}}_h^n \cdot \nabla)\tilde{\boldsymbol{u}}_h^n, \boldsymbol{v}_h)_h \\ &+ \frac{1}{Re}(\nabla \tilde{\boldsymbol{u}}_h^{n+1}, \nabla \boldsymbol{v}_h)_h - (p_h^n, \nabla \cdot \boldsymbol{v}_h)_h = 0\end{aligned} \tag{9.27}$$

- *Projection step:*
  Find $\boldsymbol{u}_h^{n+1} \in (\mathcal{V}^{nc})^2,\ \ p_h^{n+1} \in \mathcal{Q}$ such that

$$\begin{aligned}&(\boldsymbol{u}_h^{n+1} - \tilde{\boldsymbol{u}}_h^n, \boldsymbol{v}_h) = (p_h^{n+1} - p_h^n, \nabla \cdot \boldsymbol{v}_h)_h, \quad \forall \boldsymbol{v}_h \in (\mathcal{V}^{nc})^2 \\ &(\nabla \cdot \boldsymbol{u}_h^{n+1}, q_h)_h = 0, \quad \forall q_h \in \mathcal{Q}\end{aligned} \tag{9.28}$$

Variant B:
Conforming FEs at the prediction step are used to reduce the computational complexity. The accuracy of the velocities is of optimal order subject to the "cross-grid mesh" condition [27].

- *Prediction step:*
  Find $\tilde{\boldsymbol{u}}_h^{n+1} \in (\mathcal{V}^{c})^2$ such that for all $\boldsymbol{v}_h \in (\mathcal{V}^{c})^2$

$$\begin{aligned}\left(\frac{\tilde{\boldsymbol{u}}_h^{n+1} - \boldsymbol{u}_h^n}{\Delta t}, \boldsymbol{v}_h\right) &+ ((\tilde{\boldsymbol{u}}_h^n \cdot \nabla)\tilde{\boldsymbol{u}}_h^n, \boldsymbol{v}_h) \\ &+ \frac{1}{Re}(\nabla \tilde{\boldsymbol{u}}_h^{n+1}, \nabla \boldsymbol{v}_h) - (p_h^n, \nabla \cdot \boldsymbol{v}_h) = 0\end{aligned} \tag{9.29}$$

- *Projection step:*
  Find $\boldsymbol{u}_h^{n+1} \in (\mathcal{V}^{nc})^2,\ \ p_h^{n+1} \in \mathcal{Q}$ such that

$$\begin{aligned}&(\boldsymbol{u}_h^{n+1} - \tilde{\boldsymbol{u}}_h^n, \boldsymbol{v}_h) = (p_h^{n+1} - p_h^n, \nabla \cdot \boldsymbol{v}_h)_h, \quad \forall \boldsymbol{v}_h \in (\mathcal{V}^{nc})^2 \\ &(\nabla \cdot \boldsymbol{u}_h^{n+1}, q_h)_h = 0, \quad \forall q_h \in \mathcal{Q}\end{aligned} \tag{9.30}$$

The advantage of Variant A is the inf-sup stability while Variant B is computationally cheaper due to the three times reduction of the unknowns at the prediction step. In both variants, the projection scheme splits the nonlinear Navier–Stokes system to some of the following linear elliptic problems.

- **Variant A** – *Prediction step*:
  For a given $f \in L_2(\Omega)$ find $u_h \in \mathcal{V}^{nc}$, satisfying

$$(\nabla u_h, \nabla v_h)_h + \frac{1}{\Delta t}(u_h, v_h) = (f, v_h) \qquad \forall v_h \in \mathcal{V}^{nc}. \tag{9.31}$$

- **Variant B** – *Prediction step*:
  For a given $f \in L_2(\Omega)$ find $u_h \in \mathcal{V}^{c}$, satisfying

$$(\nabla u_h, \nabla v_h) + \frac{1}{\Delta t}(u_h, v_h) = (f, v_h) \qquad \forall v_h \in \mathcal{V}^{c}. \tag{9.32}$$

- **Variants A, B** – *Projection step*:
  For a given $f \in (L_2(\Omega))^2$ find $(u_h, p_h) \in (\mathcal{V}^{nc}, \mathcal{Q})$, satisfying

$$\begin{aligned} (u_h, v_h) - (p_h, \nabla \cdot v_h)_h &= (f, v_h) \qquad && \forall v_h \in \mathcal{V}^{nc}, \\ (\nabla \cdot u_h, q_h)_h &= 0 && \forall q_h \in \mathcal{Q}. \end{aligned} \tag{9.33}$$

Optimal order robust AMLI preconditioners for problems (9.31) and (9.32) can be constructed following directly the techniques presented in Chapter 3 and Chapter 4. One could expect that the AMLI methods are not applicable to the saddle point system which arises from the mixed FE problem (9.33). The last part of this section is devoted to this question (including the positive answer) how this can be done efficiently.

9.2.2 AMLI preconditioning of the mixed FE system: weighted graph-Laplacian

The mixed FE problem to be solved at the projection step leads to the system

$$\begin{bmatrix} M & & B_1 \\ & M & B_2 \\ B_1^T & B_2^T & \end{bmatrix} \mathbf{w}_h = \mathbf{b}_h. \tag{9.34}$$

It is important to note that in the 2D case the scalar mass matrix M of the Crouzeix–Raviart FEs is diagonal. This follows directly from the fact, that the quadrature formula on a triangle with nodes in the midpoints of the edges is exact for second degree polynomials, see e.g. [89]. For the saddle point system (9.34) we can either eliminate the pressure unknowns and end up with a system for $\mathbf{u}_h$

that corresponds to a divergence-free basis or, since the mass matrix is diagonal, we can eliminate the velocity unknowns and derive a system for the pressure with a symmetric and positive semidefinite matrix $B_1^T M^{-1} B_1 + B_2^T M^{-1} B_2$. Here we discuss the second approach and present an optimal order AMLI preconditioner based on a properly constructed hierarchical splitting. Let us remind that the problem is definite due to the condition $\int_\Omega p_h = 0$.

The structure of the reduced (Schur complement) matrix is the same as of the graph-Laplacians considered at the end of Chapter 8. It will be referred to as *weighted graph-Laplacian.* In the case of a uniform rectangular mesh the weighted graph-Laplacian corresponds to the T-shaped four point stencil shown in Figure 9.3.

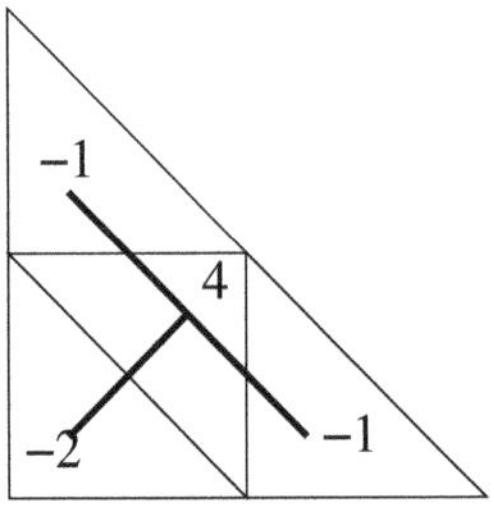

Figure 9.3: Schur complement four point stencil for the pressure

The following AMLI algorithm is a generalization of the approach used in Section 8.4 where the graph-Laplacians appear in the multilevel preconditioning of discontinuous Galerkin problems. The presentation here is based on results which can be found in [34].

Let us consider two consecutive triangulations $\mathcal{T}_H \subset \mathcal{T}_h$ and a decomposition of the weighted graph-Laplacian

$$A_h = \sum_{E \in \mathcal{E}} R_E^T A_E R_E$$

as a sum of local matrices associated with the set of edges $\mathcal{E}$ of the triangles $T \in \mathcal{T}_H$. We will analyze the model 2D problem with rectangular polygonal domain Ω covered by a uniform mesh $\mathcal{T}_H$ composed of square elements. Let the refined mesh be obtained by dividing the current triangles into four congruent ones connecting the midpoints of the sides. Following the numbering from Figure 9.4, we introduce the local (macroelement) matrix $A_E = A_E^{(H)}$, corresponding to a hypotenuse, in the form

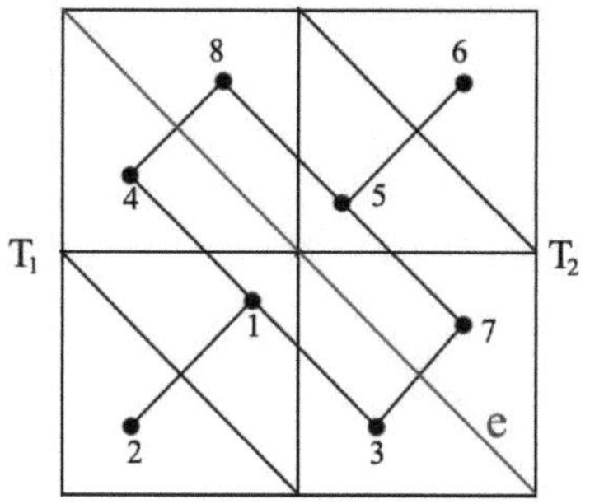

Figure 9.4: Macroelement of two adjacent triangles from $\mathcal{T}_H$ with a common hypotenuse

$$A_E^{(H)} = \left[\begin{array}{cccc|cccc}
t+1 & -2t & \frac{t-1}{2} & \frac{t-1}{2} & & & & \\
-2t & 2t & & & & & & \\
\frac{t-1}{2} & & \frac{5-t}{2} & & & & -2 & \\
\frac{t-1}{2} & & & \frac{5-t}{2} & & & & -2 \\
\hline
 & & & & t+1 & -2t & \frac{t-1}{2} & \frac{t-1}{2} \\
 & & & & -2t & 2t & & \\
 & & -2 & & \frac{t-1}{2} & & \frac{5-t}{2} & \\
 & & & -2 & \frac{t-1}{2} & & & \frac{5-t}{2}
\end{array}\right]. \tag{9.35}$$

The corresponding local matrix $A_E = A_E^{(C)}$ in the case of a common cathetus is as follows

$$A_E^{(C)} = \left[\begin{array}{cccc|cccc}
\frac{3-t}{2} & t-1 & \frac{t-1}{2} & -t & & & & \\
t-1 & 2-t & & & & & & -1 \\
\frac{t-1}{2} & & \frac{3-t}{2} & & & -1 & & \\
-t & & & t & & & & \\
\hline
 & & & & \frac{3-t}{2} & t-1 & -t & \frac{t-1}{2} \\
 & & -1 & & t-1 & 2-t & & \\
 & & & & -t & & t & \\
 & -1 & & & \frac{t-1}{2} & & & \frac{3-t}{2}
\end{array}\right]. \tag{9.36}$$

Similarly to the construction from Section 8.4, the role of the weight parameter $t \in (0, 1)$ is to distribute properly the contributions of the links between the interior nodes among the (macroelement) edge matrices of the current coarse triangles $T_i \in \mathcal{T}_H$, $i \in \{1, 2\}$ (see Figure 9.3).

We define now the (macroelement) local transformation matrix J_E^T as

$$J_E^T = \begin{bmatrix} 1 & p & q & q & & & & \\ 1 & q & p & q & & & & \\ 1 & q & q & p & & & & \\ & & & & 1 & p & q & q \\ & & & & 1 & q & p & q \\ & & & & 1 & q & q & p \\ r & r & r & r & & & & \\ & & & & r & r & r & r \end{bmatrix}, \tag{9.37}$$

where p, q are parameters to be determined later, and r is a scaling factor. Then the hierarchical basis local matrices are obtained as follows

$$\tilde{A}_E = J_E^T A_E J_E = \begin{bmatrix} \tilde{A}_{E:11} & \tilde{A}_{E:12} \\ \tilde{A}_{E:21} & \tilde{A}_{E:22} \end{bmatrix}. \tag{9.38}$$

The second diagonal block $\tilde{A}_{E:22}$ is a 2×2 matrix corresponding to the aggregated (coarse-grid) basis functions. The hierarchical basis global stiffness matrix is assembled by the edge matrices $\tilde{A}_E$, i.e.,

$$\tilde{A}_h = \sum_{E \in \mathcal{E}} R_E^T \tilde{A}_E R_E,$$

where the global two-by-two splitting

$$\tilde{A}_h = \begin{bmatrix} \tilde{A}_{h:11} & \tilde{A}_{h:12} \\ \tilde{A}_{h:21} & \tilde{A}_{h:22} \end{bmatrix} \tag{9.39}$$

is naturally induced by the local representation (9.38).

Lemma 9.10. *Consider the hierarchical basis defined by the local transformation (9.37). Then*

$$\tilde{A}_{h:22} = A_H \quad \textit{if and only if} \quad r = \frac{\sqrt{2}}{2}.$$

Proof. The definition of the last two terms in the local hierarchical basis ensures that $\tilde{A}_{E:22}$ has row-sums/column-sums equal to zero. Then, the equivalence statements

$$\tilde{A}_{E:22} = \tilde{A}_{E:22}^{(H)} = \begin{bmatrix} 2 & -2 \\ -2 & 2 \end{bmatrix}$$

and

$$\tilde{A}_{E:22} = \tilde{A}^{(C)}_{E:22} = \begin{bmatrix} 1 & -1 \\ -1 & 1 \end{bmatrix}$$

simply follow from the equations

$$\tilde{A}_{E:22}(1,1) = r^2 \sum_{i,j=1}^{4} A_E(i,j)$$

and

$$\tilde{A}_{E:22}(2,2) = r^2 \sum_{i,j=5}^{8} A_E(i,j).$$

□

Following the general theory (see Lemma 2.4) we estimate the related CBS constant locally, namely

$$\gamma^2 \le \gamma_E^2 = 1 - \lambda,$$

where λ is the eigenvalue (which is unique in this particular case) of the eigenproblem

$$\tilde{S}_E \mathbf{v} = \lambda \tilde{A}_{E:22} \mathbf{v}\,, \quad \mathbf{v} \neq \mathbf{c},$$

where $\tilde{S}_E = \tilde{A}_{E:22} - \tilde{A}_{E:21}(\tilde{A}_{E:11})^{-1}\tilde{A}_{E:12}$.

For the considered model problem, two local problems are to be solved to complete the local analysis. They correspond to the cases when $E \in \mathcal{E}$ is either a hypotenuse or a cathetus. Let us assume that $r = \frac{\sqrt{2}}{2}$. Then due to Lemma 9.10, the two-level estimates of the CBS constant are applicable in the multilevel setting, i.e. they are uniform with respect to the level of uniform refinement.

Varying the parameters (p, q, t) we get a family of hierarchical splittings. The following two lemmas are derived via simple direct computations.

Lemma 9.11 ([34]). *Let us consider the hierarchical splitting of the weighted graph-Laplacian for the model 2D problem with parameters $r = \frac{\sqrt{2}}{2}$, $p = 1$, $q = -0.5$ and $t = 0.5$. Then the CBS constant is uniformly bounded by*

$$\gamma^2 \le 0.73.$$

Let us remember that this parameter setting for (p, q, t) was introduced in Section 8.4 for the case of graph-Laplacians, see also [83]. The parameters in the next lemma are obtained by a local optimization with respect to the CBS constant estimate.

Lemma 9.12 ([34]). *Let us consider the hierarchical splitting of the weighted graph-Laplacian for the model 2D problem with parameters* $r = \frac{\sqrt{2}}{2}$, $p = 1$, $q = -0.1$ *and* $t = 0.75$. *Then the related CBS constant is uniformly bounded by*

$$\gamma^2 \leq 0.58.$$

We consider now the sequence of ℓ nested uniform refinements of the initial triangulation $\mathcal{T}_0 \subset \mathcal{T}_1 \subset \ldots \subset \mathcal{T}_\ell$, and the related weighted graph-Laplacians $A^{(0)}, A^{(1)}, \ldots, A^{(\ell)}$. The above three lemmas lead to the following theorem.

Theorem 9.13. *The AMLI algorithms for the weighted graph-Laplacian* $A = A^{(\ell)}$, *with hierarchical splittings defined by the parameter settings form Lemma 9.11 and Lemma 9.12, have optimal computational complexity if* $\upsilon \in \{2, 3\}$.

Remark 9.14. The development of robust AMLI preconditioners for the weighted graph-Laplacians in the case of a general triangulation (mesh anisotropy) is still an open problem.

Now let us summarize the results of the considerations in this section. Two locally conservative (divergence-free) projection schemes (Variant A and Variant B) for stable discretization of the initial-boundary value problem for the Navier–Stokes equations are considered. They both have discretization accuracy of optimal order [27]. As we showed, optimal order AMLI preconditioners can be successfully applied to the decoupled scalar elliptic problems at the prediction step as well as to the mixed FE problem at the projection step. As a result, the related composite time-stepping solution methods have a total computational complexity of optimal order.

10 Practical issues

The final chapter of this book provides the basic algorithms for linear and nonlinear algebraic multilevel iteration methods. We will also comment on some implementation aspects in each case. Finally, we will present an example for integrating AMLI methods for the solution of one more complex problem.

10.1 Linear AMLI algorithm

In order to solve a system

$$M^{(k)}\mathbf{v}^{(k)} = \mathbf{d}^{(k)}, \quad k = \ell, \ell-1, \dots, 1,$$

for an unknown vector $\mathbf{v}^{(k)}$ where $M^{(k)}$ denotes the linear AMLI preconditioner (defined in (2.39)) at level k, i.e.,

$$M^{(k)} = \begin{bmatrix} C_{11}^{(k)} & 0 \\ \tilde{A}_{21}^{(k)} & Z^{(k-1)} \end{bmatrix} \begin{bmatrix} I & C_{11}^{(k)^{-1}} \tilde{A}_{12}^{(k)} \\ 0 & I \end{bmatrix} \tag{10.1}$$

and $\mathbf{d}^{(k)}$ is a given right-hand side vector we need the solution of a system

$$M^{(k-1)}\mathbf{v}^{(k-1)} = \mathbf{d}^{(k-1)}$$

with some right-hand side $\mathbf{d}^{(k-1)}$. Let

$$\mathbf{d}^{(k)} = \begin{bmatrix} \mathbf{d}_1^{(k)} \\ \mathbf{d}_2^{(k)} \end{bmatrix}, \quad \mathbf{v}^{(k)} = \begin{bmatrix} \mathbf{v}_1^{(k)} \\ \mathbf{v}_2^{(k)} \end{bmatrix},$$

and

$$\mathbf{u}^{(k)} = \begin{bmatrix} \mathbf{u}_1^{(k)} \\ \mathbf{u}_2^{(k)} \end{bmatrix} = \begin{bmatrix} I & C_{11}^{(k)^{-1}} \tilde{A}_{12}^{(k)} \\ 0 & I \end{bmatrix} \begin{bmatrix} \mathbf{v}_1^{(k)} \\ \mathbf{v}_2^{(k)} \end{bmatrix},$$

then from (10.1) it follows that

$$\begin{aligned} \mathbf{u}_1^{(k)} &= C_{11}^{(k)^{-1}} \mathbf{d}_1^{(k)} \\ Z^{(k-1)}\mathbf{u}_2^{(k)} &= \mathbf{d}_2^{(k)} - \tilde{A}_{21}^{(k)}\mathbf{u}_1^{(k)} =: \mathbf{w}^{(k-1)}. \end{aligned} \tag{10.2}$$

Using (2.44) we write (10.2) in the form

$$M^{(k-1)}\mathbf{u}_2^{(k)} = Q^{(k)}(A^{(k-1)}M^{(k-1)^{-1}})\mathbf{w}^{(k-1)} \tag{10.3}$$

where $Q^{(k)}(t) = (1 - P^{(k)}(t))/t = q_0^{(k)} + q_1^{(k)} t + \ldots + q_{\upsilon_k - 1}^{(k)} t^{\upsilon_k - 1}$. Hence

$$\begin{aligned} \mathbf{d}^{(k-1)} &= Q^{(k)}(A^{(k-1)} M^{(k-1)^{-1}}) \mathbf{w}^{(k-1)}, \\ \mathbf{v}^{(k-1)} &= \mathbf{u}_2^{(k)}. \end{aligned}$$

Applying the linear AMLI method, in order to solve a linear system (at level ℓ associated with the finest mesh), in this book is understood as using the PCG method (Algorithm 1.20) with the linear AMLI preconditioner (Algorithm 10.1) implemented in step (1.94). The following algorithm computes the solution of

$$M^{(\ell)} \mathbf{v}^{(\ell)} = \mathbf{d}^{(\ell)}. \tag{10.4}$$

Algorithm 10.1 (Linear AMLI, cf. [17]).

---

	for $k = 1$ **to** ℓ set $\sigma_k := 0$	
	$k := \ell$	
forward:	$\sigma_k := \sigma_k + 1$	
	if $\sigma_k = 1$	
	$\quad \mathbf{d}^{(k)} := (J^{(k)})^T \mathbf{d}^{(k)}$	(10.5)
	$\quad \mathbf{v}_1^{(k)} := (C_{11}^{(k)})^{-1} \mathbf{d}_1^{(k)}$	
	$\quad \mathbf{w}^{(k-1)} := \mathbf{d}_2^{(k)} - \tilde{A}_{21}^{(k)} \mathbf{v}_1^{(k)}$	
	$\quad \mathbf{d}^{(k-1)} := q_{\upsilon_k - 1}^{(k)} \mathbf{w}^{(k-1)}$	
	else	
	$\quad \mathbf{d}^{(k-1)} := A^{(k-1)} \mathbf{v}^{(k-1)} + q_{\upsilon_k - \sigma_k}^{(k)} \mathbf{d}^{(k-1)}$	(10.6)
	end	
	$k := k - 1$	
	if $k > 0$ **goto forward**	
	solve $A^{(0)} \mathbf{v}^{(0)} = \mathbf{d}^{(0)}$ for $\mathbf{v}^{(0)}$	(10.7)
backward:	$k := k + 1$	
	$\mathbf{v}_2^{(k)} := \mathbf{v}^{(k-1)}$	
	if $\sigma_k < \upsilon_k$ **goto forward**	
	$\mathbf{v}_1^{(k)} := \mathbf{v}_1^{(k)} - (C_{11}^{(k)})^{-1} \tilde{A}_{12}^{(k)} \mathbf{v}_2^{(k)}$	
	$\mathbf{v}^{(k)} := J^{(k)} \mathbf{v}^{(k)}$	(10.8)
	$\sigma_k := 0$	
	if $k < \ell$ **goto backward**	

Here the vector $\boldsymbol{v} = [v_1, v_2, \ldots, v_\ell]^T$ defines the cycle, i.e., $v_k = 1$ for $1 \le k \le \ell$ corresponds to the V-cycle, and $v_k = 2$ for $1 \le k < \ell$, $v_\ell = 1$, corresponds to the classical W-cycle. Higher-order stabilization or mixed cycles with varying polynomial degree are possible and sometimes preferable from a computational point of view.

The above algorithm is based on the multiplicative two-level preconditioner (2.47). It uses the approximations $C_{11}^{(k)^{-1}}$ for $A_{11}^{(k)^{-1}}$, and $Z^{(k-1)^{-1}}$ for the inverse of the Schur complement at level k, see (2.44). The action of the matrix polynomial $Q^{(k)}(A^{(k-1)}M^{(k-1)^{-1}})$ in (2.44) on a vector is computed via Horner's rule, where the coefficients of $Q^{(k)}(t)$ are given by $q^{(k)}_{v_k-\sigma_k}$, $1 \le \sigma_k \le v_k$.

The classical V-cycle is obtained for $Q^{(k)}(t) = Q(t) := 1$ for all $k = 1, 2, \ldots, \ell$, cf. (2.59). The (linear) AMLI W-cycle, based on the shifted and scaled Chebyshev polynomial (2.57), is obtained for $Q^{(k)}(t) = Q(t) := (\lambda + 1)/\lambda - 1/\lambda t$ where we assume an approximation property of the form (2.48). In this case λ is an upper bound for the largest eigenvalue of the preconditioned matrix and all its eigenvalues will be greater or equal to 1 by construction, i.e.,

$$\lambda_i(M^{(\ell)^{-1}}A^{(\ell)}) \in [1, \lambda],$$

cf. Chapter 2.4. However, if in the approximation property (2.48) the lower bound 1 does not hold, i.e., we allow the smallest eigenvalue $\lambda_{\min}$ to be less than 1, the interval with respect to which we want to minimize the maximum of $t\, Q(t)$ has to be shifted accordingly in order to optimize the performance of the method. Note that for instance by choosing $Q^{(k)}(t) = Q(t) := 2 - t$, which corresponds to $P^{(k)}(t) = P_2(t) := (1-t)^2$, the method will converge if the largest eigenvalue is bounded by 2, i.e., $\lambda_{\max} \le \lambda = 2$, which can always be achieved by using a proper(ly scaled) approximation of the pivot block. In this case, in order to derive uniform condition number estimates, the smallest eigenvalue $\lambda_{\min}$ has to be bounded away from 0 uniformly; The derivation follows again similar arguments as were presented in Chapter 2.4. The latter choice, however, has the advantage that there are no estimates or bounds on the spectrum of the two-level preconditioner or on the CBS constant involved in the construction of the stabilization polynomial. Different (and other) choices of the stabilization polynomial have been discussed in [16, 17].

The matrices $\tilde{A}_{12}^{(k)}$ and $\tilde{A}_{21}^{(k)}$ are the off-diagonal blocks of the two-level hierarchical basis matrix $\tilde{A}^{(k)}$ at level k, and $A^{(k-1)}$ is the matrix associated with the coarse grid (with respect to the coarse-grid nodal basis). Note that for the Schur complement based multilevel preconditioner proposed in Chapter 5 the matrix $A^{(k-1)}$ is assembled from the local Schur complements, cf. Section 5.4, and no hierarchical basis transformation is involved but only a renumbering according to the fine-coarse partitioning of DOF is needed – the approximate Schur complement is associated with the coarse DOF then. In this case $\tilde{A}_{12}^{(k)}$ and $\tilde{A}_{21}^{(k)}$ are the off-diagonal blocks of the reordered nodal basis stiffness matrix at level k, and the forward and backward transformation (10.5) and (10.8) reduce to a permutation and its inverse permutation, respectively.

However, if the AMLI algorithm is based on the recursive two-level transformation $J^{(k)}$, which is the standard situation in this book, we have

$$A^{(k-1)} := \tilde{A}_{22}^{(k)}. \tag{10.9}$$

In the particular situation when $J^{(k)}$ has the form

$$J^{(k)} := \begin{bmatrix} I & J_{12}^{(k)} \\ 0 & I \end{bmatrix}, \tag{10.10}$$

which is the standard construction in the case of conforming linear finite elements (cf. Chapter 3), we have

$$A^{(k-1)} = (J_{12}^{(k)})^T A_{11}^{(k)} J_{12}^{(k)} + A_{21}^{(k)} J_{12}^{(k)} + (J_{12}^{(k)})^T A_{12}^{(k)} + A_{22}^{(k)}, \tag{10.11}$$

and

$$\tilde{A}_{12}^{(k)} = A_{11}^{(k)} J_{12}^{(k)} + A_{12}^{(k)}, \quad \tilde{A}_{21}^{(k)} = (J_{12}^{(k)})^T A_{11}^{(k)} + A_{21}^{(k)}. \tag{10.12}$$

If the coarse-level matrix $A^{(k-1)}$ can be obtained via assembling one typically avoids its computation based on (10.11) from the fine-level stiffness matrix. Moreover, since we need only the action of $A^{(k-1)}$ on a vector in (10.6) the matrix-vector product can also be computed without assembling $A^{(k-1)}$. If (10.12) holds, the actions of $\tilde{A}_{12}^{(k)}$ and $\tilde{A}_{21}^{(k)}$ on a vector can be implemented without explicit computation of the HB matrices, even by using only the standard nodal basis stiffness matrices on macro-element level.[1] If we have an additive representation of ${C_{11}^{(k)}}^{-1}$, based on local matrices, this provides us with the opportunity of implementing a *matrix-free* AMLI algorithm, which is especially well suited for parallel computer architectures. The solution of the system on the coarsest level ($k = 0$) in step (10.7) is typically performed by using a direct method, e.g., by Gaussian elimination.

When using nonnested finite element spaces, e.g., for the nonconforming finite element systems that we were discussing in Chapters 4 and 7, the transformation $J^{(k)}$ not necessarily has to be of the form (10.10). In such a situation the whole

[1] Note that the multiplication with the global transformation matrices $J_{12}^{(k)}$ and $(J_{12}^{(k)})^T$ can also be performed locally.

AMLI procedure can be based on the two-level hierarchical basis matrices $\tilde{A}^{(k)}$, thereby explicitly constructing $\tilde{A}_{12}^{(k)}$ and $\tilde{A}_{21}^{(k)}$, and defining $A^{(k-1)}$ according to (10.9), the latter being associated with the coarse grid again.

The additive variant of linear AMLI, which is based on the block-diagonal two-level preconditioner (2.27) substituting $C_{11} = C_{11}^{(k)}$ and $C_{22} = Z^{(k-1)}$ is obtained from Algorithm 10.1 by simply skipping all multiplications with $\tilde{A}_{12}^{(k)}$ and $\tilde{A}_{21}^{(k)}$. It therefore has a considerably smaller arithmetic cost of each outer iteration. In the computations presented in this book the additive cycle was typically cheaper by a factor close to two, which is due to the fact that each visit of any level k, either in the forward elimination or in the backward substitution loop, involves one application of $C_{11}^{(k)^{-1}}$ and one application of either $\tilde{A}_{12}^{(k)}$ or $\tilde{A}_{21}^{(k)}$. These matrix operations typically dominate the overall cost. If the cost of applying $C_{11}^{(k)^{-1}}$ is comparable to that of one multiplication with an off-diagonal block of the k-th level HB matrix, skipping the latter will approximately halve the total number of arithmetic operations that accumulate in one outer iteration. At the same the condition number of the additive preconditioner typically is less than four times bigger than the condition number of the corresponding multiplicative preconditioner, which in case of the two-level methods with $C_{11}^{(k)} = A_{11}^{(k)}$ can be seen by comparing the respective upper bounds, i.e.,

$$\frac{1+\gamma}{1-\gamma} = (1+\gamma)^2 \frac{1}{1-\gamma^2} \leq 4 \frac{1}{1-\gamma^2},$$

cf. (2.19) and (2.20). This means that the number of iterations required to achieve a certain accuracy with the additive method typically is less than two times the number of iterations performed by the multiplicative algorithm. Finally, this often results in almost the same total solution time for both algorithms, as we observed in Chapter 7.5, see e.g., Table 7.6 and Figure 7.8.

Moreover, the additive algorithm, i.e., the block-diagonal preconditioner, is clearly favorable for parallel implementation on multiprocessor systems. That is why it often is a true alternative for the multiplicative AMLI. The situation is similar in case of nonlinear AMLI.

10.2 Nonlinear AMLI algorithm

As already pointed out in Chapter 2 the linear AMLI algorithm crucially depends on a proper choice of the stabilization polynomial $Q(t)$, cf. (2.45), to be used in the construction of the approximation of the inverse of the Schur complement, see (2.44). By contrast, the following parameterfree nonlinear AMLI algorithm uses inner iterations for the action of $Z^{(k-1)^{-1}}$ providing the coarse-grid correction.

Algorithm 10.2 (Nonlinear AMLI, cf. [19, 73]).

---

for $k = 1$ **to** ℓ set $\sigma_k := 0$; $\mathbf{x}^{(k)} := \mathbf{0}$

$k := \ell$; $\mathbf{d}^{(\ell)} := \tilde{\mathbf{b}} = (J^{(\ell)})^T \mathbf{b}$; $\mathbf{r}^{(\ell)} := \mathbf{d}^{(\ell)}$

while (termination criterion is false) **do** { (10.13)

forward: $\sigma_k := \sigma_k + 1$

if $(\sigma_k = 1$ && $k < \ell)$

  $\mathbf{x}^{(k)} := \mathbf{0}$; $\mathbf{r}^{(k)} := \mathbf{d}^{(k)}$

$\mathbf{p}^{(k)}_{1(\sigma_k)} := (C^{(k)}_{11})^{-1} \mathbf{r}^{(k)}_1$

$\mathbf{d}^{(k-1)} := \mathbf{r}^{(k)}_2 - \tilde{A}^{(k)}_{21} \mathbf{p}^{(k)}_{1(\sigma_k)}$

$k := k - 1$

if $k > 0$

  if $(\sigma_k = 0)$ $\mathbf{d}^{(k)} := (J^{(k)})^T \mathbf{d}^{(k)}$

  goto forward

solve $A^{(0)}\mathbf{x}^{(0)} = \mathbf{d}^{(0)}$ for $\mathbf{x}^{(0)}$ (10.14)

backward: $\sigma_k = 0$; $k := k + 1$; $\mathbf{p}^{(k)}_{2(\sigma_k)} := \mathbf{x}^{(k-1)}$

$\mathbf{p}^{(k)}_{1(\sigma_k)} := \mathbf{p}^{(k)}_{1(\sigma_k)} - (C^{(k)}_{11})^{-1} \tilde{A}^{(k)}_{12} \mathbf{p}^{(k)}_{2(\sigma_k)}$

if $\nu_k = 1$

  $\mathbf{x}^{(k)} := \mathbf{p}^{(k)}_{(\sigma_k)}$

else

  $\mathbf{q}^{(k)}_{(\sigma_k)} := \tilde{A}^{(k)} \mathbf{p}^{(k)}_{(\sigma_k)}$ (10.15)

  for $j = 1$ **to** $\sigma_k - 1$ (10.16)

    $\beta = (\mathbf{q}^{(k)}_{(\sigma_k)}, \mathbf{p}^{(k)}_{(j)}) / \gamma^{(k)}_{(j)}$

    $\mathbf{p}^{(k)}_{(\sigma_k)} := \mathbf{p}^{(k)}_{(\sigma_k)} - \beta \mathbf{p}^{(k)}_{(j)}$

    $\mathbf{q}^{(k)}_{(\sigma_k)} := \mathbf{q}^{(k)}_{(\sigma_k)} - \beta \mathbf{q}^{(k)}_{(j)}$

  $\gamma^{(k)}_{(\sigma_k)} = (\mathbf{q}^{(k)}_{(\sigma_k)}, \mathbf{p}^{(k)}_{(\sigma_k)})$; $\alpha = (\mathbf{r}^{(k)}, \mathbf{p}^{(k)}_{(\sigma_k)}) / \gamma^{(k)}_{(\sigma_k)}$

  $\mathbf{x}^{(k)} := \mathbf{x}^{(k)} + \alpha \mathbf{p}^{(k)}_{(\sigma_k)}$; $\mathbf{r}^{(k)} := \mathbf{r}^{(k)} - \alpha \mathbf{q}^{(k)}_{(\sigma_k)}$

if $(\sigma_k < \nu_k$ && $k < \ell)$ **goto forward**

if $k < \ell$

  $\mathbf{x}^{(k)} := J^{(k)} \mathbf{x}^{(k)}$; **goto backward**

if $(\sigma_\ell = \nu_\ell)$ $\sigma_\ell = 0$

}

$\mathbf{x}^{(\ell)} := J^{(\ell)} \mathbf{x}^{(\ell)}$

Here we denoted by

$$
\begin{array}{rcl}
C_{11}^{(k)} & \ldots & \text{a preconditioner for } A_{11}^{(k)} \\
\mathbf{d}^{(k)} & \ldots & \text{the current right-hand side at level } k \\
\mathbf{r}^{(k)} = (\mathbf{r}_1^{(k)T}, \mathbf{r}_2^{(k)T})^T & \ldots & \text{the current residual at level } k \\
\nu_k & \ldots & \text{the number of recursive calls at level } k \\
\sigma_k & \ldots & \text{a counter for the number of visits at level } k \\
\mathbf{p}_{(j)}^{(k)} = (\mathbf{p}_{1(j)}^{(k)T}, \mathbf{p}_{2(j)}^{(k)T})^T & \ldots & \text{the } j\text{-th search direction at level } k, \quad 1 \le i \le \sigma_k.
\end{array}
$$

Note that Algorithm 10.2 is a multilevel extension of Algorithm 1.22 presented in Chapter 1. Due to the nonlinearity of the preconditioner it requires the explicit orthogonalization of the search directions at every level, cf. (10.16). Here, for convenience we applied the two-level hierarchical basis transformation at level ℓ outside the loop (10.13) that implements the GCG iteration for the solution of $\tilde{A}^{(\ell)}\tilde{\mathbf{x}}^{(\ell)} = \tilde{\mathbf{b}}$ then. The reverse transformation is applied after the termination criterion for the outmost loop (10.13) is satisfied. The basis transformations at all intermediate levels $k = \ell - 1, \ell - 2, \ldots, 1$ are performed inside the loop (10.13), and only at level 0, which is associated with the coarsest mesh, a linear system in standard nodal basis is solved, cf. (10.14). This explains why the matrix-vector multiplication in step (10.15) in the present setting is performed with the two-level HB matrix $\tilde{A}^{(k)}$ at levels $k = 1, 2, \ldots, \ell$. In the end, at level ℓ, Algorithm 10.2 is equivalent to Algorithm 1.22 if in the latter the nonlinear AMLI preconditioner is used in step (1.104)[2]. We usually refer to Algorithm 10.2 as the nonlinear AMLI method.

10.3 Case study: Integrating of new AMLI solvers

In this section we demonstrate how some different techniques which were introduced and studied in previous chapters can be integrated and extended in the construction of efficient AMLI solvers for new and more complex problems.

In the considered example, the locking-free discretization of almost incompressible pure displacement 2D problems (see Section 9.1.3) is first extended to the 3D case. It is shown in addition, that for certain problems, which are in particular related to applications in numerical upscaling, the Dirichlet conditions on the whole boundary can be replaced by Dirichlet conditions for the normal displacements only.

[2]and if the linear system at the fine-grid level ℓ is transformed into a two-level hierarchical basis.

Then, a composite First Reduce (FR) AMLI preconditioner is constructed. The FR splitting was first considered in Section 4.2. Here, two separate steps of this kind are combined:

- The FR step with respect to the cubic macroelements, divided into 6 tetrahedral elements (see Figure 10.1), starts with a local elimination (static condensation) of the interior unknowns. Then an aggregation splitting is applied with respect to each of the faces of the macroelements.
- The reduced matrix has the same structure as the stiffness matrix related to the Rannacher–Turek FEs. This motivates to apply to it the FR algorithm introduced and studied in Section 7.3.

The first set of the presented numerical tests illustrate both, the locking-free approximation and the locking-free PCG convergence of the AMLI algorithm, when the Poisson ratio tends to the incompressibility limit.

The last numerical tests come from μ-FEM (micro-FEM) analysis of bone structures. The studied voxel model contains solid and fluid phases. Both of them are considered as linearly elastic bodies where the fluid phase is almost incompressible. This is a real-life large-scale problem. The complex geometry (interfaces) of the microstructure is taken from a highly resolution computer tomography (CT) image. One important issue is that the related composite material has strong coefficient jumps which are resolved at the finest level of the (voxel) discretization.

10.3.1 Crouzeix–Raviart FE discretization of 3D pure displacement elasticity problems

The weak formulation of the 3D linear elasticity problem (homogeneous boundary conditions are assumed) reads as follows:

For $\boldsymbol{f} = (f_1, f_2, f_3)^T \in (L_2(\Omega))^3$, find $\boldsymbol{u} \in (H_0^1(\Omega))^3 = \{\boldsymbol{v} \in (H^1(\Omega))^3, \boldsymbol{v} \mid_{\Gamma_D} = \boldsymbol{0}\}$ such that

$$\mathcal{A}(\boldsymbol{u}, \boldsymbol{v}) = \int_\Omega \boldsymbol{f}^T \boldsymbol{v} dx \qquad \forall \boldsymbol{v} \in (H_0^1(\Omega))^3. \tag{10.17}$$

The bilinear form $\mathcal{A}(\boldsymbol{u}, \boldsymbol{v})$ is of the form

$$\mathcal{A}(\boldsymbol{u}, \boldsymbol{v}) = \int_\Omega \lambda \operatorname{div}(\boldsymbol{u}) \operatorname{div}(\boldsymbol{v}) + 2\mu \sum_{i,j=1}^{3} \varepsilon_{ij}(\boldsymbol{u}) \varepsilon_{ij}(\boldsymbol{v}) dx = \int_\Omega \langle C \boldsymbol{d}(\boldsymbol{u}), \boldsymbol{d}(\boldsymbol{v}) \rangle dx,$$

where

$$C = \begin{bmatrix} \lambda+2\mu & 0 & 0 & 0 & \lambda & 0 & 0 & 0 & \lambda \\ 0 & \mu & 0 & \mu & 0 & 0 & 0 & 0 & 0 \\ 0 & 0 & \mu & 0 & 0 & 0 & \mu & 0 & 0 \\ 0 & \mu & 0 & \mu & 0 & 0 & 0 & 0 & 0 \\ \lambda & 0 & 0 & 0 & \lambda+2\mu & 0 & 0 & 0 & \lambda \\ 0 & 0 & 0 & 0 & 0 & \mu & 0 & \mu & 0 \\ 0 & 0 & \mu & 0 & 0 & 0 & \mu & 0 & 0 \\ 0 & 0 & 0 & 0 & 0 & \mu & 0 & \mu & 0 \\ \lambda & 0 & 0 & 0 & \lambda & 0 & 0 & 0 & \lambda+2\mu \end{bmatrix}, \tag{10.18}$$

and

$$d(u) = \left[\frac{\partial u_1}{\partial x_1}, \frac{\partial u_1}{\partial x_2}, \frac{\partial u_1}{\partial x_3}, \frac{\partial u_2}{\partial x_1}, \frac{\partial u_2}{\partial x_2}, \frac{\partial u_2}{\partial x_3}, \frac{\partial u_3}{\partial x_1}, \frac{\partial u_3}{\partial x_2}, \frac{\partial u_3}{\partial x_3}\right]^T.$$

Similarly to the 2D case, cf. Section 9.1.3, in the case of pure displacement problems, that is when $\partial\Omega = \Gamma_D$, the following (stabilized) modification of the bilinear form holds true

$$\mathcal{A}(u, v) = \int_\Omega \langle C^s d(u), d(v)\rangle dx = \mathcal{A}^s(u, v) \tag{10.19}$$

where

$$C^s = \begin{bmatrix} \lambda+2\mu & 0 & 0 & 0 & \lambda+\mu & 0 & 0 & 0 & \lambda+\mu \\ 0 & \mu & 0 & 0 & 0 & 0 & 0 & 0 & 0 \\ 0 & 0 & \mu & 0 & 0 & 0 & 0 & 0 & 0 \\ 0 & 0 & 0 & \mu & 0 & 0 & 0 & 0 & 0 \\ \lambda+\mu & 0 & 0 & 0 & \lambda+2\mu & 0 & 0 & 0 & \lambda+\mu \\ 0 & 0 & 0 & 0 & 0 & \mu & 0 & 0 & 0 \\ 0 & 0 & 0 & 0 & 0 & 0 & \mu & 0 & 0 \\ 0 & 0 & 0 & 0 & 0 & 0 & 0 & \mu & 0 \\ \lambda+\mu & 0 & 0 & 0 & \lambda+\mu & 0 & 0 & 0 & \lambda+2\mu \end{bmatrix}. \tag{10.20}$$

The equality (10.19) holds due to the pure displacement boundary conditions as for $u, v \in (H_0^1(\Omega))^3$ we have

$$\int_\Omega \frac{\partial u_i}{\partial x_j}\frac{\partial v_j}{\partial x_i} dx = \int_\Omega \frac{\partial u_i}{\partial x_i}\frac{\partial v_j}{\partial x_j} dx. \tag{10.21}$$

As in the 2D case, the matrix C^s is positive definite. It is also important that the rotations are again excluded from the kernel of the related (stabilized) Neumann boundary conditions operator. As a direct result, the nonconforming Crouzeix–Raviart FEs are straightforwardly applicable to the variational problem (10.19).

Let us remind that locking-free error estimates for the 2D pure displacement problem discretized by Crouzeix–Raviart FEs are presented, e.g., in [42, 43, 57]. The same scheme of analysis is applicable to the 3D case as well.

Let us consider the model linear elasticity problem in $\Omega = (0,1)^3$ and Dirichlet conditions on the whole boundary $\Gamma_D = \partial\Omega$. The right-hand side $\boldsymbol{f}$ of the Lamé system of elasticity corresponds to the exact solution $\boldsymbol{u}$ where $u_1 = x_1^3 + \sin(x_2 + x_3)$, $u_2 = x_2^3 + x_3^2 - \sin(x_1 - x_3)$, $u_3 = x_1^2 + x_3^3 + \sin(x_1 - x_2)$. Let also $u_{h:i}$, $1 \le i \le 3$, be the FEM numerical solution and let $\boldsymbol{u}$, $\boldsymbol{u}_h$, and $\boldsymbol{e}$ be the vectors of the nodal values (three values per node corresponding to the three displacements) of the related functions, where $\boldsymbol{e} = \boldsymbol{u} - \boldsymbol{u}_h$ is the error.

Table 10.1: Relative error stability for $\nu \to 1/2$

ν	0.4	0.49	0.499	0.4999
$\max\limits_{i\in\{1,2,3\}} \frac{\lVert e_i \rVert_{l_\infty}}{\lVert f_i \rVert_{l_\infty}}$	1.66688E-7	5.26416E-8	5.63936E-8	5.67591E-8

In the numerical tests, the Poisson ratio ν is varied, approaching the limit of one half, while the elasticity modulus E corresponds to a certain almost incompressible rubber material. The results in Table 10.1 are for discretization with mesh parameter $h = 1/32$, i.e. the FEM problem has $N = 875\,520$ degrees of freedom. The presented relative errors well illustrate the expected locking-free approximation.

10.3.2 Composite FR algorithm

Let us assume that the domain Ω is covered by a tetrahedral mesh $\mathcal{T}_1$ based on cubic (macro)elements, each of which is split into 6 tetrahedra. We also suppose that the edges of the cubes are parallel to the coordinate axes. The sequence of nested meshes $\mathcal{T}_1 \subset \mathcal{T}_2 \subset \ldots \subset \mathcal{T}_\ell = \mathcal{T}_h$ is obtained by uniform refinement of the coarser macroelements into 8 finer cubes, and then splitting again each of them into 6 similar tetrahedra.

Here we present a composite FR algorithm for AMLI preconditioning of the stiffness matrix A_h, which corresponds to the FEM discretization of the modified elasticity problem (10.19), using Crouzeix–Raviart elements defined on the finest mesh $\mathcal{T}_\ell = \mathcal{T}_h$.

The algorithm is described on a macroelement level. The global stiffness matrix A_h is written in the form

$$A_h^s = \sum_{E \in \mathcal{T}_h} R_E^T A_E^s R_E,$$

where $E \in \mathcal{T}_h$ are the cubic macroelements. In what follows we use the numbering of nodes from Figure 10.1. For a better understanding of their location, the coordinates with respect to the reference macroelement are given in Table 10.2.

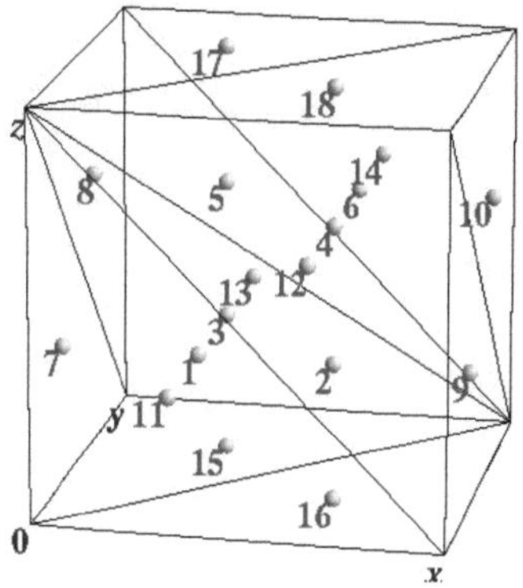

Figure 10.1: Reference macroelement of 6 Crouzeix–Raviart elements

Table 10.2: Coordinates of the nodes of the reference macroelement

node	(x_1, x_2, x_3)	node	(x_1, x_2, x_3)	node	(x_1, x_2, x_3)
1	(1/3, 1/3, 1/3)	7	(0, 1/3, 1/3)	13	(1/3, 1, 1/3)
2	(2/3, 1/3, 1/3)	8	(0, 2/3, 2/3)	14	(2/3, 1, 2/3)
3	(1/3, 2/3, 1/3)	9	(1, 1/3, 1/3)	15	(1/3, 2/3, 0)
4	(2/3, 1/3, 2/3)	10	(1, 2/3, 2/3)	16	(2/3, 1/3, 0)
5	(1/3, 2/3, 2/3)	11	(1/3, 0, 1/3)	17	(1/3, 2/3, 1)
6	(2/3, 2/3, 2/3)	12	(2/3, 0, 2/3)	18	(2/3, 1/3, 1)

Let $\phi_1, \ldots, \phi_{18}$ be the standard (scalar) nonconforming linear finite element nodal basis functions on the macroelement E. Then for the 3D elasticity problem we use the basis functions $\phi_i^{(1)} = (\phi_i, 0, 0)^T$, $\phi_i^{(2)} = (0, \phi_i, 0)^T$, and $\phi_i^{(3)} = (0, 0, \phi_i)^T$, $i = 1, \ldots, 18$. The vector of the macroelement basis functions

$$\varphi_E = \{\Phi_i\}_{i=1}^{54} = \left(\phi_1^{(1)}, \phi_1^{(2)}, \phi_1^{(3)}, \phi_2^{(1)}, \phi_2^{(2)}, \phi_2^{(3)}, \ldots, \phi_{18}^{(1)}, \phi_{18}^{(2)}, \phi_{18}^{(3)}\right)^T$$

is transformed into a vector of new hierarchical basis functions $\tilde{\varphi}_E = \{\tilde{\Phi}_i\}_{i=1}^{54}$, where $\tilde{\Phi} = J_E \Phi$. Following the FR procedure, we consider a transformation matrix J_E corresponding to the splitting $\mathcal{V}(E) = \{\Phi_i\}_{i=1}^{54} = \tilde{\mathcal{V}}_1(E) \oplus \tilde{\mathcal{V}}_2(E) \oplus \tilde{\mathcal{V}}_3(E)$,

$$\begin{aligned}
\tilde{\mathcal{V}}_1(E) &= \operatorname{span}\{\tilde{\Phi}_i\}_{i=1}^{18} \\
&= \operatorname{span}\{\phi_1^{(k)}, \phi_2^{(k)}, \phi_3^{(k)}, \phi_4^{(k)}\, \phi_5^{(k)}\, \phi_6^{(k)}\}_{k=1}^{3} \\
\tilde{\mathcal{V}}_2(E) &= \operatorname{span}\{\tilde{\Phi}_i\}_{i=19}^{36} \\
&= \operatorname{span}\{\phi_8^{(k)} - \phi_7^{(k)}, \phi_{10}^{(k)} - \phi_9^{(k)}, \phi_{12}^{(k)} - \phi_{11}^{(k)}, \\
&\qquad \phi_{14}^{(k)} - \phi_{13}^{(k)}, \phi_{16}^{(k)} - \phi_{15}^{(k)}, \phi_{18}^{(k)} - \phi_{17}^{(k)}\}_{k=1}^{3} \\
\tilde{\mathcal{V}}_3(E) &= \operatorname{span}\{\tilde{\Phi}_i\}_{i=37}^{54} \\
&= \operatorname{span}\{\phi_8^{(k)} + \phi_7^{(k)}, \phi_{10}^{(k)} + \phi_9^{(k)}, \phi_{12}^{(k)} + \phi_{11}^{(k)}, \\
&\qquad \phi_{14}^{(k)} + \phi_{13}^{(k)}, \phi_{16}^{(k)} + \phi_{15}^{(k)}, \phi_{18}^{(k)} + \phi_{17}^{(k)}\}_{k=1}^{3}.
\end{aligned}$$

Accordingly, J_E transforms the macroelement stiffness matrix A_E^s into the hierarchical form $\tilde{A}_E^s = J_E^T A_E^s J_E$,

$$\tilde{A}_E^s = \begin{bmatrix} \tilde{A}_{E:11}^s & \tilde{A}_{E:12}^s & \tilde{A}_{E:13}^s \\ \tilde{A}_{E:21}^s & \tilde{A}_{E:22}^s & \tilde{A}_{E:23}^s \\ \tilde{A}_{E:31}^s & \tilde{A}_{E:32}^s & \tilde{A}_{E:33}^s \end{bmatrix} \begin{matrix} \tilde{\phi}_i \in \tilde{\mathcal{V}}_1(E) \\ \tilde{\phi}_i \in \tilde{\mathcal{V}}_2(E) \\ \tilde{\phi}_i \in \tilde{\mathcal{V}}_3(E) \end{matrix} .$$

The corresponding global stiffness matrix

$$\tilde{A}_h^s = \sum_{E \in \mathcal{T}_h} R_E^T \tilde{A}_E^s R_E$$

is again decomposed into 3×3 blocks

$$\tilde{A}_h^s = \begin{bmatrix} \tilde{A}_{h:11}^s & \tilde{A}_{h:12}^s & \tilde{A}_{h:13}^s \\ \tilde{A}_{h:21}^s & \tilde{A}_{h:22}^s & \tilde{A}_{h:23}^s \\ \tilde{A}_{h:31}^s & \tilde{A}_{h:32}^s & \tilde{A}_{h:33}^s \end{bmatrix}, \tag{10.22}$$

which are induced by the decomposition on macroelement level. The block $\tilde{A}_{h:11}^s$ corresponds to the interior nodal unknowns with respect to the macroelements $E \in \mathcal{T}_h$. The matrices $\tilde{A}_{h:22}^s$ and $\tilde{A}_{h:33}^s$ correspond to certain differences and aggregates of nodal unknowns (basis functions) related to the sides of E. As we know, the first step of the FR algorithm is to eliminate locally (static condensation) the first block of the unknowns. For this purpose we factor $\tilde{A}_h$, and get the Schur complement $\tilde{B}_h^s$ in the form

$$\tilde{B}_h^s = \begin{bmatrix} \tilde{B}_{h:11}^s & \tilde{B}_{h:12}^s \\ \tilde{B}_{h:21}^s & \tilde{B}_{h:22}^s \end{bmatrix}, \tag{10.23}$$

where its first block corresponds to the differences of the (two) basis functions corresponding to each macroelement face. The matrix $\tilde{B}^s_{h:22}$ corresponds to the half-sum (average) of the (same) basis functions from each macroelement face, and thus is associated with the coarse grid. Here, "coarse grid" means the grid of cubic elements associated with $\mathcal{T}_h$. After applying a two-level method to (10.23) the problem is reduced to a system with the coarse level matrix $\tilde{B}^s_{h:22}$. This is the end of the first step of our composite algorithm.

The next observation to note is that $\tilde{B}^s_{h:22}$ has the same structure as the related Rannacher–Turek FE stiffness matrix. This allows us to apply the FR method from Section 7.3 as a second step of the algorithm.

As we know, the convergence of the AMLI algorithm depends on, and is controllable by, the related CBS constants. The next figure shows the multilevel behavior of the locally estimated CBS constant varying the Poisson ratio.

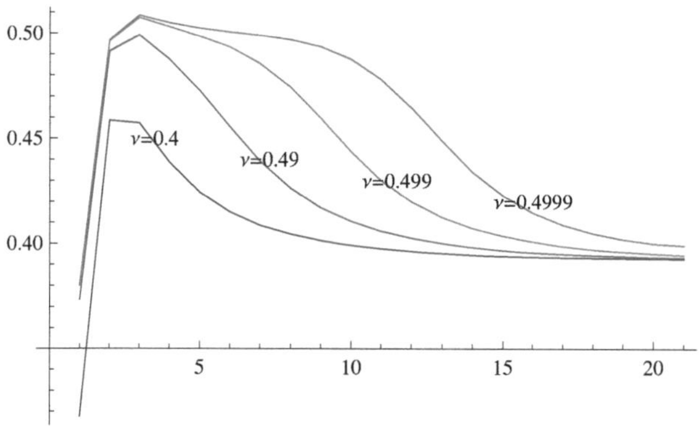

Figure 10.2: Numerically computed γ_E^2 as a function of the refinement level $k = 1, 2, \ldots, \ell$

The first important observation is that γ_E is uniformly bounded when the Poisson ratio tends to the incompressibility limit of $\nu = 1/2$.

We see also that in the multilevel setting, the local CBS constant at the first step of the composite FR algorithm is considerably smaller. Then there is a steep (but bounded by 0.52) increase of γ_E^2, corresponding to the first aggregation step of the second stage (where the Rannacher–Turek FR construction is applied) of the composite algorithm. The peak at the beginning (first two levels) is followed by a monotonically decreasing part of the graphics of $\gamma_E^2(k)$, which tends to a certain fixed value (close to 0.4), uniformly with respect to the Poisson ratio.

Based on the general theory stated in Section 2.6, the presented analysis of the CBS constant leads to the following conclusions:

- The multiplicative variant of the composite AMLI algorithm fulfills the optimality condition (2.78) for degrees of the stabilization matrix polynomial $2 \leq \upsilon < 8$.
- For the additive AMLI, see (2.79), the optimality conditions hold strictly for $3 \leq \upsilon < 8$. As we will see from the numerical tests in the next section, the additive AMLI stabilizes even for $\upsilon = 2$.

10.3.3 Numerical tests: Towards μFEM analysis of bone structures

Two test problems are included in this section.

Problem 1. We consider the pure displacement linear elasticity problem with constant coefficients E and ν in the unit cube $\Omega = (0,1)^3$. The robustness of the composite AMLI algorithm is studied varying the mesh size h and for a Poisson ratio ν approaching $1/2$.

Problem 2. Here we consider a composite material representing a bone microstructure. What we see in Figure 10.3 is the solid phase of a bone specimen at a millimeter scale which is obtained by a computer tomography (CT) image at a micron scale. The geometry of the solid phase is described in terms of voxels the size of which corresponds to the resolution of the CT. Such problems are also referred to as μFEM analysis of voxel structures.

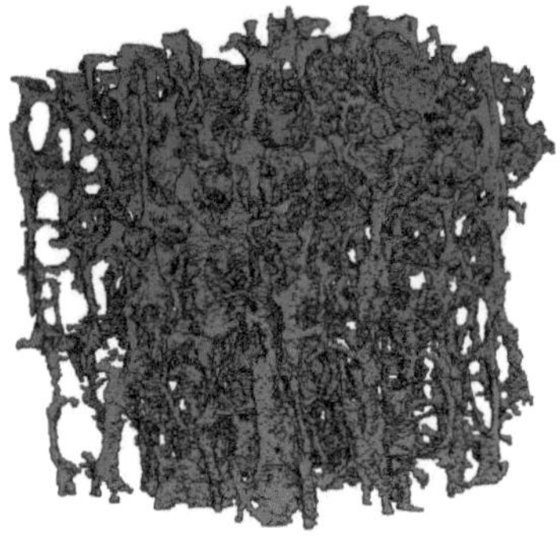

Figure 10.3: CT image of a trabecular bone specimen

The bone specimen is a composite material of two phases, namely solid and fluid. The related material properties are described in terms of the bulk moduli $k_s = 14\ GPa$ and $k_f = 2.3\ GPa$, and the Poisson ratios $\nu_s = 0.325$ and $\nu_f \approx 0.5$. In our tests we vary the Poisson ratio ν_f for the fluid phase, i.e., we consider the cases $\nu_f \in \{0.4, 0.45, 0.49\}$. The related elasticity moduli are computed by the relation $k = E/(1-2\nu)$.

Dirichlet boundary conditions with respect to the normal displacements only are imposed. This setting comes from the related numerical homogenization (numerical upscaling) scheme if (at macro level) the homogenized properties of the elastic material are isotropic. It is easy to check that the equality (10.21) holds true for this kind of boundary conditions. However, for problems with jumping coefficients, some additional smoothness conditions on the interfaces are needed to get (10.21). Let us note that we apply the FEM discretization based on (10.19) in the context of numerical homogenization only.

The presented results are for the second step of the composite algorithm, i.e., the solution of the linear systems with the coarse level matrix $\tilde{B}^s_{h:22}$ in (10.23) for which we use different variants of the AMLI algorithm. Note that the first step of the solution procedure results in an optimal order process if the two-level preconditioner for (10.23) has a uniformly bounded condition number. We will comment on this issue at the end of this section. If we implement Dirichlet boundary conditions without elimination of variables, e.g., by equating the corresponding rows and columns with the identity matrix, the dimension of the full system to be solved is $N_T \times N_T$ where $N_T = 18n^2(1 + 2n)$ and $n = 1/h$. The number of interior DOF on voxel level is $N_I = 18n^3$. Thus the size of the condensed matrix (10.23) is $N_C \times N_C$, $N_C = 18n^2(1 + n)$, and, finally, its lower-right block, to which we apply the recursive multilevel method, yields a linear system of dimension $N = 9n^2(1 + n)$.

In the tables below we list the number of outer iterations that are required to reduce the A-norm of the initial error by a factor 10^6. The right-hand side vector is the vector of all zeros and the iteration is initialized with a random initial guess. The approximate inverses of the pivot blocks we realize in all cases by static condensation of the interior DOF on macro element level followed by an incomplete factorization without any additional fill-in, i.e., ILU(0) applied to the Schur complement, cf. Section 7.5.2.

The numerical results for Problem 1 are presented in Tables 10.3–10.5, those for Problem 2 are collected in Tables 10.6 and 10.7.

We observe that if the material is homogeneous, as for Problem 1, our method is completely robust with respect to the Poisson ratio ν; The convergence even becomes faster when ν approaches $1/2$. In case of the V-cycle method the number of PCG iterations increases moderately according to the increase of the condition number of the preconditioner when adding levels of approximate factorization. The W-cycle method in both cases, linear and nonlinear AMLI, stabilizes the number of outer iterations and thus yields an optimal order solution process.

However, the situation is not that favorable for inhomogeneous materials as the bone micro structure in Problem 2. There the jump of the PDE coefficients introduced by the phase change produces an additional difficulty, which the imple-

Table 10.3: Convergence results for AMLI V-cycle: Problem 1

# voxels	$\nu = 0.49$	$\nu = 0.4999$	$\nu = 0.499999$
16^3	12 (18)	10 (14)	7 (11)
32^3	15 (24)	12 (19)	8 (14)
64^3	19 (31)	15 (25)	9 (17)

Table 10.4: Convergence results for linear AMLI W-cycle: Problem 1

# voxels	$\nu = 0.49$	$\nu = 0.4999$	$\nu = 0.499999$
16^3	9 (14)	7 (11)	6 (9)
32^3	10 (15)	8 (13)	6 (10)
64^3	10 (16)	8 (14)	6 (10)

Table 10.5: Convergence results for nonlinear AMLI W-cycle: Problem 1

# voxels	$\nu = 0.49$	$\nu = 0.4999$	$\nu = 0.499999$
16^3	8 (13)	7 (11)	5 (9)
32^3	8 (14)	7 (12)	5 (9)
64^3	8 (15)	7 (12)	5 (9)

Table 10.6: Convergence results for AMLI V-cycle: Problem 2

# voxels	$\nu_f = 0.4$	$\nu_f = 0.45$	$\nu_f = 0.49$
16^3	18 (28)	25 (37)	52 (71)
32^3	23 (37)	32 (46)	68 (92)
64^3	28 (45)	37 (57)	78 (109)

Table 10.7: Convergence results for nonlinear AMLI W-cycle: Problem 2

# voxels	$\nu_f = 0.4$	$\nu_f = 0.45$	$\nu_f = 0.49$
16^3	14 (22)	19 (30)	46 (68)
32^3	14 (23)	20 (32)	50 (72)
64^3	14 (23)	21 (32)	53 (72)

mented method can only cope with satisfactorily in case of moderate jumps on the finest mesh. However, this can be expected from the numerical experiments we performed for the 3D scalar elliptic problems discretized by rotated trilinear elements with a similar method, cf. Chapter 7.5.2. Note that for $\nu_f = 0.49$ we have $E_s/E_f > 10^2$ and $\mu_s/\mu_f > 10^2$, which means that both the modulus of elasticity E and the Lamé constant μ exhibit a jump of more than two orders of magnitude on the interfaces of the solid and the fluid phase of the bone. We come to the conclusion that for problems with highly oscillatory coefficients on the finest mesh the hierarchical splitting – in particular the fine-coarse partitioning and in the present context also the aggregation of unknowns – has to be adapted. One possible direction to go has been described in the previous chapter, see Section 8.3, for the case of DG discretizations.

Remark 10.3. For the case of homogeneous materials the first block $\tilde{B}^s_{h:11}$ of the Schur complement (10.23) as well as the pivot blocks of the recursively computed coarse-level matrices after static condensation of the interior unknowns are well-conditioned with a condition number bound that is uniform with respect to the Poisson ratio $\nu \in (0, 1/2)$. This can be shown by a local analysis on (macro) element level.

Remark 10.4. The presented composite algorithm can also be used to solve the system of linear algebraic equations arising from mixed finite element discretization of the Stokes problem efficiently. This is due to the fact that by applying the augmented Lagrangian method to this indefinite problem the resulting reduced, nearly singular problem is equivalent to the linear elasticity problem for (almost) incompressible materials. For details see [58, 84].

Bibliography

[1] M. Abramowitz and I.A. Stegun, *Handbook of Mathematical Functions*. Dover Publications, New York, 1965, Ninth Printing 1972.

[2] B. Achchab and J.F. Maitre, *Estimate of the constant in two strengthened CBS inequalities for FEM systems of 2D elasticity: Application to multilevel methods and a posteriori error estimators*, Numer. Lin. Alg. Appl. 3(2) (1996), pp. 147–160.

[3] D.N. Arnold, F. Brezzi, B. Cockburn, and L.D. Marini, *Unified analysis of discontinuous Galerkin methods for elliptic problems*, SIAM J. Numer. Anal. 39 (2002), pp. 1749–1779.

[4] O. Axelsson, *A generalized conjugate gradient, least square method*, Numer. Math. 51 (1987), pp. 209–227.

[5] ———, *Iterative Solution Methods*. Cambridge University Press, 1994.

[6] ———, *Stabilization of algebraic multilevel iteration methods; Additive methods*, Numerical Algorithms 21 (1999), pp. 23–47.

[7] O. Axelsson and V.A. Barker, *Finite Element Solution of Boundary Value Problems: Theory and Computations*. Academic Press, Orlando, 1984.

[8] O. Axelsson and R. Blaheta, *Two simple derivations of universal bounds for the C.B.S. inequality constant*, Appl. Math. 49 (2004), pp. 57–72.

[9] O. Axelsson, R. Blaheta, and M. Neytcheva, *Preconditioning of boundary value problems using elementwise Schur complements*, Uppsala University, Information Technology, Report, 2006. No. 2006-048.

[10] O. Axelsson and V. Eijkhout, *The nested recursive two-level factorization method for nine-point difference matrices*, SIAM J. Sci. Stat. Comput. 12 (1991), pp. 1373–1400.

[11] O. Axelsson and I. Gustafsson, *Preconditioning and two-level multigrid methods of arbitrary degree of approximations*, Math. Comp. 40 (1983), pp. 219–242.

[12] O. Axelsson and G. Lindskog, *On the eigenvalue distribution of a class of preconditioning methods*, Numer. Math. 48 (1986), pp. 479–498.

[13] O. Axelsson and G. Lindskog, *On the rate of convergence of the preconditioned conjugate gradient method*, Numer. Math. 48 (1986), pp. 499–523.

[14] O. Axelsson and S. Margenov, *On multilevel preconditioners which are optimal with respect to both problem and discretization parameters*, Computational Methods in Applied Mathematics 3(1) (2003), pp. 6–22.

[15] O. Axelsson and A. Padiy, *On the additive version of the algebraic multilevel iteration method for anisotropic elliptic problems*, SIAM J. Sci. Comput. 20 (1999), pp. 1807–1830.

[16] O. Axelsson and P.S. Vassilevski, *Algebraic multilevel preconditioning methods I*, Numer. Math. 56 (1989), pp. 157–177.

[17] ———, *Algebraic multilevel preconditioning methods II*, SIAM J. Numer. Anal. 27 (1990), pp. 1569–1590.

[18] ———, *A black box generalized conjugate gradient solver with inner iterations and variable-step preconditioning*, SIAM J. Matrix Anal. Appl. 12 (1991), pp. 625–644.

[19] ———, *Variable-step multilevel preconditioning methods, I: Self-adjoint and positive definite elliptic problems*, Numer. Lin. Alg. Appl. 1 (1994), pp. 75–101.

[20] O. Axelsson B. Achchab, L. Laayouni, and A. Souissi, *Strengthened Cauchy-Bunyakowski-Schwarz inequality for a three dimensional elasticity system*, Numer. Lin. Alg. Appl. 8(3) (2001), pp. 191–205.

[21] I. Babuška and M. Zlamal, *Nonconforming elements in the finite element method with penalty*, SIAM J. Numer. Anal. 10 (1973), pp. 863–875.

[22] E. Bängtsson and M. Neytcheva, *An agglomerate multilevel preconditioner for linear isostasy saddle point problems*, pp. 113–120. Springer LNCS 3743, 2006.

[23] R. Bank, *Hierarchical bases and the finite element method*, Acta Numerica 5 (1996), pp. 1–43.

[24] R. Bank and T. Dupont, *An optimal order process for solving finite element equations*, Math. Comp. 36 (1981), pp. 427–458.

[25] R. Barret, M. Berry, T.F. Chan, J. Demmel, J. Donato, J. Dongarra, V. Eijkhout, R. Pozo, C. Romine, and H. van der Vorst, *Templates for the Solution of Linear Systems: Building Blocks for Iterative Methods*. SIAM, Society for Industrial and Applied Mathematics, Philadelphia, 1994.

[26] C.E. Baumann and J.T. Oden, *A discontinuous hp finite element method for convection-diffusion problems*, Computer Methods in Applied Mechanics and Engineering 175 (1999), pp. 311–341.

[27] B. Bejanov, J.-L. Guermond, and P. Minev, *A locally* div*-free projection scheme for incompressible flows based on non-conforming finite elements*, Int. J. Numer. Meth. Fluids 49 (2005), pp. 239–258.

[28] R. Blaheta, *Nested tetrahedral grids and strengthened C.B.S. inequality*, Numer. Lin. Alg. Appl. 10 (2003), pp. 619–637.

[29] ———, *Algebraic multilevel methods with aggregations: An overview*, pp. 3–14. Springer LNCS 3743, 2006.

[30] ———, *Application of hierarchical decomposition: Preconditioners and error estimates for conforming and nonconforming FEM*. Large Scale Scientific Computing (I. Lirkov, S. Margenov, and J. Wasniewski, eds.), pp. 78–85. Springer LNCS 4818, 2008.

[31] R. Blaheta, S. Margenov, and M. Neytcheva, *Uniform estimate of the constant in the strengthened CBS inequality for anisotropic non-conforming FEM systems*, Numer. Lin. Alg. Appl. 11 (2004), pp. 309–326.

[32] ______, *Robust optimal multilevel preconditioners for non-conforming finite element systems*, Numer. Lin. Alg. Appl. 12 (2005), pp. 495–514.

[33] ______, *Aggregation-based multilevel preconditioning of non-conforming FEM elasticity problems*, pp. 847–856. Springer LNCS 3732, 2006.

[34] P. Boyanova and S. Margenov, *Multilevel splitting of weighted graph-Laplacian arising in non-conforming mixed FEM elliptic problems*. Numerical Analysis and Applications. Proceedings of the 4th NAA conference, Lozenetz, Bulgaria, June 16–20, 2008 (L. Vulkov et al., ed.), to appear. Springer LNCS, Singapore, 2009.

[35] D. Braess, *Finite Elements: Theory, Fast Solvers, and Applications in Solid Mechanics*. Cambridge University Press, 2001, Second Edition.

[36] A. Brandt, *Algebraic multigrid theory: The symmetric case*, Appl. Math. Comput. 19 (1986), pp. 23–56.

[37] ______, *General highly accurate algebraic coarsening*, Electronic Transactions on Numerical Analysis 10 (2000), pp. 1–20.

[38] ______, *Multiscale Scientific Computation: Review 2001*, Multiscale and Multiresolution Methods, T. Barth, R. Haimes and T. Chan, eds., Springer-Verlag, 2001.

[39] A. Brandt, S. McCormick, and J. Ruge, *Algebraic multigrid (AMG) for automatic multigrid solutions with applications to geodetic computations*, Inst. for Computational Studies, Fort Collins, Colorado, Report, 1982.

[40] ______, *Algebraic multigrid (AMG) for sparse matrix equations*, Sparsity and Its Applications, D.J. Evans, ed., Cambridge University Press, Cambridge, 1984.

[41] S. Brenner and C. Carstensen, *Finite Element Methods*, Encyclopedia of Computational Mechanics (T.J.R. Hughes E. Stein., R. de Borst, ed.), J. Wiley, 2004, pp. 73–118.

[42] S. Brenner and L. Scott, *The Mathematical Theory of Finite Element Methods*, Texts in Applied Mathematics 15. Springer-Verlag, 1994.

[43] S. Brenner and L. Sung, *Linear finite element methods for planar linear elasticity*, Math. Comp. 59 (1992), pp. 321–338.

[44] M. Brezina, A. Cleary, R. Falgout, V. Henson, J. Jones, T. Manteuffel, S. McCormick, and J. Ruge, *Algebraic multigrid based on element interpolation (AMGe)*, SIAM J. Sci. Comput. 22 (2000), pp. 1570–1592.

[45] M. Brezina, R. Falgout, S. MacLachlan, T. Manteuffel, S. McCormick, and J. Ruge, *Adaptive smoothed aggregation (αSA)*, SIAM J. Sci. Comput. 25 (2004), pp. 1896–1920.

[46] ______, *Adaptive algebraic multigrid*, SIAM J. Sci. Comput. 27 (2006), pp. 1261–1286 (electronic).

[47] P.G. Ciarlet, *The Finite Element Method for Elliptic Problems*. North Holland, 1978.

[48] A. Cleary, R. Falgout, V. Henson, J. Jones, T. Manteuffel, S. McCormick, G. Miranda, and J. Ruge, *Robustness and scalability of algebraic multigrid*, SIAM J. Sci. Stat. Comput. 21 (2000), pp. 1886–1908.

[49] M. Crouzeix and P.-A. Raviart, *Conforming and non-conforming finite element methods for solving the stationary Stokes equations*, RAIRO, Anal. Num. 3 (1973).

[50] V.A. Dobrev, R.D. Lazarov, P.S. Vassilevski, and L.T. Zikatanov, *Two-level preconditioning of discontinuous Galerkin approximations of second order elliptic equations*, Numer. Lin. Alg. Appl. 13 (6) (2006), pp. 753–770.

[51] T. Dupont, R.P. Kendall, and H.H. Rachford, *An approximate factorization procedure for solving self-adjoint elliptic difference equations*, SIAM Journal on Numerical Analysis 5 (1968), pp. 559–573.

[52] V. Eijkhout, *On the existence problem of incomplete factorisation methods*, Computer Science Department, University of Tennessee, Report, U.S.A., 1999. Lapack working note 144, UT-CS-99-435.

[53] V. Eijkhout and P.S. Vassilevski, *The role of the strengthened Cauchy-Bunyakowski-Schwarz inequality in multilevel methods*, SIAM Review 33 (1991), pp. 405–419.

[54] R.E. Ewing, J. Wang, and Y. Yang, *A stabilized discontinuous finite element method for elliptic problems*, Numer. Lin. Alg. Appl. 10 (2003), pp. 83–104.

[55] R. Falgout and P.S. Vassilevski, *On generalizing the AMG framework*, SIAM J. Numer. Anal. 42 (2004), pp. 1669–1693.

[56] R. Falgout, P.S. Vassilevski, and L. Zikatanov, *On two-grid convergence estimates*, Numer. Lin. Alg. Appl. 12 (2005), pp. 471–494.

[57] R.S. Falk, *Nonconforming finite element methods for the equations of linear elasticity*, Math. Comp. 57 (1991), pp. 529–550.

[58] M. Fortin and R. Glowinski, *Augmented Lagrangian Methods*, Studies in Mathematics and its Applications 15. North-Holland Publishing Co., Amsterdam, 1983, Applications to the numerical solution of boundary value problems, Translated from the French by B. Hunt and D. C. Spicer.

[59] A. George and J. Lu, *Computer Solution of Large Sparse Positive Definite Systems*. Prentice-Hall, Englewood Cliffs, N.Y., 1981.

[60] I. Georgiev, J. Kraus, and S. Margenov, *Multilevel preconditioning of 2D Rannacher-Turek FE problems; Additive and multiplicative methods*. Numerical Methods and Applications (T. Boyanov et al., ed.), pp. 56–64. Springer LNCS 4310, 2007.

[61] ———, *Multilevel algorithms for Rannacher-Turek finite element approximation of 3D elliptic problems*, Computing 82 (2008), pp. 217–239.

[62] ———, *Multilevel preconditioning of rotated bilinear non-conforming FEM problems*, Comput. Math. Appl. 55 (2008), pp. 2280–2294.

[63] G.H. Golub and C.F. van Loan, *Matrix Computations*. The Johns Hopkins University Press, 1989.

[64] W. Hackbusch, *Multigrid Methods and Applications*, Springer Series in Computational Mathematics 4. Springer-Verlag, Berlin, 1985.

[65] P. Hansbo and M. Larson, *A simple nonconforming bilinear element for the elasticity problem*. Trends in Computational Structural Mechanics, pp. 317–327. CIMNE, 2001.

[66] P. Hansbo and M.G. Larson, *Discontinuous Galerkin and the Crouzeix-Raviart element: Application to elasticity*, ESAIM: Math. Model. Numer. Anal. 37(1) (2003), pp. 63–72.

[67] V. Henson and P.S. Vassilevski, *Element-free AMGe: General algorithms for computing the interpolation weights in AMG*, SIAM J. Sci. Comput. 23 (2001), pp. 629–650.

[68] C. Johnson, *Numerical Solutions of Partial Differential Equations by the Finite Element Method*. Cambridge University Press, Cambridge, 1987.

[69] J. Jones and P.S. Vassilevski, *AMGe based on element agglomeration*, SIAM J. Sci. Comput. 23 (2001), pp. 109–133.

[70] J. Douglas Jr. and T. Dupont, *Interior penalty procedures for elliptic and parabolic Galerkin methods*. Lecture notes in Phys. 58. Springer-Verlag, 1976.

[71] Tz. Kolev and S. Margenov, *Two-level preconditioning of pure displacement nonconforming FEM systems*, Numer. Lin. Alg. Appl. 6 (1999), pp. 533–555.

[72] J. Kraus, *An algebraic preconditioning method for M-matrices: Linear versus nonlinear multilevel iteration*, Numer. Lin. Alg. Appl. 9 (2002), pp. 599–618.

[73] ———, *Algebraic multilevel preconditioning of finite element matrices using local Schur complements*, Numer. Lin. Alg. Appl. 13 (2006), pp. 49–70.

[74] ———, *Algebraic multigrid based on computational molecules, II: Linear elasticity problems*, SIAM J. Sci. Comput. 30 (2008), pp. 505–524.

[75] J. Kraus and C. Brand, *Condition numbers of approximate Schur complements in two- and three-dimensional discretizations on hierarchically ordered grids*, Computing 65 (2000), pp. 135–154.

[76] J. Kraus and S. Margenov, *Generalized hierarchical bases for discontinuous Galerkin discretizations*, Numer. Lin. Alg. Appl. (submitted).

[77] J. Kraus, S. Margenov, and J. Synka, *On the multilevel preconditioning of Crouzeix-Raviart elliptic problems*, Numer. Lin. Alg. Appl. 15 (2008), pp. 395–416.

[78] J. Kraus and J. Schicho, *Algebraic multigrid based on computational molecules, I: Scalar elliptic problems*, Computing 77 (2006), pp. 57–75.

[79] J. Kraus and J. Synka, *An agglomeration-based multilevel-topology concept with application to 3D-FE meshes*, RICAM, Report, Linz, 2004. No. 2004-08.

[80] J. Kraus and S. Tomar, *A multilevel method for discontinuous Galerkin approximation of three-dimensional anisotropic elliptic problems*, Numer. Lin. Alg. Appl. 15 (2008), pp. 417–438.

[81] ______, *Multilevel preconditioning of elliptic problems discretized by a class of discontinuous Galerkin methods*, SIAM J. Sci. Comput. 30 (2008), pp. 684–706.

[82] U. Langer and W. Queck, *On the convergence factor of Uzawa's algorithm*, J. Comput. Appl. Math. 15 (1986), pp. 191–202.

[83] R.D. Lazarov and S. Margenov, *CBS constants for graph-Laplacians and application to multilevel methods for discontinuous Galerkin systems*, Journal of Complexity 4–6 (2007), pp. 498–515.

[84] Y. Lee, J. Wu, J. Xu, and L. Zikatanov, *Robust subspace correction methods for nearly singular systems*, Mathematical Models and Methods in Applied Sciences 17 (2007), pp. 1937–1963.

[85] J.F. Maitre and F. Musy, *The contraction number of a class of two-level methods; An exact evaluation for some finite element subspaces and model problems*, pp. 535–544. Lect. Notes Math. 960, 1982.

[86] J. Mandel, M. Brezina, and P. Vaněk, *Energy optimization of algebraic multigrid bases*, Computing 62 (1999), pp. 205–228.

[87] S. Margenov, *Upper bound of the constant in the strengthened C.B.S. inequality for FEM 2D elasticity equations*, Numer. Lin. Alg. Appl. 1(1) (1994), pp. 65–74.

[88] ______, *Semicoarsening AMLI algorithms for elasticity problems*, Numer. Lin. Alg. Appl. 5 (1998), pp. 347–362.

[89] S. Margenov and P. Minev, *On a MIC(0) preconditioning of non-conforming mixed FEM elliptic problems*, Math. Comput. Simul. 76 (2007), pp. 149–154.

[90] S. Margenov and J. Synka, *Generalized aggregation-based multilevel preconditioning of Crouzeix-Raviart FEM elliptic problems*, RICAM, Report, Linz, 2006. No. 2006-23.

[91] S. Margenov and P.S. Vassilevski, *Algebraic multilevel preconditioning of anisotropic elliptic problems*, SIAM J. Sci. Comp. 15(5) (1994), pp. 1026–1037.

[92] ______, *Algebraic two-level preconditioning of non-conforming FEM systems*. Large-Scale Scientific Computations of Engineering and Environmental Problems, pp. 239–248. Notes on Numerical Fluid Mechanics, V 62, VIEWEG, 1998.

[93] Y. Notay, *Optimal V-cycle algebraic multilevel preconditioning*, Numer. Lin. Alg. Appl. 5 (1998), pp. 441–459.

[94] ______, *Using approximate inverses in algebraic multilevel methods*, Numer. Math. 80 (1998), pp. 397–417.

[95] ______, *A multilevel block incomplete factorization preconditioning*, Appl. Numer. Math. 31 (1999), pp. 209–225.

[96] ———, *Flexible conjugate gradients*, SIAM J. Sci. Comput. 22 (2000), pp. 1444–1460.

[97] ———, *Robust parameter-free algebraic multilevel preconditioning*, Numer. Lin. Alg. Appl. 9 (2002), pp. 409–428.

[98] I. Pultarova, *The strengthened CBS inequality constant for second order elliptic partial differential operator and for hierarchical bilinear finite element functions*, Applications of Mathematics 50 (2005), pp. 323–329.

[99] R. Rannacher and S. Turek, *Simple non-conforming quadrilateral Stokes Element*, Numerical Methods for Partial Differential Equations 8(2) (1992), pp. 97–112.

[100] B. Riviere, M.F. Wheeler, and V. Girault, *Improved energy estimates for interior penalty, constrained and discontinuous Galerkin methods for elliptic problems I*, Comput. Geosci. 3 (1999), pp. 337–360.

[101] J. Ruge and K. Stüben, *Efficient solution of finite difference and finite element equations by algebraic multigrid (AMG)*. Multigrid Methods for Integral and Differential Equations, D.J. Paddon and H. Holstein, eds., pp. 169–212. The Institute of Mathematics and Its Applications Conference Series, Clarendon Press, Oxford, 1985.

[102] J. Ruge and K. Stüben, *Algebraic Multigrid*, Multigrid Methods, vol. 3 of Frontiers in Applied Mathematics (S.F. McCormick, ed.), SIAM, Philadelphia, 1987, pp. 73–130.

[103] Y. Saad, *Iterative Methods for Sparse Linear Systems*. PWS Publishing Company, 1996.

[104] K. Shahbazi, *An explicit expression for the penalty parameter of the interior penalty method*, Journal of Computational Physics 205 (2005), pp. 401–407.

[105] J. Shewchuk, *What is a good linear finite element? – Interpolation, conditioning, anisotropy, and quality measures*. Eleventh international meshing roundtable, pp. 115–126, 2002.

[106] K. Stüben, *Algebraic multigrid (AMG): Experiences and comparisons*, Appl. Math. Comput. 13 (1983), pp. 419–452.

[107] J. Synka, *The effect of a minimum angle condition on the preconditioning of the pivot block arising from 2-level-splittings of Crouzeix-Raviart FE-spaces*. Large Scale Scientific Computing (I. Lirkov, S. Margenov, and J. Wasniewski, eds.), pp. 105–112. Springer LNCS 4818, 2008.

[108] P. Vaněk, M. Brezina, and J. Mandel, *Convergence of algebraic multigrid based on smoothed aggregation*, Numer. Math. 88 (2001), pp. 559–579.

[109] P. Vaněk, J. Mandel, and M. Brezina, *Algebraic multigrid based on smoothed aggregation for second and fourth order problems*, Computing 56 (1996), pp. 179–196.

[110] R.S. Varga, *Matrix Iterative Analysis*. Prentice-Hall, 1962.

[111] P.S. Vassilevski, *Nearly optimal iterative methods for solving finite element elliptic equations based on multilevel splitting of the matrix*, Institute for Scientific Computation, University of Wyoming, Report, Laramie, U.S.A., 1989. No. 1989-09.

[112] ______, *Hybrid V-cycle algebraic multilevel preconditioners*, Math. Comp. 58 (1992), pp. 489–512.

[113] ______, *On two ways of stabilizing the hierarchical basis multilevel methods*, Siam Review 39 (1997), pp. 18–53.

[114] ______, *Multilevel Block Factorization Preconditioners*. Springer-Verlag, 2008.

[115] J. Xu and L. Zikatanov, *On an energy minimizing basis for algebraic multigrid methods*, Computing and Visualization in Science 7 (2004), pp. 121–127.

[116] D. Young, *Iterative Solution of Large Linear Systems*. Academic Press, 1971.

[117] H. Yserentant, *On the multilevel splitting of finite element spaces*, Numer. Math. 49 (1986), pp. 379–412.

[118] L. Zikatanov, *Two-sided bounds on the convergence rate of two-level methods*, Numer. Linear Algebra Appl. 15 (2008), pp. 439–454.

Index

Bei Fragen zur Produktsicherheit wenden Sie sich bitte an:
If you have any questions regarding product safety,
please contact:

Walter de Gruyter GmbH
Genthiner Straße 13
10785 Berlin
productsafety@degruyterbrill.com